SELF-ORGANIZED 3D INTEGRATED OPTICAL INTERCONNECTS

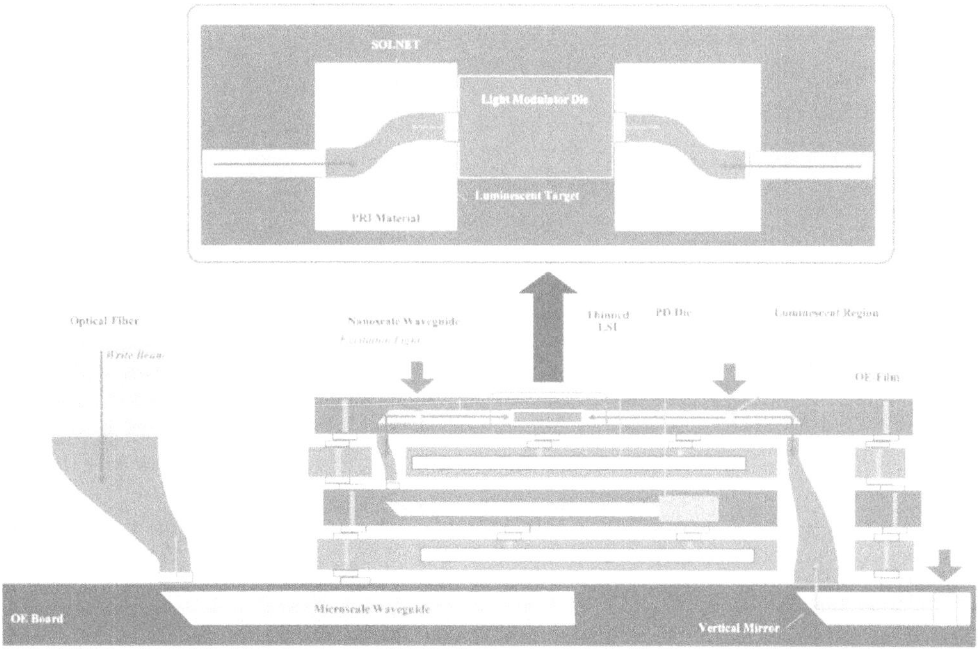

SELF-ORGANIZED 3D INTEGRATED OPTICAL INTERCONNECTS

with All-Photolithographic Heterogeneous Integration

Tetsuzo Yoshimura

CRC Press
Taylor & Francis Group
Boca Raton London New York

CRC Press is an imprint of the
Taylor & Francis Group, an **informa** business

Self-Organized 3D Integrated Optical Interconnects with All-Photolithographic Heterogeneous Integration

First published 2021 by Jenny Stanford Publishing Pte. Ltd.

Published 2023 by CRC Press
Taylor & Francis Group
6000 Broken Sound Parkway NW, Suite 300
Boca Raton, FL 33487-2742

ISBN 13: 978-981-4877-04-6 (hbk)

Visit the Taylor & Francis Web site at
http://www.taylorandfrancis.com

and the CRC Press Web site at
http://www.crcpress.com

British Library Cataloguing-in-Publication Data
A catalogue record for this book is available from the British Library.

To the memory of my wife, Yoriko

Contents

Preface

Lightwaves have been introduced into long-haul communications, metro/local-area networks, and inter-box optical interconnects. Currently, lightwaves are ready to come into boxes of computers, especially in high-performance computing systems like data centers and supercomputers to realize intra-box optical interconnects. Lightwaves are also required in massive switching systems.

In inter-box optical interconnects, lightwaves have successfully been introduced by optoelectronic (OE) modules, in which discrete bulk-chip OE and electronic devices are assembled using the flip-chip-bonding-based packaging technology called *jisso gijutsu* in Japanese. The OE modules, however, are not applicable to intra-box optical interconnects because of the following reason. Intra-box optical interconnects involve "very short line distances of cm–mm order" and "very large line counts of hundreds-thousands." This causes *optics excess*, namely, excess components, materials, spaces for components, fabrication efforts for packaging, and design efforts. *Optics excess* raises sizes and costs of intra-box optical interconnects enormously if they are built using *jisso gijutsu*.

In the present book, the concept of self-organized three-dimensional (3D) integrated optical interconnects is proposed to show the possibility of the *optics excess* reduction, and the strategy to reach the goal is described.

Self-organized 3D integrated optical interconnects have a 3D architecture with 3D optical wiring backbones built by stacking films, and are fabricated using the resource-saving all-photolithographic heterogeneous integration of small/thin OE dies, which enables us to replace *jisso gijutsu* with a sophisticated fabrication process for large-scale integrated circuits (LSIs). The implementation of the self-organized lightwave network (SOLNET) reduces the efforts of OE device alignments in the assembling process and makes it easy to construct vertical optical waveguides for inter-film optical links in 3D optical wiring backbones. These advantages are brought by two abilities of SOLNETs: (i) self-aligned optical couplings between misaligned OE devices with different core sizes, and (ii) 3D waveguide construction in free spaces. Thus, self-organized 3D integrated optical interconnects are expected to reduce *optics excess*, achieving reductions in the sizes and costs of intra-box optical interconnects.

The core technologies of self-organized 3D integrated optical interconnects will contribute to ultrahigh-speed future optical communication systems, too.

The present book is intended to be read by scientists, engineers, and graduate students who study optical interconnects within computers and optical switching systems. It would help the readers to produce new ideas or strategies to reduce sizes and costs of intra-box optical interconnects as well ultrahigh-speed future optical communication systems. Readers with an interest in solar energy conversion systems would also get hints to expand their ideas.

The author published a book titled *Optical Electronics: Self-Organized Integration and Applications* with Jenny Stanford Publishing in 2012. The present book is a kind of second edition. A good deal of content from the previous book has been used in the present one. The chapters are rearranged to make the story line clear. The chapter on the SOLNET is fully rewritten using latest data. Embodiments disclosed in our patents are newly included for integrated OE multichip modules (MCMs), 3D optical interconnects, micro optical link modules, and active optical sheets/boards/connectors.

The author would like to thank Professor Kunihiko Asama of Tokyo University of Technology for his instruction and encouragement, General Managers Isao Fudemoto and Osamu Takahashi of Fujitsu Laboratories Ltd. for giving the author the great opportunity to work in Silicon Valley, and General Managers Kohei Kiyota, Masao Tanaka, Akira Ihaya, Hideo Takahashi, Masaaki Kobayashi, Hirofumi Okuyama of Fujitsu Laboratories Ltd., and President Yoshio Honda of Fujitsu Computer Packaging Technologies Inc. (FCPT), San Jose, California, for their encouragement and support in the research. The author would also like to thank Professor Michael Haney of

University of Delaware and Professor Eugenio Fazio of Sapienza University of Rome for leading the author to the international scene of optical interconnects, Drs. Hideyuki Nawata, Tetsuo Sato, and Juro Oshima of Nissan Chemical Industries Ltd. for supporting the SOLNET research for a long time, colleagues at FCPT and Fujitsu Laboratories, and students of Yoshimura Laboratory in Tokyo University of Technology for their collaboration in research work that contributed to the accomplishment of this work. Finally, the author would like to express sincere gratitude to Stanford Chong, Director and Publisher, and Jenny Rompas, Director and Publisher, of Jenny Stanford Publishing Pte. Ltd. for giving me the great opportunity to write this book, and Sarabjeet Garcha, Senior Editorial Manager, for editing and completing it.

Tetsuzo Yoshimura

Chapter 1

Introduction

Since lightwaves were introduced into long-haul communication systems, optics penetration into electronics have progressed as summarized in Fig. 1.1. Lightwaves have been introduced into shorter-distance systems in the order of metro networks, local-area networks (LANs), and inter-box optical interconnects in high-performance computing systems such as data centers and supercomputers. Currently, lightwaves are ready to come into boxes of computers to realize the intra-box optical interconnects.

Within boxes, increases in the clock frequency and wiring density cause problems such as heat generation (power dissipation) and noise generation. Lightwave implementation will solve these problems. Lightwaves will also be required in massive switching systems to treat the rapidly increasing data rates and volume in networking. Recent development of nano-photonics has dramatically improved the light modulation and optical switching functions, opening the way toward the intra-box optical interconnects, and massive optical switching systems.

In the inter-box optical interconnects, lightwaves have successfully been introduced by optoelectronic (OE) modules, in which discrete bulk-chip OE devices and electronic devices are assembled using conventional flip-chip-bonding-based packaging

Self-Organized 3D Integrated Optical Interconnects with All-Photolithographic Heterogeneous Integration
Tetsuzo Yoshimura

ISBN 978-981-4877-04-6 (Hardcover), 978-1-003-04305-8 (eBook)
www.jennystanford.com

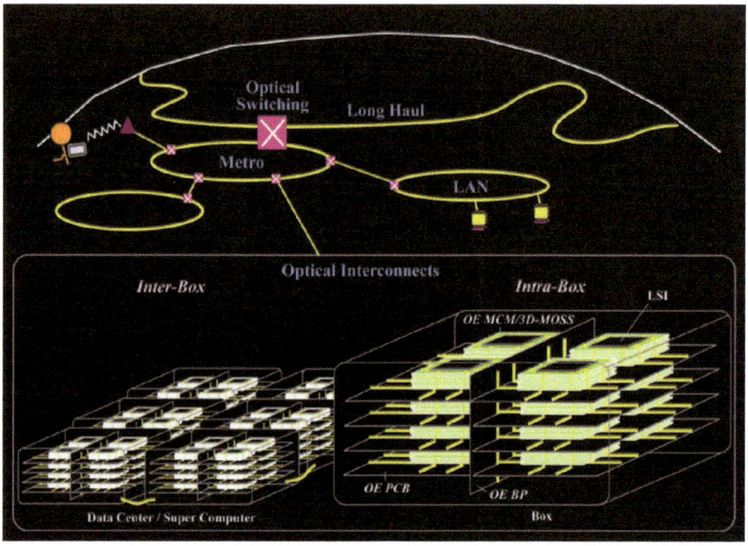

Figure 1.1 Optics penetration into electronics.

processes called *jisso gijutsu* in Japanese. The OE-module-based technology, however, does not seem applicable to the intra-box optical interconnects due to two remarkable features of the intra-box optical interconnects, being distinguished from the inter-box optical interconnects.

(1) Very short line distances of cm–mm order
(2) Very large line counts of hundreds to thousands

These features imply that OE devices of hundreds to thousands should be distributed in the board/chip areas within boxes with high density to achieve the electrical-to-optical (E-O) signal conversion and the optical-to-electrical (O-E) signal conversion. In the case that the OE modules fabricated by the *jisso gijutsu* are used for the E-O/O-E conversion, the following *optics excess* will become serious.

- Excess components and materials
- Excess spaces for components
- Excess fabrication efforts for packaging
- Excess design efforts

The *optics excess* enormously raises sizes and costs of the intra-box optical interconnects.

To realize the intra-box optical interconnects, the *optics excess* should be reduced. In the present book, *a concept of self-organized three-dimensional (3D) integrated optical interconnects* is proposed to show the possibility of reducing the *optics excess*, and the strategy to reach the goal is presented.

The self-organized 3D integrated optical interconnects have 3D architecture with 3D optical wiring backbones, where electronic devices like large-scale integrated circuits (LSIs) for processors, memories, and switches are distributed, and small/thin OE dies for the E-O/O-E signal conversion are embedded. Sometimes optical switching and wavelength division multiplexing (WDM) functions are involved.

The 3D architecture is constructed utilizing scalable film optical link modules (S-FOLMs), which provide various types of 3D OE platforms, including OE multichip modules (MCMs), 3D micro optical switching systems (3D-MOSSs), OE printed circuit boards (PCBs), and OE backplanes (BPs) as depicted in Fig. 1.1. S-FOLMs are fabricated using the resource-saving all-photolithographic heterogeneous integration process called *photolithographic packaging with selectively occupied repeated transfer* (PL-Pack with SORT). The 3D structures are built by stacking optical waveguide films, in which small/thin OE dies are embedded, using the Film/Z-connection technology.

In the 3D integrated optical interconnects, due to the inherent lightwave nature of "Go Straight," the optical coupling problem is caused. That is, to construct the 3D optical wiring backbones, where lightwaves should be transmitted through a lot of OE devices, including optical waveguides, light modulators, optical switches, vertical cavity surface-emitting lasers (VCSELs), photodetectors (PDs), and so on, strict positional/angular control of the OE devices is necessary. This raises the amount of efforts for OE device assembling, and consequently, the system costs rise, too. Another problem is that not only horizontal waveguides for planar optical links but also vertical waveguides for inter-film optical links are required for 3D optical wiring backbones. The implementation of the self-organized lightwave network (SOLNET) solves these two problems simultaneously to realize self-organized 3D integrated optical interconnects because of the following capabilities.

- SOLNET enables self-aligned optical couplings between misaligned optical devices with different core sizes.
- SOLNET enables 3D waveguide construction in free spaces.

The contents of the present book are as follows. In Chapter 2, guidelines toward the goal of the self-organized 3D integrated optical interconnects are proposed, and required core technologies are summarized.

In Chapters 3–7, details of the core technologies are explained. In Chapter 3 the concept of the S-FOLM, which provides 3D OE platforms for the 3D integrated optical interconnects, is presented. In Chapter 4, fabrication processes for optical waveguide films with vertical mirrors, which are the elementary units of S-FOLMs, and experimental demonstrations of the optical waveguide films and 3D optical circuits built by stacking the optical waveguide films are described. In Chapter 5 the resource-saving all-photolithographic heterogeneous integration process, PL-Pack with SORT, which is used for embedding small/thin OE dies in films, is proposed and the experimental demonstrations are presented. In Chapter 6, examples of light modulators and optical switches for the 3D integrated optical interconnects are described, emphasizing the waveguide prism deflector (WPD) optical switches, which we precisely designed to predict the modulation/switching performance by computer simulations. In Chapter 7 the concept and proof-of-concepts of SOLNETs, which realize the self-organization of the 3D integrated optical interconnects, are presented.

Then, in Chapter 8, precise models of the self-organized 3D integrated optical interconnects are proposed. In Chapters 9, precise models of the 3D-MOSS are proposed, and the effects of S-FOLMs and SOLNETs on their performance are predicted. In Chapter 10, as a related application, self-organized integrated solar energy conversion systems are proposed. In Chapter 11, embodiments of structures and fabrication processes for the self-organized 3D integrated optical interconnects are explained based on our US patents.

Finally, in Chapter 12, as future challenges, possibilities of polymer multiple quantum dots (polymer MQDs) as electro-optic (EO) materials for the light modulators and optical switches are discussed. Growth of the polymer MQDs is experimentally demonstrated by the molecular layer deposition (MLD), indicating that the MLD is expected to be a promising method for producing high-performance EO polymer MQDs in the future.

Chapter 2

Guidelines toward Self-Organized 3D Integrated Optical Interconnects

Although the lightwave implementation into boxes of computers brings us advantages, it causes *optics excess* that increases system sizes and costs. To reduce the *optics excess*, the lightwave-implemented computers should be simple; in other words, structures, spaces and material consumption for components, fabrication processes, and design procedures in the lightwave-implemented computers should be just like those in pure electronic computers. Such system simplification is achieved by self-organized 3D integrated optical interconnects.

In the present chapter, first, after advantages of lightwave implementation into boxes of computers are reviewed, the concept of the integrated optical interconnects is explained. The integrated optical interconnects simplify the lightwave-implemented computers and enable us to treat them just like the pure electronic computers, reducing the *optics excess*. Next, various types of structures for E-O/O-E signal conversion, which is a critical issue in the integrated optical interconnects, are compared. It is revealed that the E-O/O-E signal conversion using S-FOLMs has advantages over that using conventional OE modules made from assembled bulk chips. Then,

Self-Organized 3D Integrated Optical Interconnects with All-Photolithographic Heterogeneous Integration
Tetsuzo Yoshimura

ISBN 978-981-4877-04-6 (Hardcover), 978-1-003-04305-8 (eBook)
www.jennystanford.com

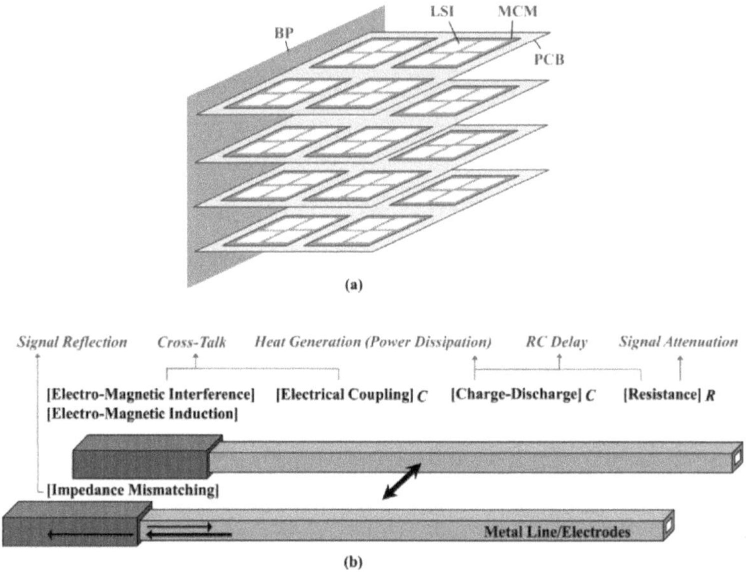

Figure 2.1 (a) Structural hierarchy in high-performance computing systems, and (b) problems in the electrical wiring.

the concept of the self-organized 3D integrated optical interconnects is explained. They are constructed by implementing SOLNETs into S-FOLM-based 3D integrated optical interconnects, enabling us to construct 3D optical wiring backbones in self-organized manners. Finally, core technologies required to reach the goal of the self-organized 3D integrated optical interconnects are summarized.

2.1 Advantages of Lightwave Implementation into Boxes of Computers

An example of the hierarchy for high-performance computing systems is shown in Fig. 2.1a. LSIs are placed on MCMs, which are placed on PCBs. The PCBs are placed on BPs. Typical sizes of the LSI, MCM, PCB, and BP are respectively ~1 cm, ~10 cm, a few 10 cm, and ~100 cm.

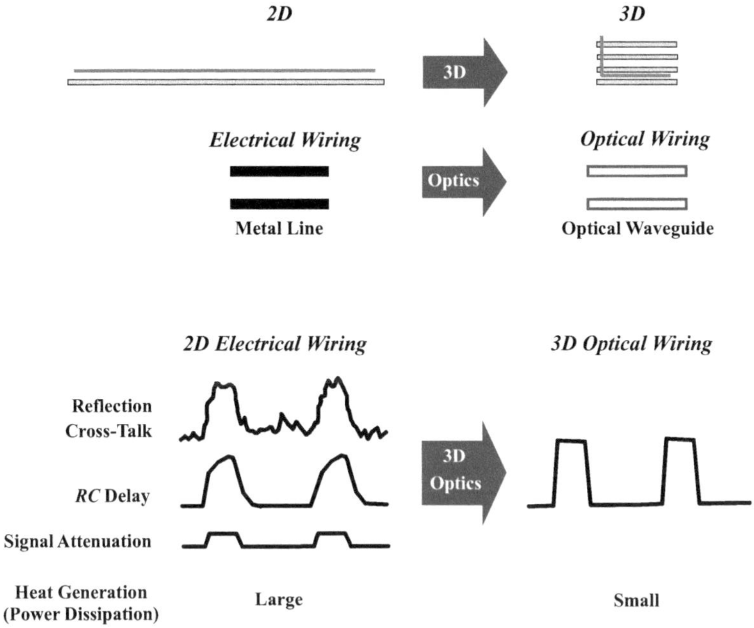

Figure 2.2 Advantages of 3D optical wiring.

In each level in the hierarchy, as Fig. 2.1b shows, cross-talk, heat generation (power dissipation), *RC* delay, signal attenuation, and signal reflection are caused. The cross-talk arises from electromagnetic interference, electromagnetic induction, and electrical couplings between metal lines/electrodes due to capacitance, *C*. The heat generation and *RC* delay mainly arise from charge–discharge of metal lines/electrodes through electrical resistance, *R*, of transistors in driver circuits. The signal attenuation arises from electrical resistance, *R*, of the lines/electrodes including the skin effect. The signal reflection arises from impedance mismatching in electrical circuits. Since the influence of the electromagnetic interference, electromagnetic induction, capacitance, and electrical resistance increases with increasing clock frequency and wiring density, the problems become serious with increases in clock frequency and wiring density in computers. The same situation occurs in switching systems.

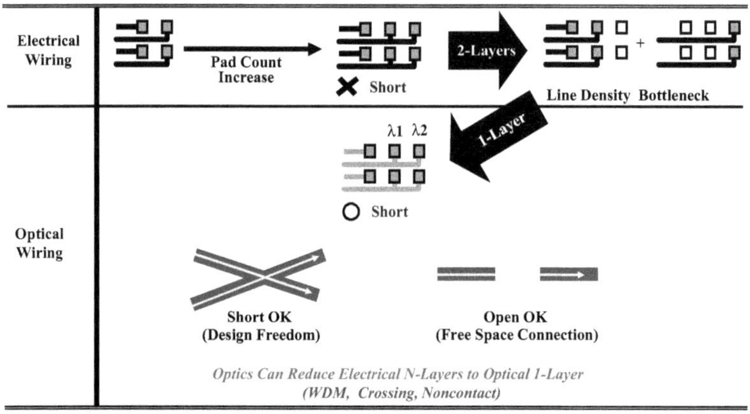

Figure 2.3 Additional advantages of optical wiring.

There are two ways to overcome the problems; 3D wiring and optical wiring [1–3] as summarized in Fig. 2.2. By introducing 3D wiring, wiring distances are reduced, resulting in reduction of the electromagnetic interference, electromagnetic induction, and capacitance/electrical resistance originated from the metal lines/electrodes. Actually, research and development on the 3D electrical wiring in LSIs and boards is under way.

By replacing metal lines/electrodes with optical waveguides, lightwaves are implemented into computers. Optical wiring can dramatically reduce the electromagnetic interference, electromagnetic induction, capacitance/electrical resistance, and impedance mismatching. The removal of the skin effect realizes frequency-independent loss of propagated signals. Therefore, it is expected that 3D optical wiring will achieve data transmission with high-S/N-ratio and small-delay signals. The 3D optical wiring is also expected to reduce the heat generation (power dissipation).

Optical wiring has additional advantages of "Short OK" and "Open OK" characteristics as depicted in Fig. 2.3. For the "Short OK" characteristics, when optical waveguides cross each other, almost no cross-talk appears due to "Go Straight" characteristics of lightwaves, enabling an increase in wiring design freedom. By using the "Short OK" characteristics, optics can reduce layer counts for wiring; for example, electrical 2-layers to optical 1-layer, by using

WDM. Furthermore, by increasing the number of the wavelengths in the WDM, a large number of optical signals can be transmitted in a single optical waveguide. This realizes a drastic increase in transmission data rates as well as reduction in layer counts for wiring, for example, electrical N-layers to optical 1-layer.

Meanwhile, when optical waveguides are separated with some distances, most of the lightwaves from one optical waveguide can be transmitted into the other optical waveguide due to their free-space-propagation characteristics. The "Open OK" characteristics enable "free space connections." This property is useful to construct 3D optical circuits, in which lightwave transmission along vertical directions can be easily carried out without vertical waveguides.

As described, the 3D optical wiring improves the performance of computers.

2.2 Integrated Optical Interconnects

In Fig. 2.4, a model for optical interconnects is compared with that for electrical interconnects. In electrical interconnects, electronic units such as LSIs, MCMs, and PCBs are simply connected by metal lines. In optical interconnects, on the other hand, electrical signals from output terminals of an electronic unit are converted to optical signals by light modulators or VCSELs. The optical signals are

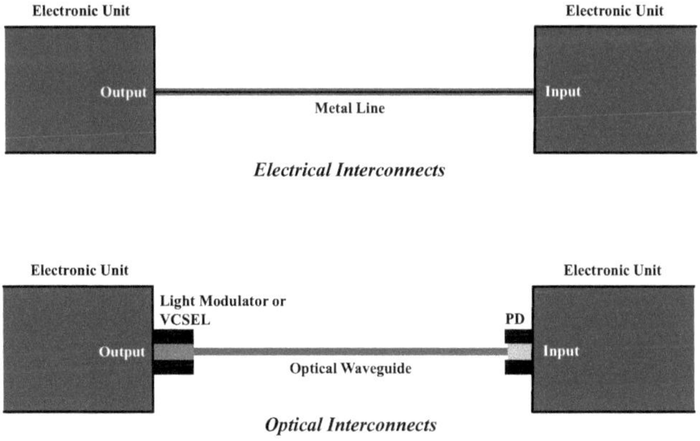

Figure 2.4 Models for optical interconnects and electrical interconnects.

transmitted through optical waveguides to PDs to be converted to electrical signals. The electrical signals are provided into input terminals of another electronic unit.

It is found from Fig. 2.4 that optical interconnects cause the following *optics excess*: components, materials and spaces for the components, fabrication efforts, and design efforts. The excess increases system sizes and costs. Since the optical interconnects within boxes of computers involve a lot of electronic units, an enormous number of components need to be distributed with high density for the E-O/O-E signal conversion, increasing the *optics excess* unacceptably.

The problem can be solved by the integrated optical interconnects, in which optics can be treated like electronics. The general concept is shown in Fig. 2.5 together with the concept of conventional optical interconnects. In the conventional optical interconnects, discrete bulk-chip OE devices and driver/amplifier (D/A) integrated circuits (ICs) are used, and consequently, spaces for the OE devices and the D/A ICs are necessary in addition to the spaces for LSIs. In integrated optical interconnects, on the other

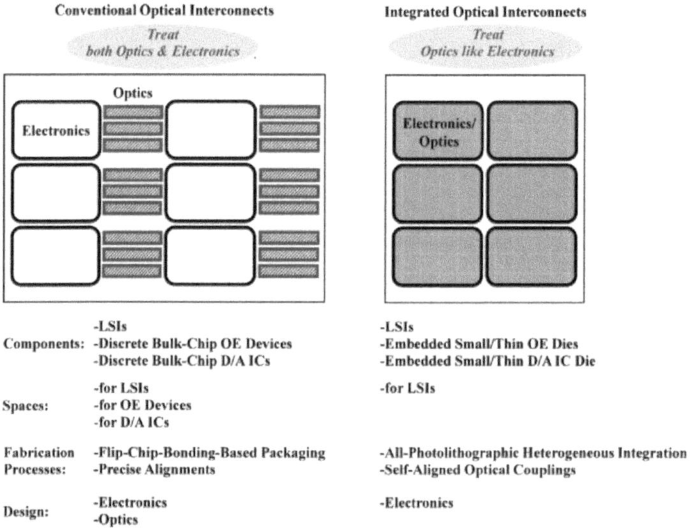

Figure 2.5 Concepts of integrated optical interconnects and conventional optical interconnects.

hand, small/thin dies of OE devices are embedded in the LSI, MCM, or PCB areas together with small/thin dies of D/A ICs. Thus, the *optics excess* related to the components, namely, the excess material consumption and spaces, can be reduced to a level close to that in electrical interconnects.

For fabrication processes, in conventional optical interconnects, typically, packaging based on the flip-chip bonding, in which assembling with precise alignments of individual OE devices is necessary, is used. In integrated optical interconnects, on the other hand, heterogeneous integration based on the all-photolithographic process, which is similar to the process used in the LSI industries, is used for embedding the small/thin OE dies. This reduces the die assembling efforts drastically. And furthermore, assembling with rough alignments of the dies is permitted if optical couplings are performed in self-aligned manners using SOLNETs, as described in section 2.3. This contributes to reduction in the fabrication efforts.

Design in conventional optical interconnects is carried out by considering both electronics and optics. In integrated optical interconnects, the design is carried out by considering only electronics because standardized-interface characteristics are given by embedding interface ICs underneath the LSIs. This contributes to reduction in the design efforts.

From these arguments, it is concluded that integrated optical interconnects enable us to simplify lightwave-implemented computers and treat them just like pure electronic computers.

Considering the trend of "from 2D wiring to 3D wiring," integrated optical interconnects should be built in 3D structures. 3D integrated optical interconnects can be built using S-FOLMs. S-FOLMs provide all levels of optical networking units, including 3D OE platforms by stacking OE-Films, which contain optical waveguides, small/thin OE dies, thinned LSIs, etc.

2.3 Self-Organization of 3D Integrated Optical Interconnects

In 3D integrated optical interconnects, there are two issues, namely, an issue related to optical couplings and an issue related to

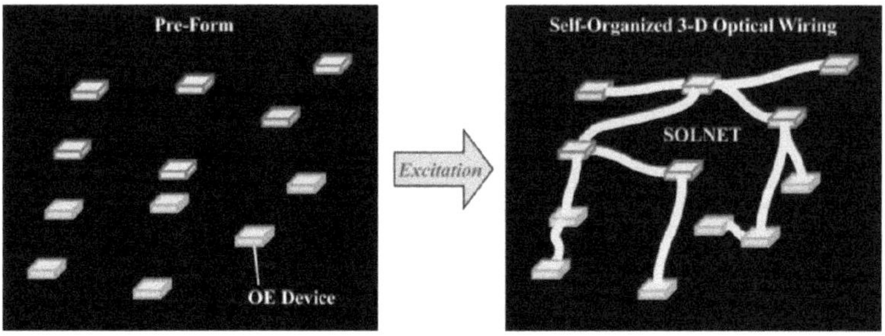

Figure 2.6 Concept of self-organized 3D optical wiring built by the SOLNET.

construction of 3D optical wiring backbones. In 3D structures a large number of optical couplings between OE devices are involved. In such cases, enormous alignment efforts with micron or submicron accuracy are required, raising system fabrication costs. It is also required to form vertical waveguides for the 3D optical wiring in 3D structures. These issues are solved by SOLNETs, which can construct 3D optical wiring backbones in self-organized manners.

The concept of self-organized 3D optical wiring built by SOLNET is schematically illustrated in Fig. 2.6. After a pre-form with connection points of OE devices distributed three-dimensionally is prepared, optical waveguides of SOLNETs are formed between the connection points automatically by some excitation from outside to construct self-organized 3D optical wiring backbones. By combining S-FOLMs and SOLNETs, self-organized 3D integrated optical interconnects are realized. The details of S-FOLMs and SOLNETs are respectively described in Chapters 3 and 7.

2.4 E-O and O-E Signal Conversion in Integrated Optical Interconnects

E-O/O-E signal conversion is a critical function in integrated optical interconnects. There are various types of E-O and O-E conversion structures as classified in Fig. 2.7. In the OE module-type structure, bulk chips of OE devices like VCSELs and PDs, and bulk chips of D/A ICs are assembled. These chips are discretely mounted on a

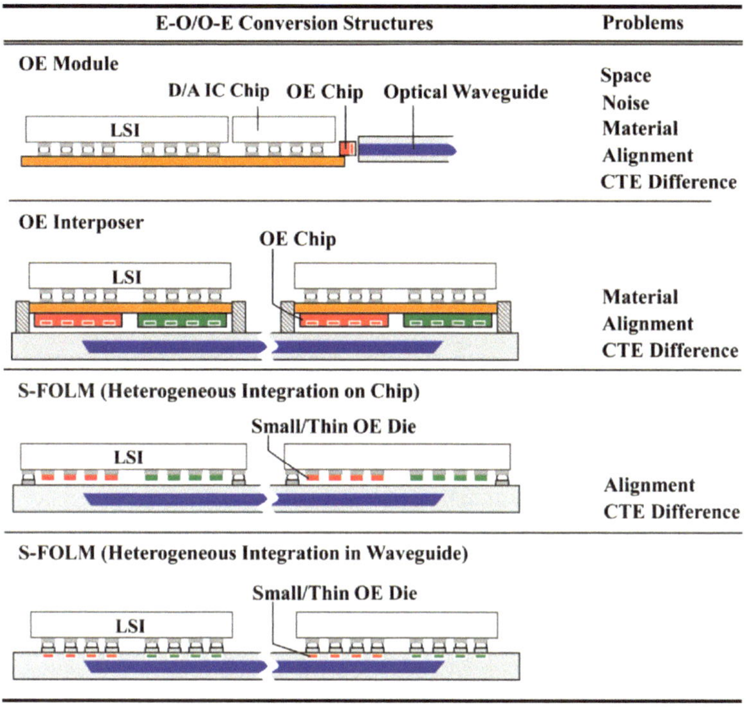

Figure 2.7 Classification of E-O and O-E conversion structures.

substrate together with LSI chips by the flip-chip-bonding-based packaging processes. OE devices are precisely aligned to optical waveguides. Although the OE module-type structure is already available, it has several problems:

- Large excess spaces for bulk chips are necessary.
- Large noise is generated due to long metal lines between LSIs and OE devices.
- Large excess material consumption is caused for OE chips.
- Large efforts for precise alignments between OE devices and optical waveguides are necessary.
- Large relative position changes between OE devices and optical waveguides are caused due to the difference in the coefficient of thermal expansion (CTE).

In the OE interposer-type structure, bulk chips of OE devices are assembled together with D/A IC chips in an interposer [4]. The interposer is inserted between LSIs and optical waveguides. The OE devices are precisely aligned to vertical mirrors in the optical waveguides to achieve optical couplings. The assembling processes are carried out by flip-chip-bonding-based packaging processes. In the OE interposer-type structure, because the OE devices are located under the LSIs and the metal line length between the LSIs and the OE devices is shortened, the problems of large excess spaces and large noise are solved. However, the following three problems still remain: large excess material consumption for bulk OE chips, large efforts for precise alignments between the OE devices and the optical waveguides, and the temperature-change-induced relative position changes between the OE devices and the optical waveguides due to their CTE mismatch.

In the S-FOLM-type structure, heterogeneous integration processes utilizing the epitaxial lift-off (ELO) [5–7] is applied. A typical ELO process, invented by Yablonovitch, is shown in Fig. 2.8.

(1) An epitaxial film is grown on a growth substrate.
(2) The epitaxial film is separated into small/thin dies in an array by etching.
(3) A polymer film is attached onto the surface.
(4) The growth substrate is removed to transfer small/thin dies on the polymer film.
(5) The small/thin dies are transferred onto LSIs by the pick-and-place method.

In the S-FOLM (heterogeneous integration on chip)-type structure, as shown in Fig. 2.7, small/thin dies of OE devices like VCSELs and PDs are mounted on output and input electrodes of LSIs by ELO. LSIs with the small/thin OE dies are mounted on optical waveguides by flip-chip-bonding-based packaging processes. In this case, material consumption for the OE devices is drastically reduced because small/thin OE dies are used. However, large efforts for precise alignments between the OE devices and the optical waveguides are still required in the packaging processes. The problem of the temperature-change-induced relative position

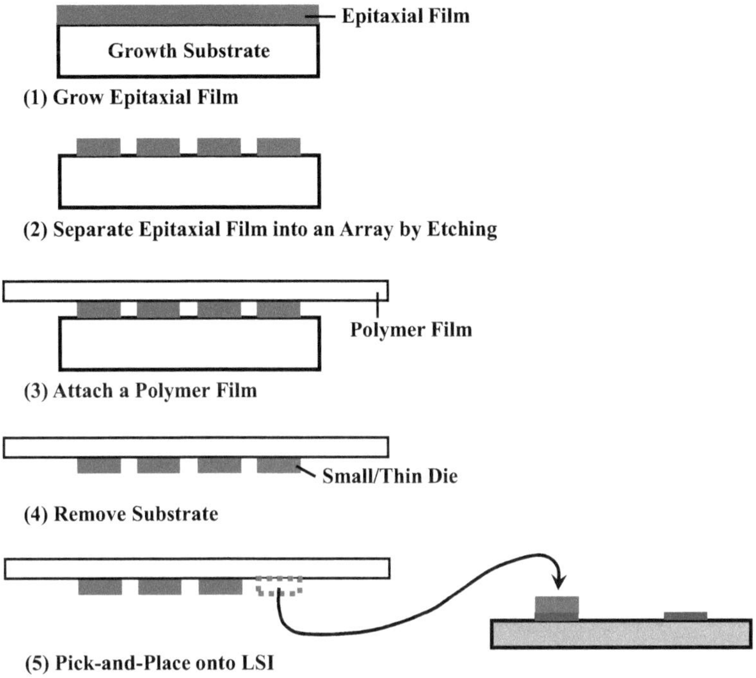

Figure 2.8 Epitaxial lift-off process.

change also remains due to the CTE mismatch between LSIs and optical waveguide layers.

In the S-FOLM (heterogeneous integration in waveguide)-type structure [8–18], small/thin OE dies are embedded into an optical waveguide film or substrate. By using PL-Pack with SORT described in Chapter 5, small/thin OE dies are embedded by the all-photolithographic process similar to the LSI fabrication process. In other words, the flip-chip-bonding-based packaging is not necessary. Consequently, the problem of the precise alignments between OE devices and optical waveguides is solved. Furthermore, the CTE mismatch problem is also solved because each of small/thin OE dies is placed just on the vertical mirror site in the optical waveguide film or substrate as an isolated tiny island. The detail is described in Chapter 5. It should be noted that interface films containing small/thin D/A IC dies is typically inserted between the

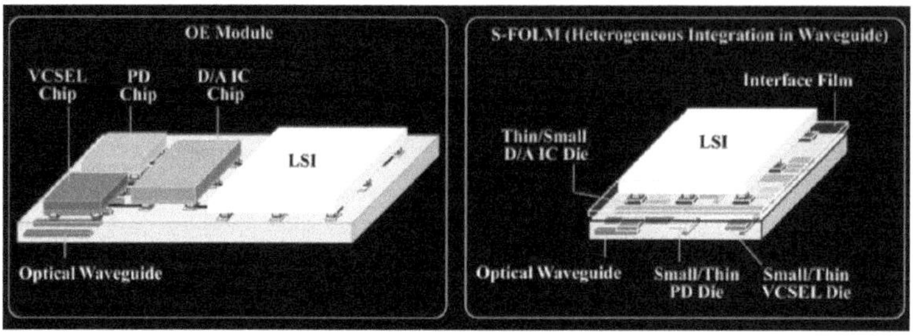

Figure 2.9 Optical transceivers built by the S-FOLM (heterogeneous integration in waveguide)-type E-O/O-E conversion structure and built by the OE module-type structure.

LSIs and the small/thin OE dies although they are not drawn in the figure for simplicity.

In Fig. 2.9, an optical transceiver built by the S-FOLM (heterogeneous integration in waveguide)-type E-O/O-E conversion structure is compared with that built by the OE module-type structure. In the optical transceiver with the OE module-type structure, OE bulk chips of VCSELs/PDs and D/A ICs are mounted beside LSIs on a substrate. In the optical transceiver with the S-FOLM (heterogeneous integration in waveguide)-type structure, small/thin OE dies of VCSELs/PDs are embedded just under the LSIs corresponding to individual pad positions. An interface film containing small/thin D/A IC dies is inserted between the LSIs and the small/thin OE dies. In this case, as mentioned above, all the problems listed in Fig. 2.7 are solved. Thus, the S-FOLM (heterogeneous integration in waveguide)-type optical transceiver reduces the excess spaces on the surface, the epitaxial-material consumption, the metal line lengths reaching the VCSELs/PDs, the noise generated from the lines, the alignment efforts, and the CTE mismatch effects to provide small-size, low-cost, and low-noise optical transceivers.

From the above-mentioned arguments, it is concluded that the S-FOLM (heterogeneous integration in waveguide)-type structure is the best solution for the E-O and O-E signal conversion, reducing the *optics excess*.

Table 2.1 Features for the self-organized 3D integrated optical interconnects model and for the conventional optical interconnects model

	Conventional optical interconnects	Self-organized 3D integrated optical interconnects
Backbone	2D optical wiring	3D optical wiring
E-O/O-E interface	Discrete bulk-chip OE devices	Embedded small/thin OE dies
Fabrication process	Flip-chip-bonding-based packaging Assembling with precise alignments	All-photolithographic heterogeneous integration Assembling with rough alignments

2.5 Core Technologies for Self-Organized 3D Integrated Optical Interconnects

Features of the proposed model for self-organized 3D integrated optical interconnects are summarized in Table 2.1 with those of the model for conventional optical interconnects. In the proposed model the backbone is 3D optical wiring instead of 2D optical wiring, and the E-O/O-E interface is made from embedded small/thin OE dies instead of discrete bulk-chip OE devices. The structures of the proposed model are fabricated by the all-photolithographic heterogeneous integration process instead of the flip-chip-bonding-based packaging process. The self-aligned optical coupling capability of SOLNETs allows us to be free from the OE device assembling with precise alignments. The 3D structures are constructed by 3D OE platforms of S-FOLMs.

A structural model for the self-organized 3D integrated optical interconnects is depicted in Fig. 2.10. Optical waveguide films with vertical mirrors, in which small/thin dies of high-speed/small-size light modulators/optical switches are embedded, and films with thinned LSIs are stacked to construct 3D structures. Small/thin OE dies are embedded into the films by the all-photolithographic heterogeneous integration process using PL-Pack with SORT. Optical Z-connections for inter-film optical links are fabricated by vertical waveguides of SOLNETs. Optical solder for self-aligned optical couplings between optical waveguides and embedded small/thin OE

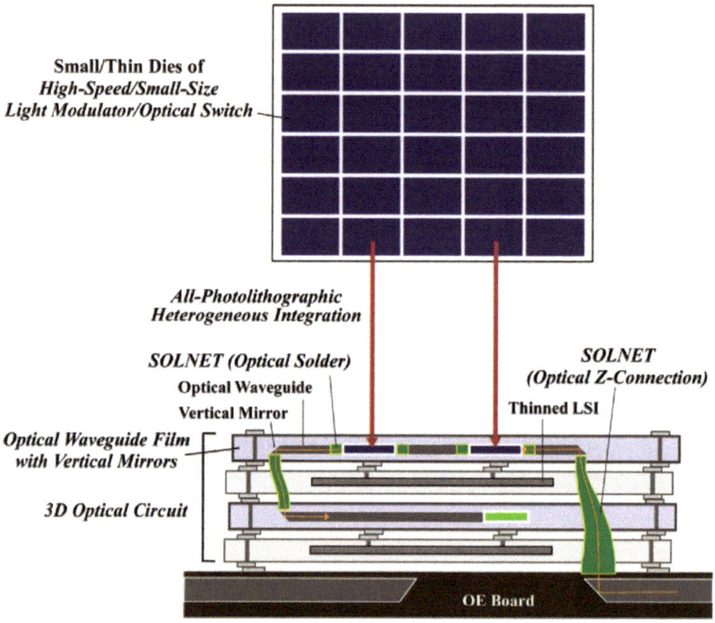

Figure 2.10 Structural model for self-organized 3D integrated optical interconnects.

dies is also realized by SOLNETs to form self-organized 3D optical wiring backbones.

A process flow example for fabrication of the self-organized 3D integrated optical interconnects using the above-described core technologies is summarized in Fig. 2.11.

(1) Optical waveguide films with vertical mirrors are prepared. Electrodes, pads, lines, vias, etc., are formed on/in the films.
(2) Thin/small OE dies for high-speed/small-size light modulators/optical switches and PDs, which have been grown on substrates, are transferred onto supporting substrates by the ELO.
(3) The small/thin OE dies on the supporting substrates are transferred onto the optical waveguide films prepared in step 1 by the all-photolithographic heterogeneous integration process of PL-Pack with SORT to embed small/thin OE dies into optical waveguide films.

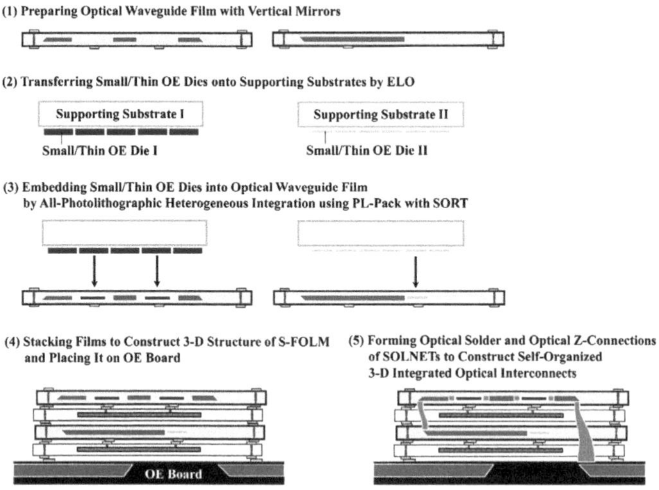

Figure 2.11 Process flow example for fabrication of self-organized 3D integrated optical interconnects.

(4) The optical waveguide films with the embedded small/thin OE dies are stacked together with films containing thinned LSIs to construct a 3D structure, and the 3D structure is placed on an OE board.
(5) In the 3D structure, optical solder and optical Z-connections of SOLNETs are formed to construct self-organized 3D integrated optical interconnects.

As described above, self-organized 3D integrated optical interconnects involve a variety of technologies related to optics and electronics. Most of them have already been established. However, the following core technologies still need to be developed.

- Optical waveguide films with vertical mirrors
- 3D optical wiring backbones constructed by stacking the optical waveguide films
- All-photolithographic heterogeneous integration processes
- High-speed/small-size light modulators and optical switches
- SOLNETs for the optical solder and the optical Z-connection

The details of these core technologies are presented in Chapters 4–7.

References

1. M. W. Haney, H. Thienpont, and T. Yoshimura, "Introduction to the issue on optical interconnects," *IEEE J. Sel. Top. Quantum Electron.* **9**, 347–349 (2003).
2. T. Gu, L. Kimerling, A. Seyedi, T. Schrans, J. Michel, D. Dai, and P. O'Brien, "Guest editorial: special issue on optical interconnects," *J. Lightwave Technol.* **38**, 3319–3321 (2020).
3. D. A. B. Miller, "How large a system can we build without optics?" *Workshop Notes, 8th Annual Workshop on Interconnections Within High Speed Digital Systems*, Lecture 1.2 (1997).
4. T. Mikawa, M. Kinoshita, K. Hiruma, T. Ishitsuka, M. Okabe, S. Hiramatsu, H. Furuyama, T. Matsui, K. Kumai, O. Ibaragi, and M. Bonkohara, "Implementation of active interposer for high-speed and low-cost chip level optical interconnects," *IEEE J. Sel. Top. Quantum Electron.* **9**, 452–459 (2003).
5. R. Pu, C. Duan, and C. Wilmsen, "Hybrid integration of VCSEL's to CMOS integrated circuits," *IEEE J. Sel. Top. Quantum Electron.* **5**, 201–208 (1999).
6. N. M. Jokerst, M. A. Brooke, S. Cho, S. Wilkinson, M. Vrazel, S. Fike, J. Tabler, Y. J. Joo, S. Seo, D. S. Wils, and A. Brown, "The heterogeneous integration of optical interconnections into integrated microsystems," *IEEE J. Sel. Top. Quantum Electron.* **9**, 350–360 (2003).
7. E. Yablonovitch, T. Gmitter, J.P. Harbison, and R. Bhat, "Extreme sensitivity in the lift-off of epitaxial GaAs films," *Appl. Phys. Lett.* **51**, 2222–2224 (1987).
8. T. Yoshimura, J. Roman, Y. Takahashi, S. Beilin, W. Wang, and M. Inao, *4th International Conference on Organic Nonlinear Optics (ICONO'4), Extended Abstracts*, Chitose, Japan, 278 (1998).
9. T. Yoshimura, J. Roman, Y. Takahashi, S. I. Beilin, W. V. Wang, and M. Inao, "Optoelectronic amplifier/driver-less substrate <OE-ADLES> for polymer-waveguide-based board-level interconnection-calculation of delay and power dissipation-," *Nonlinear Opt.* **22**, 453–456 (1999).
10. T. Yoshimura, J. Roman, Y. Takahashi, M. Lee, B. Chou, S. I. Beilin, W. V. Wang, and M. Inao, "Proposal of optoelectronic substrate with Film/Z-connection based on OE-Film," *Proc. 3rd IEMT/IMC Symposium*, 140–145 (1999).
11. T. Yoshimura, J. Roman, Y. Takahashi, W. V. Wang, M. Inao, T. Ishitsuka, K. Tsukamoto, K. Motoyoshi, and W. Sotoyama, "Self-organizing waveguide

coupling method "SOLNET" and its application to film optical circuit substrates," *Proc. 50th Electronic Components & Technology Conference (ECTC)*, 962–969 (2000).

12. T. Yoshimura, Y. Takahashi, M. Inao, M. Lee, W. Chou, S. Beilin, W.-C. Wang, J. Roman, and T. Massingill, "Systems based on opto-electronic substrates with electrical and optical interconnections and methods for making," U.S. Patent 6,343,171 B1 (2002).
13. T. Yoshimura, Y. Takahashi, J. Roman, M. T. McCormack, S. Beilin, W.-C. Wang, and M. Inao, "Single and multilayer waveguides and fabrication process," U.S. Patent 6,785,447 B2 (2004).
14. T. Yoshimura, M. Ojima, Y. Arai, and K. Asama, "Three-dimensional self-organized microoptoelectronic systems for board-level reconfigurable optical interconnects-performance modeling and simulation," *IEEE J. Sel. Top. Quantum Electron.* **9**, 492–511 (2003).
15. T. Yoshimura, T. Inoguchi, T. Yamamoto, S. Moriya, Y. Teramoto, Y. Arai, T. Namiki, and K. Asama, "Self-organized lightwave network based on waveguide films for three-dimensional optical wiring within boxes," *J. Lightwave Technol.* **22**, 2091–2100 (2004).
16. T. Yoshimura, M. Miyazaki, Y. Miyamoto, N. Shimoda, A. Hori, and K. Asama, "Three-dimensional optical circuits consisting of waveguide films and optical Z-connections," *J. Lightwave Technol.* **24**, 4345–4352 (2006).
17. T. Yoshimura, Y. Suzuki, N. Shimoda, T. Kofudo, K. Okada, Y. Arai, and K. Asama, "Three-dimensional chip-scale optical interconnects and switches with self-organized wiring based on device-embedded waveguide films and molecular nanotechnologies," *Proc. SPIE* **6126**, from Photonics West 2006, San Jose, California, 612609-1-15 (2006).
18. T. Yoshimura, *Optical Electronics: Self-Organized Integration and Applications*, Pan Stanford Publishing, Singapore (2012).

Chapter 3

Scalable Film Optical Link Modules

3.1 Concept of S-FOLM

As mentioned in section 2.2, lightwave implementation within boxes of computers causes *optics excess*. The S-FOLM is expected to be a solution for overcoming the *optics excess* problem [1–19]. The concept of the S-FOLM is shown in Fig. 3.1. Firstly, packaging elements such as electrical substrates and OE-Films are prepared. The packaging elements are denoted as follows.

- *S*: Electrical substrates
- *W*: Optical waveguide films
- *D*: Films, in which small/thin OE dies for light modulators, optical switches (SWs), optical filters, VCSELs, PDs, and so on are embedded
- *DW*: Optical waveguide films, in which small/thin OE dies are embedded
- *C*: Films, in which thinned electronic chips and small/thin electronic dies such as LSIs, ICs, capacitors, and so on are embedded

Then, the packaging elements are combined by the Film/Z-connection technology [20] to build products of the S-FOLM. Here,

Self-Organized 3D Integrated Optical Interconnects with All-Photolithographic Heterogeneous Integration
Tetsuzo Yoshimura

ISBN 978-981-4877-04-6 (Hardcover), 978-1-003-04305-8 (eBook)
www.jennystanford.com

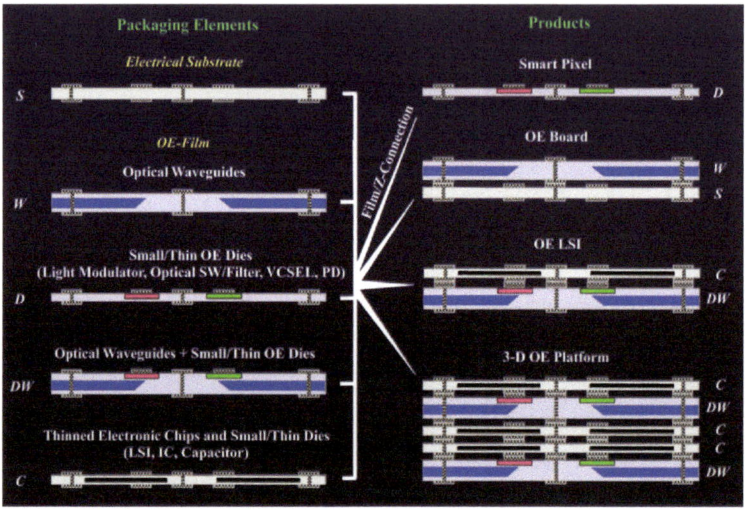

Figure 3.1 Concept of the S-FOLM.

the Z-connections act as both electrical joints and mechanical joints between films. The S-FOLM realizes a maximized variety of products built from minimized types of packaging elements, providing all levels of optical networking units. For example, an OE-Film (*D*) with VCSEL/PD dies itself is a smart pixel. When an OE-Film (*W*) is attached onto an electrical substrate (*S*), an OE board is constructed. A combination of an OE-Film (*DW*) and an OE-Film (*C*) with thinned LSIs provides an OE LSI. By stacking multiple OE-Films (*DW*) and OE-Films (*C*), a 3D OE platform is obtained. These examples indicate that S-FOLM has a great packaging scalability.

In the S-FOLM, small/thin divided OE dies are embedded just under the individual pad positions of LSIs/ICs. This gives the S-FOLM the following advantages to remove *optics excess* and reduce system cost.

- Material consumption and assembling efforts for OE components are minimized by using PL-Pack with SORT [6, 10, 21, 22] (see Chapter 5).
- Excess spaces for OE components are not necessary on the surface.

- Design efforts are minimized by inserting thin-interface ICs for defining interface specifications. This enables us to treat optics as if a transparent agent for interconnect-spec improvement and to provide standardized interface capabilities.
- Bit error rates (BERs) and signal delays are reduced by replacing long metal lines between the components with short vertical metal lines.

In addition, the S-FOLM has the remarkable property of flexibility in structures and positions. The S-FOLM can be attached at any places with any shapes/sizes on substrates.

As can be seen from the arguments mentioned above, in the S-FOLM, semiconductor materials exist only at the places where they are necessary. The other area is occupied by polymer or glass. This is cost-effective due to epitaxial-material saving. In some cases, a cost reduction of one order of magnitude or more is expected, as discussed in Chapter 5.

3.2 3D Integrated Optical Interconnects Built by S-FOLMs

3.2.1 3D OE Platforms

Figure 3.2 shows examples of 3D OE platforms of S-FOLMs for 3D integrated optical interconnects within boxes of computers and optical switching systems [3, 5–11, 14–19]. The multilayer OE MCM shown in Fig. 3.2a consists of a plurality of OE-Films (*W*) with vertical mirrors, an OE-Film (*DW*) with embedded small/thin light modulator/PD dies, and an OE-Film (*C*). Vertical optical links between optical waveguide films are carried out by optical Z-connections consisting of vertical waveguides and vertical mirrors. The second layer (OE-Film (*DW*)) is for the E-O/O-E signal conversion. The third and fourth layers (OE-Films (*W*)) are for the multilayer optical signal routing. The optical signal link between a plurality of 3D OE platforms are performed through the OE-Film (*W*) of the OE PCB by backside connections.

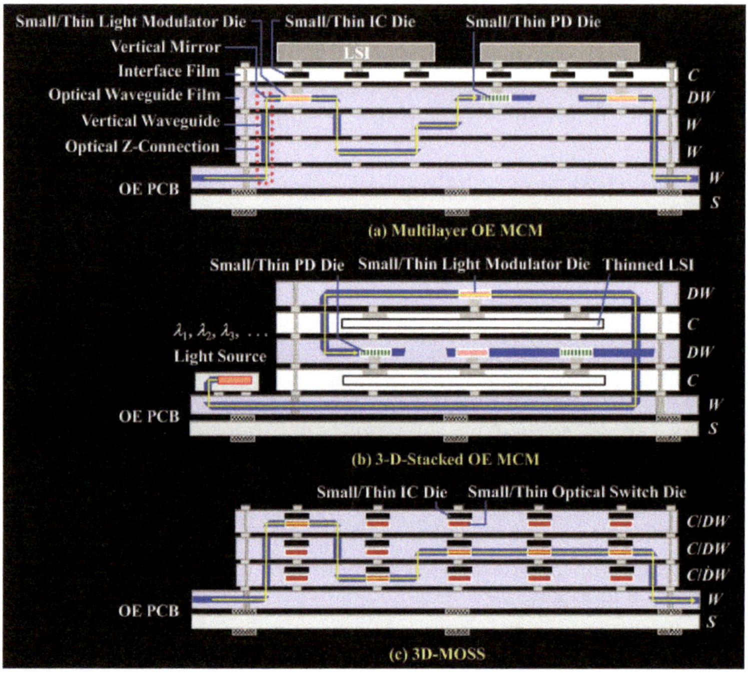

Figure 3.2 3D OE platforms of S-FOLMs for 3D integrated optical interconnects within boxes of computers and optical switching systems.

The first layer (OE-Film (*C*)), in which small/thin-interface IC dies are embedded, is an interface film, on which LSIs are mounted. Via the interface film, electrical signals from LSI outputs are converted into optical signals by light modulators. The optical signals are guided to PDs, where they are converted to electrical signals for LSI inputs via the interface film. The interface film has functions such as driver/amplifier, multiplexing (MUX)/demultiplexing (DEMUX), error correction, timing controlling, coding, and so on. This interface approach gives the 3D OE platforms a standardized-interface capability, and permits computer designers to build electrical architectures without the knowledge of optics.

When light modulators can directly be driven by LSI outputs and PDs can directly generate electrical signals for LSI inputs as in the case of OE amplifier/driver-less substrates (OE-ADLES) [1, 2,

18, 19], which is described in section 3.4, the interface film is not necessary.

In the 3D-stacked OE MCM shown in Fig. 3.2b, OE-Films (*DW*) containing embedded light modulators and PDs are stacked together with OE-Films (*C*) containing thinned LSIs. Optical signals, which are generated at E-O signal conversion sites in the OE-Films (*DW*) by LSI outputs, propagate to O-E signal conversion sites, being converted into electrical signals for LSI inputs. In some cases, the optical signals generated in the 3D-stacked OE MCM are coupled to optical waveguides in the OE-Film (*W*) of the OE PCB by backside connections to be transmitted to other OE units, and vice versa.

In the 3D-stacked OE MCM, heat generation is a serious problem. By placing light sources outside of the stacked structure and using OE-ADLES that uses the external light sources of cw or pulse trains, reduction of the heat generation in the OE MCM is achieved. By using external light sources with a plurality of wavelengths and by replacing vertical mirrors with wavelength filters including tunable filters, WDM interconnects can be performed, which is the final goal to realize flexible optical networking in computers with minimized connection line counts.

In the 3D micro optical switching system (3D-MOSS) shown in Fig. 3.2c, OE-Films (*DW*), in which small/thin optical switch die arrays are embedded, are stacked with coupled OE-Films (*C*), where small/thin-interface IC dies for optical switch drivers are embedded. Optical signals are successively switched by cascaded optical switches through 3D optical routing to be output. The 3D architecture is built by dividing conventional planar switching networks into a plurality of blocks and stacking them. The 3D structure is effective to reduce system sizes and wiring lengths, and consequently, insertion loss, as discussed in Chapter 9.

In 3D OE platforms, a large number of optical couplings in 3D optical circuits are involved. In such cases, enormous alignment efforts with micron or submicron accuracy are required, raising fabrication costs. Vertical waveguide formation for 3D optical wiring backbones is another issue. As mentioned in section 2.3, these problems are expected to be solved by self-organized 3D optical wiring of SOLNETs, the details of which are described in Chapter 7.

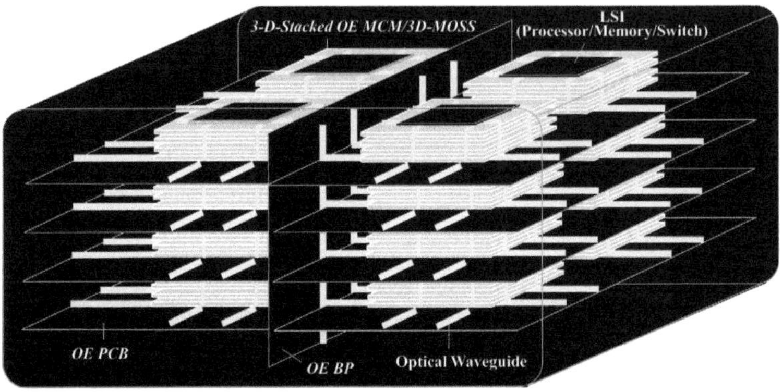

Figure 3.3 Expected structure within boxes of high-performance computers, in which 3D integrated optical interconnects are implemented.

3.2.2 Structures within Boxes of Computers

Figure 3.3 shows an expected structure within boxes of high-performance computers, in which 3D integrated optical interconnects are implemented. LSIs for processors, memories, and switches are involved in 3D stacked structures. E-O and O-E signal conversions are respectively performed at the output and input terminals of the LSIs. Optical signals are transmitted from LSI outputs, through optical waveguides in 3D-stacked OE MCMs, OE PCBs, and OE BPs to LSI inputs. 3D-MOSSs are distributed in the boxes for switching optical signals directly without the E-O/O-E signal conversion.

For extremely high-speed machine cycles, the signal propagation lengths during one clock period get close to the LSI chip size. In this case, to reduce the signal propagation delay, 3D-stacked OE MCMs are especially preferable because stacked LSI chips are connected through short vertical waveguides in 3D-stacked structures.

3.3 Various OE Structures Built by S-FOLMs

Figure 3.4 shows conceptual illustrations of the Film/Z-connection technology. It was originally developed for stacking film-type electrical substrates at Fujitsu Computer Packaging Technologies

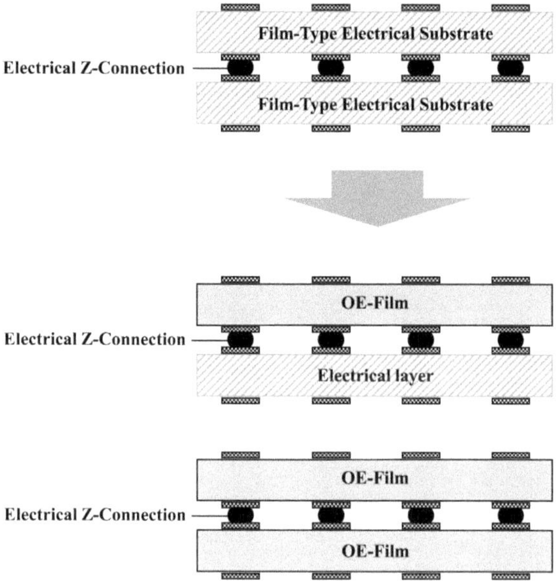

Figure 3.4 Conceptual illustration of the Film/Z-connection technology.

Inc. (FCPT), San Jose, California [20]. First, fine metal lines and pads are deposited on both sides of films, and vias are formed in the films to prepare film-type electrical substrates (in the figure, metal lines and vias are not drawn). Then, the film-type electrical substrates are stacked by Z-connections to construct multilayer electrical substrates.

By replacing the film-type electrical substrates with OE-Films, as shown in Fig. 3.4, multilayer OE substrates, in which electrical and optical interconnects coexist, are constructed. This brings us the concept of the S-FOLM described in section 3.1. In comparison with the conventional build-up technology, the Film/Z-connection technology has the advantage that it can construct multilayer OE substrates with large freedoms in substrate structures and fabrication processes. This advantage spreads the substrate design freedom, enabling us to introduce a lot of functions into multilayer OE substrates by combining various types of OE-Films. Each type of OE-Film can be optimized independently of other types of OE-Films. Thus, versatility improvement is expected.

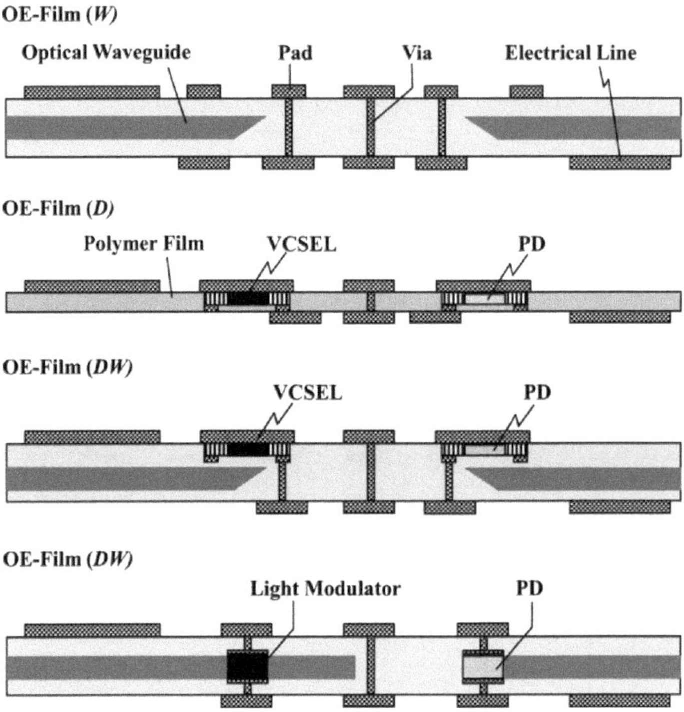

Figure 3.5 Examples of detailed structures of OE-Films.

Figure 3.5 shows examples of detailed structures of OE-Films.

- OE-Film (*W*): Optical waveguide film with metallization
- OE-Film (*D*): Small/thin OE dies are embedded in a polymer/glass film with metallization
- OE-Film (*DW*): Small/thin OE dies are embedded in an optical waveguide film with metallization

It should be noted that the OE-Films contain metal lines, pads, and vias, meaning that OE-Films also have electrical substrate functions. By combining these OE-Films with film-type electrical substrates, all levels of OE substrates for optical interconnects can be provided according to system requirements, as mentioned in section 3.1. Various types of OE structures built by S-FOLMs are presented below.

3.3.1 OE-Film/Electrical Substrate Stack

Figure 3.6a shows an example of an OE PCB and an OE BP which are built from S-FOLMs using the Film/Z-connection technology. The OE PCB is constructed by stacking an OE-Film (*DW*) on an electrical substrate (*S*). Thin/small OE dies and optical waveguides

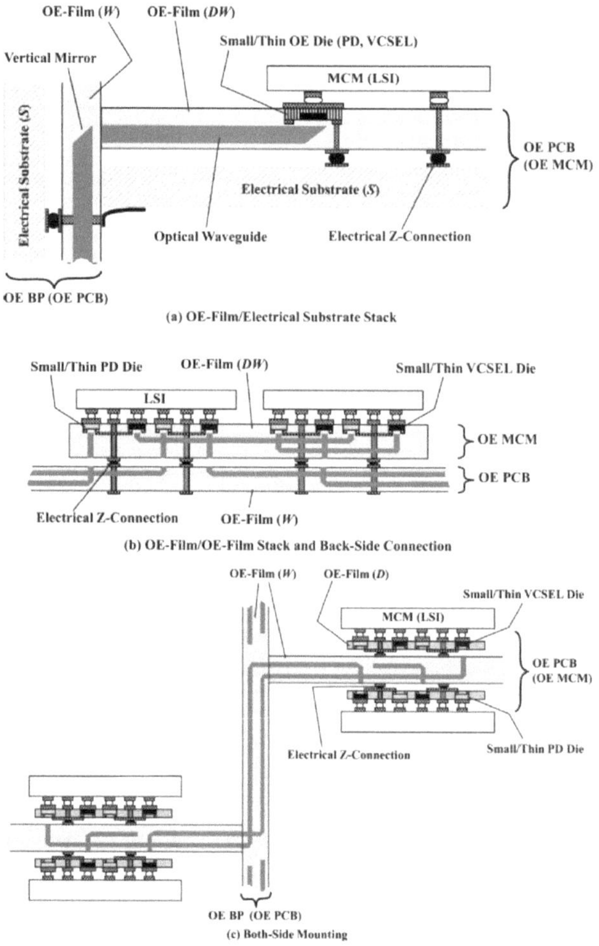

Figure 3.6 Examples of OE substrate structures built from the S-FOLM using the Film/Z-connection technology. Reprinted with permission from Yoshimura et al. [6].

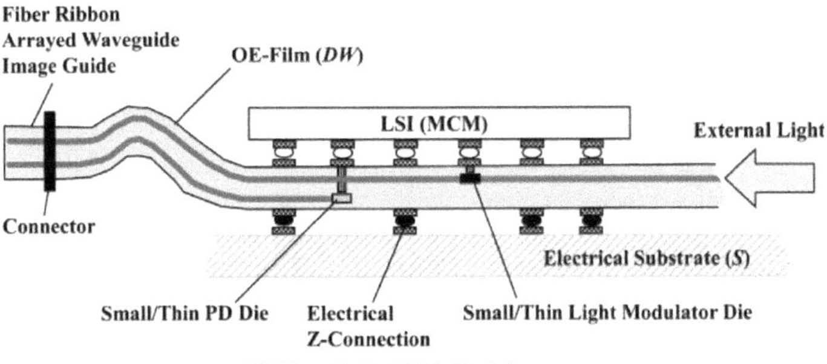

(d) Micro Optical Link Module

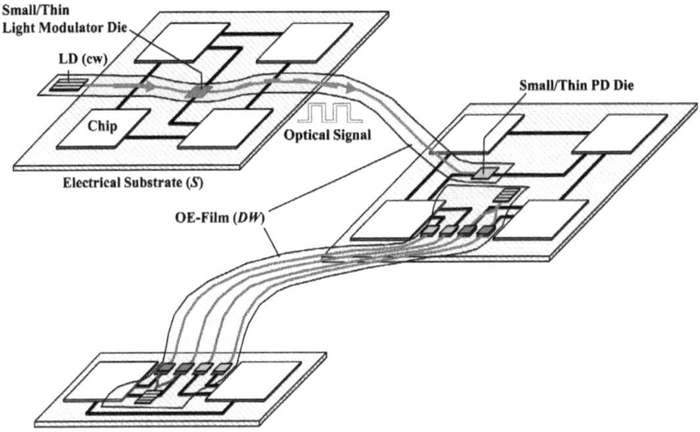

(e) OE Tap Guide

Figure 3.6 (*Continued*)

are coupled by vertical mirrors in optical waveguides. The OE BP is constructed by stacking an OE-Film (*W*) on an electrical substrate (*S*). The OE BP and OE PCB are coupled by vertical mirrors in optical waveguides. Z-connections have two roles, namely, electrical connections and mechanical supports between the OE-Films and the electrical substrates. The gaps between the films and the substrates are filled by dielectric materials, if necessary. In this model, the OE MCM, the OE PCB and the OE BP can be an LSI, an OE MCM and an OE PCB, respectively. The OE-Films and the electrical substrates are fabricated separately. Therefore, materials and processes for OE-Films can independently be optimized, not being restricted by the electrical substrate condition.

3.3.2 OE-Film/OE-Film Stack and Backside Connection

Figure 3.6b shows an example of an OE-Film/OE-Film stack for an OE MCM/PCB structure. The OE MCM is an OE-Film (*DW*), and the OE PCB is an OE-Film (*W*). They are coupled by vertical mirrors. By electrical output signals from LSIs, VCSELs in the OE MCM is driven. The emitted light beams from the VCSELs are coupled to optical waveguides in the OE MCM to transmit optical signals to PDs in the same OE MCM. PDs convert optical signals to electrical input signals into LSIs. When the optical signal transmission is performed between outputs and inputs of different LSIs, inter-LSI optical interconnects are achieved. When the optical signal transmission is performed between outputs and inputs of the same LSI, intra-LSI optical interconnects are achieved.

In some cases the optical signals go out to the OE PCB through the backside connection to be guided to other OE MCMs or other OE units. Only OE-Film-based technology can realize such backside connections.

3.3.3 Both-Side Mounting

Figure 3.6c shows an example of both-side mounting. OE-Films (*D*) are stacked on both sides of an OE-Film (*W*) to construct a both-side OE PCB. OE PCBs on the opposite sides can also be connected by a both-side OE BP consisting of an OE-Film (*W*). In this model, the OE PCB and the OE BP can be an OE MCM and an OE PCB, respectively. The both-side mounting enables a butterfly structure, which is important in multiprocessor systems. The both-side mounting availability is a significant advantage of the Film/Z-connection approach because it is difficult in OE substrates constructed by the conventional build-up technology.

Smart pixels, which consist of VCSEL/PD-array chips bonded onto CMOS ICs, have received much attention. OE-Films (*D*) play the same role as VCSEL/PD-array chips. OE-Films (*D*) have the following advantages over VCSEL/PD-array chips, that is, semiconductor materials exist only at the places where they are necessary in OE-Films (*D*). The other area is occupied by polymer or glass films. This is cost-effective from a viewpoint of expensive semiconductor material consumption. Furthermore, processability for via/pad/metal line formation is very high. This indicates that OE-Films (*D*) can contribute to realizing low-cost smart pixels.

3.3.4 Micro Optical Link Module

Figure 3.6d shows an example of micro optical link modules that enable direct jump from LSI chips. In this example, an OE-Film (*DW*) is stacked on an electrical substrate (*S*). Light modulators act as transmitters. External light (cw or pulse train) is modulated by outputs of LSIs through light modulators to generate optical signals. If very high-power external light is applied, the high-speed and low-heat-generation OE-ADLES (see section 3.4) is available. Instead of the OE-Film (*DW*), a combination of an OE-Film (*D*) and an OE-Film (*W*) can also be used for the module. By a micro optical link module, optical signals can directly be extracted from LSIs and transmitted to any levels of OE units, including to LSI, to board, to shelf, and to fiber. LSIs can be replaced with MCMs. The micro optical link module is suitable for optical co-packaging.

3.3.5 OE Tap Guide

Figure 3.6e shows examples of OE tap guides using the OE-ADLES. An OE-Film (*DW*) is attached to electrodes/pads/lines on an electrical substrate to connect the electrodes/pads/lines to small/thin light modulator dies embedded in the OE-Film (*DW*). Cw or pulse train light beams from LDs are modulated by the electrical signals from electrodes/pads/lines through light modulators to generate optical signals. The optical signals are transmitted through optical waveguides to small/thin PD dies and are converted to electrical signals to be introduced into the electrodes/pads/lines of other places. By OE tap guides, optical signals can directly be extracted from electrical substrates and brought to any levels of OE units, and to any places in the same substrate or other substrates. The OE tap guide is also suitable for optical co-packaging.

3.3.6 WDM Transceiver and WDM Inter-PCB Connect

Figure 3.7 shows examples of OE substrate structures built from S-FOLMs for WDM optical interconnects. The WDM transceiver illustrated in Fig. 3.7a can be used in massive parallel optical links and OE BPs. An OE-Film (*DW*), in which small/thin VCSEL/PD dies for wavelength λ_1 and λ_2 are embedded together with optical waveguides, is attached onto an electrical substrate (*S*) by electrical Z-connections. On the OE-Film (*DW*), interface films containing thinned interface ICs with transceiver circuits for the

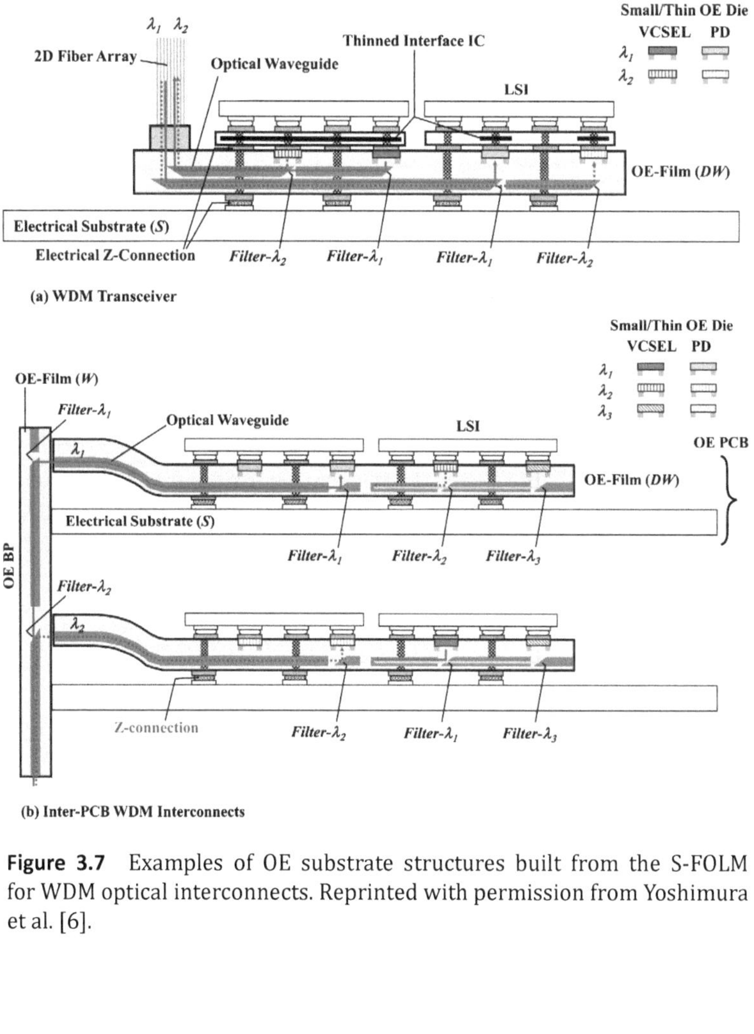

Figure 3.7 Examples of OE substrate structures built from the S-FOLM for WDM optical interconnects. Reprinted with permission from Yoshimura et al. [6].

VCSEL/PD are attached. The optical waveguides in the OE-Film are connected to a 2D optical fiber array (or image guide). Power/GND and electrical signals are supplied from the electrical substrate through the electrical vias/pads/electrodes of the OE-Film. Surface-normal couplers with wavelength-filter functions, which enable wavelength MUX and DEMUX, carry out optical couplings between small/thin OE dies and optical waveguides. For single wavelength use, surface-normal couplers can be vertical mirrors.

Interface films give standardized-interface characteristics to transceivers. Namely, by embedding appropriate thinned interface ICs into films, the S-FOLM I/O specs can freely be adjusted to the I/O specs of LSIs to be mounted on S-FOLMs. The interface films can involve many functions such as MUX/DEMUX, error correction, timing controlling, and so on. Thinned interface ICs have through vias or are divided into small/thin dies for individual pad connections.

Another remarkable feature of the WDM transceiver utilizing S-FOLMs is flexibility in structures and positions. S-FOLMs can be attached at any places with any shapes/sizes on a substrate.

Figure 3.7b shows an example of S-FOLMs for inter-PCB WDM interconnects. An OE-Film (*DW*), in which small/thin VCSEL/PD dies for wavelength λ_1, λ_2, and λ_3 are embedded, is attached onto an electrical substrate (*S*) by electrical Z-connections to form an OE PCB (for simplicity, interface films are neglected in the figure). Surface-normal couplers of wavelength filters in an OE BP optically connect the OE BP to the S-FOLM of the OE PCB. The couplers act as wavelength MUX/DEMUX.

3.3.7 3D Optical Circuits for WDM

Figure 3.8 shows an example of 3D optical circuits constructed by stacking optical waveguide films. Optical waveguide films are inserted into the gaps in Z-connection units, in which vertical waveguides are formed.

Figure 3.9 shows examples of the 3D structures for MUX and DEMUX in Z-connection units. Vertical wavelength filters are

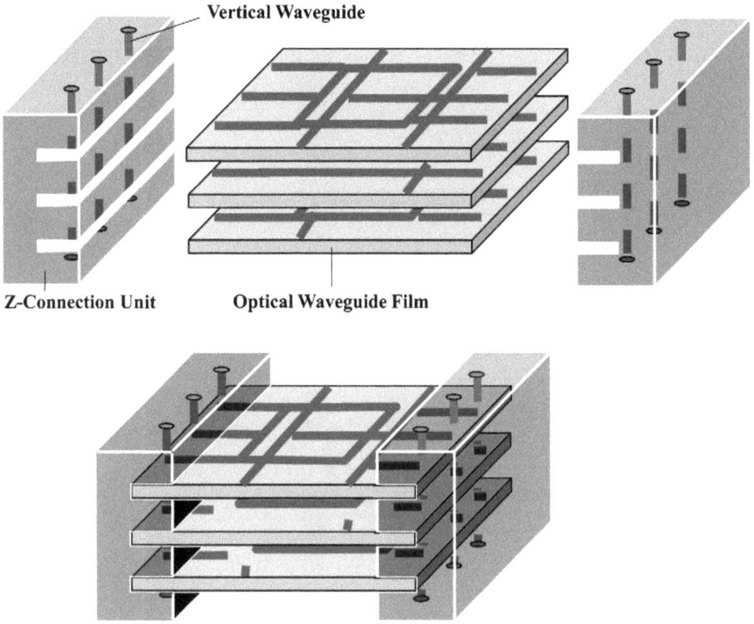

Figure 3.8 3D optical circuits constructed by Z-connection units for WDM optical interconnects.

embedded in optical waveguide films. Figure 3.10 shows examples of signal flows in WDM optical interconnects with Z-connection units. At cross-points, where wavelength filters are formed in the optical waveguide films, MUX and DEMUX are performed to achieve wavelength-dependent designated routings.

By embedding semitransparent mirrors or gratings, 1:N fan-out or N:1 coupling can be achieved. In the case of through, holes are placed instead of wavelength filters, mirrors, and gratings.

Z-connection units can be fabricated by constructing vertical waveguides using the build-up process like cladding/core (patterning)/cladding/. . ., followed by forming gaps by, for example, the laser cut or the dicing saw cut. As Fig. 3.11 shows, Z-connection units can also be fabricated using SOLNETs. In this case, for the vertical waveguide materials, photo-induced refractive-index increase (PRI) materials are used. By exposing the

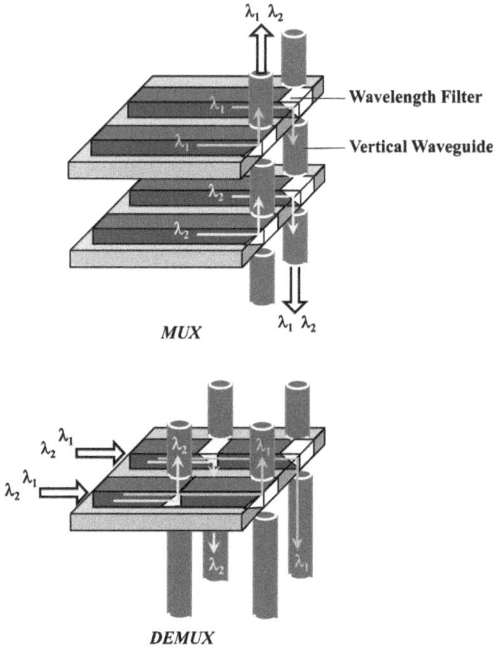

Figure 3.9 MUX and DEMUX in the Z-connection units.

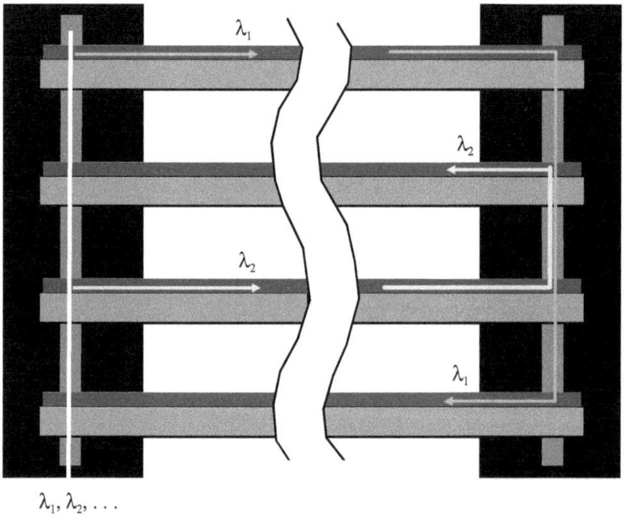

Figure 3.10 Examples of signal flows in WDM optical interconnects with Z-connection units.

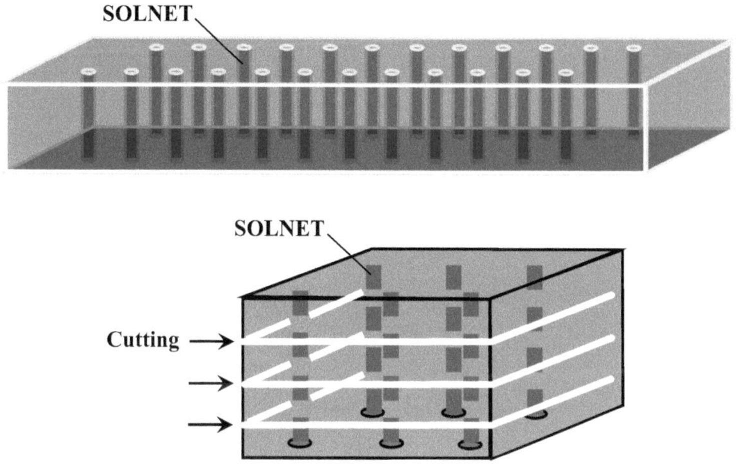

Figure 3.11 Fabrication of Z-connection units using SOLNETs.

PRI material to write beams coming through a mask with a hole array pattern, due to the self-organization, a vertical waveguide array is automatically constructed along the light-propagating direction toward a luminescent target array or mirror array formed at the bottom part (see Chapter 7).

3.4 Optoelectronic Amplifier/Driver-Less Substrate

3.4.1 Concept of OE-ADLES

Although optical interconnects have the potential to overcome the problems in electrical interconnects as mentioned in section 2.1, there is a large barrier which prevents optics from getting into boxes of computers. It is ascribed to the *optics excess* explained in Chapters 1 and 2.

In Fig. 3.12, the electrical interconnects, conventional optical interconnects, and integrated optical interconnects based on the OE-ADLES are compared for an inter-LSI interconnection. In conventional optical interconnects using bulk-chip-based OE modules, chips of VCSELs, PDs, VCSEL drivers, and PD amplifiers, which

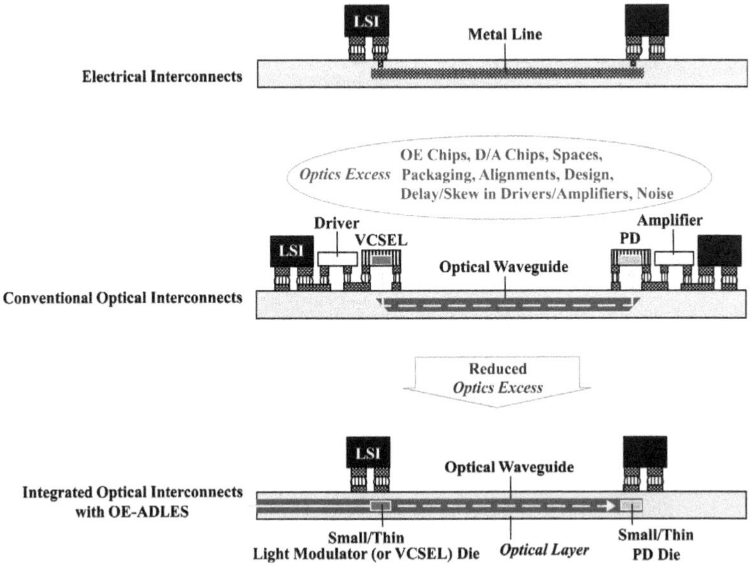

Figure 3.12 Comparison of electrical interconnects, conventional optical interconnects, and integrated optical interconnects with the OE-ADLES.

are unnecessary in electrical interconnects, are required as well as spaces for them. Packaging, precise OE device alignments, and design considering optics and electronics simultaneously are also required to fabricate OE modules. Signal delay/skew in drivers and amplifiers for the E-O/O-E signal conversion is caused, and the noise generation due to relatively long metal lines for LSI-driver-VCSEL and PD-amplifier-LSI connections is increased.

In integrated optical interconnects with the OE-ADLES [1, 2, 18, 19], the *optics excess* is reduced as shown in Fig. 3.12. The E-O/O-E signal conversion functions achieved by small/thin light modulator and PD dies are completely buried in an optical layer by integrating dies together with optical waveguides. Optical linking is performed by directly connecting LSI outputs to the electrodes of light modulators and LSI inputs to the electrodes of PDs. Thus, a simple interconnect configuration just like electrical interconnects can be obtained. In some cases, VCSELs with high quantum efficiency can be used instead of light modulators.

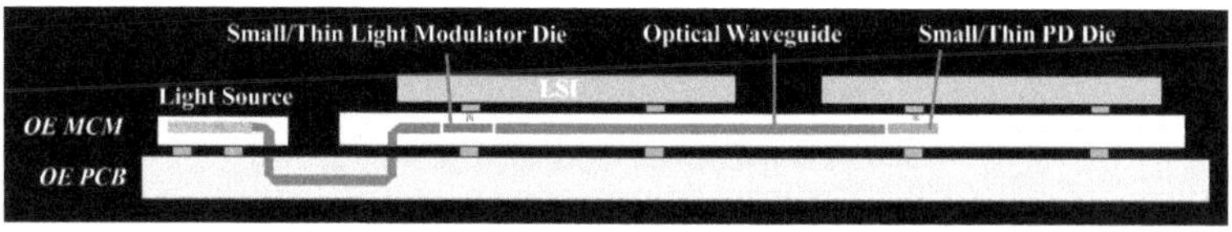

Figure 3.13 OE-ADLES for OE MCM. Reprinted with permission from Yoshimura et al. [18].

Figure 3.13 shows an OE-ADLES for OE MCMs. Strong light beams from external light sources are directly modulated by LSI outputs through small/thin light modulator dies without drivers. The generated optical signals are transmitted through optical waveguides to small/thin PD dies, where the optical signals are converted back to electrical signals for LSI inputs without amplifiers. The OE-ADLES is expected to have low power dissipation and high data rate due to the following reasons.

- Light sources are put outside, reducing power dissipation within the interconnect units.
- Light modulators are driven directly by LSI output electrical signals, reducing the driving power and the rising/falling time for optical signal generation.
- Strong light beams are available from external light sources, typically, high-power LD and mode-locked LD, achieving fast PD response.

3.4.2 Power Dissipation and *RC* Delay in OE-ADLES

In order to estimate power dissipation and *RC* delay in the OE-ADLES, models shown in Fig. 3.14 are considered. For electrical substrates, data transmission is treated as charge/discharge processes of capacitance in a buffer driver, a transmission line, and a receiver through gate resistance of the buffer driver. For OE-ADLESs, data transmission is treated as charge/discharge processes of capacitance in a buffer driver and a light modulator through gate resistance of the buffer driver, and charge/discharge processes of capacitance in a PD and a receiver by currents generated in the PD.

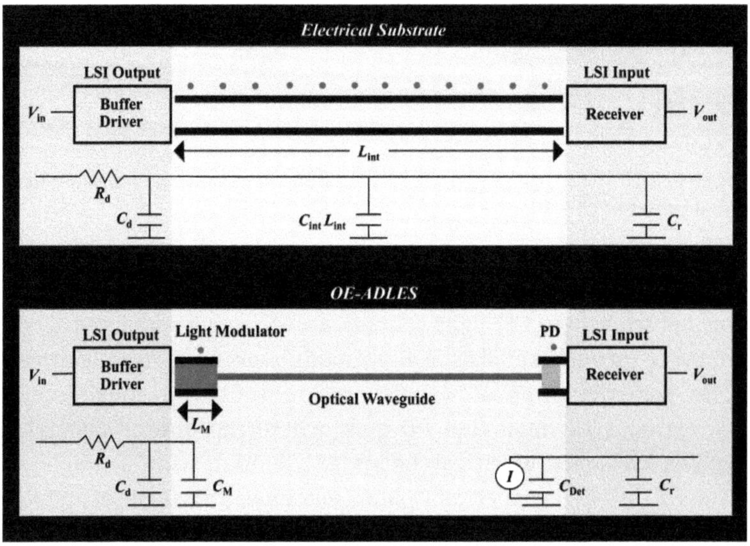

Figure 3.14 Models for electrical substrates and OE-ADLESs. Reprinted with permission from Yoshimura et al. [18].

In Table 3.1, expressions for *RC* delay and power dissipation are summarized as well as parameters for the calculations. Attenuation along transmission lines is neglected. External power dissipation at external light sources is not included. The WPD optical switch described in Chapter 6 is assumed as the light modulator for the transmitter.

In Fig. 3.15, the *RC* delay and power dissipation of an electrical substrate and an OE-ADLES are compared. Lead lanthanum zirconate titanate (PLZT) and polymer MQD, described in Chapter 6, are assumed as EO materials for WPD optical switches. Here, the polymer MQD is an artificial material consisting of polymer wires with MQDs, whose EO coefficient was calculated by using the molecular orbital method. The overview is described in Chapter 12.

For *RC* delay, the OE-ADLES with the polymer MQD has an advantage over the electrical substrate in a line length range longer than 12 mm for light beam power of 3 mW. By increasing the light beam power, the response time becomes faster, moving the break-even point toward shorter line length of a few mm ranges.

Table 3.1 Expressions for RC delay and power dissipation, and parameters for calculations [18]

Electrical substrate	RC delay	$\tau = R_d(C_{int}L_{int} + C_d + C_r)$
	Power dissipation	$P = f(C_{int}L_{int} + C_d + C_r)V_{out}^2$ $\quad C_{int} = \varepsilon_0\varepsilon_i w/d$
OE-ADLES	RC delay	$\tau = R_d(C_M + C_d) + V_{out}(C_{Det} + C_r)/(\eta_c P_{Det}) + \tau_P$
	Power dissipation	$P = f[(C_M + C_d)V_{in}^2 + (C_{Det} + C_r)V_{out}^2]$
		$C_M = \varepsilon_0\varepsilon_r A_M/d$

		Electrical substrate	OE-ADLES	
			PLZT	Polymer MQDs (Calc.)
Dielectric constant	ε_r	3	300	4
Refractive index	n		2.48	2
EO coefficient	R, r		$R = 5.7 \times 10^{-16}$ m^2/V^2	$r = 1.35 \times 10^{-9}$ m/V

f	Data rate	d	Gap between electrodes: 1 μm
V_{in}	Power supply voltage: 3 V	A_M	Light modulator area: 16 × 16 μm^2
V_{out}	Received detector voltage: 3 V	C_M	Light modulator capacitance
C_{int}	Transmission line capacitance per cm	C_{Det}	Photodetector capacitance: 0.01 pF
R_d	Driver gate resistance: 50 Ω	η_c	Responsibility of photodetector: 0.8 A/W
C_d	Driver capacitance: 0.02 pF	τ_p	Photodetector internal propagation delay: 25 ps
C_r	Receiver capacitance: 0.02 pF	P_{Det}	Incident power to photodetector
L_{int}	Transmission line length	λ	Wavelength: 1.3 μm
w	Electrode width: 4 μm		

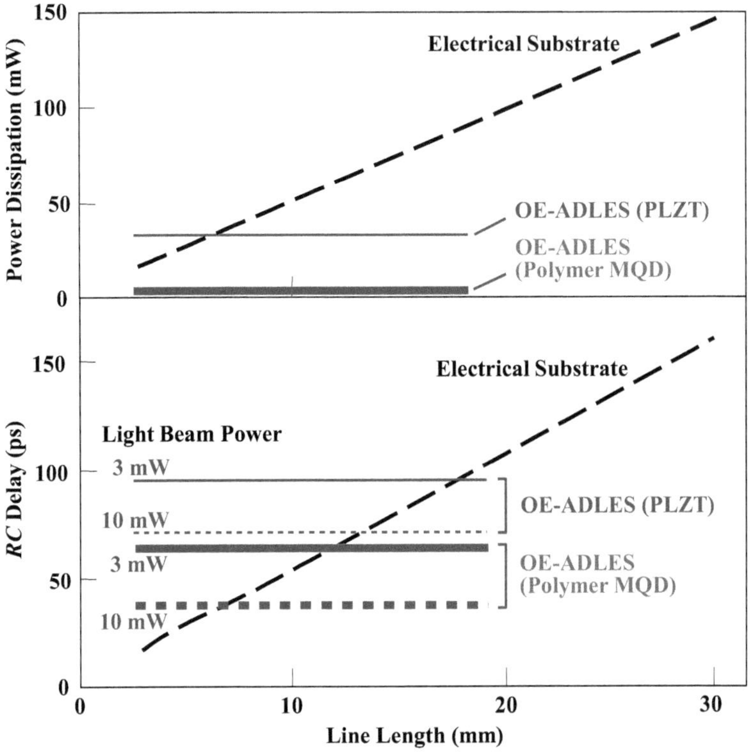

Figure 3.15 Results of calculations for delay and power dissipation at 5 Gbps in an electrical substrate and an OE-ADLES. Reprinted with permission from Yoshimura et al. [18].

For power dissipation, drastic reduction can be seen in the OE-ADLES with the polymer MQD. The break-even point is located at line length of sub-mm ranges. The break-even point in the OE-ADLES with PLZT is around 6 mm. In the OE-ADLES with PLZT, compared to that with the polymer MQD, the break-even point is located at longer line length regions for both *RC* delay and power dissipation. This comes from the large dielectric constant of PLZT.

The superiority of the OE-ADLES in the calculated *RC* delay and power dissipation is attributed to the capacitance reduction. That is to say, in the electrical substrate a large amount of charge is required to charge up the electrical transmission line due to its large line capacitance, while in the OE-ADLES only a small amount of charge is required to charge up the electrodes of the light modulator and PD having a small capacitance.

A key issue in the OE-ADLES is EO material availability with a large EO coefficient and a small dielectric constant. The polymer MQD seems promising for the EO material, and is expected to be grown by molecular layer deposition (MLD), as mentioned in Chapter 12, although the experimental demonstration is a future challenge. Size reduction of light modulators and PDs is another important issue. When they are miniaturized by one order of magnitude for each dimension, that is, the device thickness is reduced by one order of magnitude to 100 nm and the device area by two orders of magnitude to $1.6 \times 1.6\ \mu m^2$, their capacitance as well as the PD internal propagation delay will be reduced by one order of magnitude in principle. This will move the break-even points to shorter line length regions for both *RC* delay and power dissipation.

References

1. T. Yoshimura, J. Roman, Y. Takahashi, S. Beilin, W. Wang, and M. Inao, *4th International Conference on Organic Nonlinear Optics (ICONO'4), Extended Abstracts*, Chitose, Japan, 278 (1998).
2. T. Yoshimura, J. Roman, Y. Takahashi, S. I. Beilin, W. V. Wang, and M. Inao, "Optoelectronic amplifier/driver-less substrate <OE-ADLES> for polymer-waveguide-based board-level interconnection-calculation of delay and power dissipation-," *Nonlinear Opt.* **22**, 453–456 (1999).
3. T. Yoshimura, J. Roman, Y. Takahashi, M. Lee, B. Chou, S. I. Beilin, W. V. Wang, and M. Inao, "Proposal of optoelectronic substrate with Film/Z-connection based on OE-film," *Proc. 3rd IEMT/IMC Symposium*, 140–145 (1999).
4. T. Yoshimura, J. Roman, P. Reilly, Y. Takahashi, M. Lee, B. Chou, S. I. Beilin, W. V. Wang, and M. Inao, "Proposal of film optical link module (FOLM) and beveled cut process for the high density waveguide couplers," *Proc. 1999 IEEE LEOS Annual Meeting Conference*, San Francisco, California, 27–28 (1999).
5. T. Yoshimura, J. Roman, Y. Takahashi, W. V. Wang, M. Inao, T. Ishitsuka, K. Tsukamoto, K. Motoyoshi, and W. Sotoyama, "Self-organizing waveguide

coupling method "SOLNET" and its application to film optical circuit substrates," *Proc. 50th Electronic Components & Technology Conference (ECTC)*, 962–969 (2000).

6. T. Yoshimura, J. Roman, Y. Takahashi, M. Lee, B. Chou, S. I. Beilin, W. V. Wang, and M. Inao, "Optoelectronic scalable substrates based on Film/Z-connection and its application to film optical link module (FOLM)," *Proc. SPIE* **3952**, from Photonics West 2000, San Jose, California, 202–213 (2000).
7. T. Yoshimura, J. Roman, Y. Takahashi, W. V. Wang, M. Inao, T. Ishituka, K. Tsukamoto, K. Motoyoshi, and W. Sotoyama, "Self-organizing lightwave network (SOLNET) and its application to film optical circuit substrates," *IEEE Trans. Compon. Packag. Technol.* **24**, 500–509 (2001).
8. T. Yoshimura, Y. Takahashi, M. Inao, M. Lee, W. Chou, S. Beilin, W.-C. Wang, J. Roman, and T. Massingill, "Systems based on opto-electronic substrates with electrical and optical interconnections and methods for making," U.S. Patent 6,343,171 B1 (2002).
9. T. Yoshimura, Y. Takahashi, M. Inao, M. Lee, W. Chou, S. Beilin, W.-C. Wang, J. Roman, and T. Massingill, "Optoelectronic substrates with electrical and optical interconnections and method for making," U.S. Patent 6,611,635 B1 (2003).
10. T. Yoshimura, M. Ojima, Y. Arai, and K. Asama, "Three-dimensional self-organized microoptoelectronic systems for board-level reconfigurable optical interconnects-performance modeling and simulation," *IEEE J. Sel. Top. Quantum Electron.* **9**, 492–511 (2003).
11. T. Yoshimura, Y. Takahashi, M. Inao, M. Lee, W. Chou, S. Beilin, W.-C. Wang, J. Roman, and T. Massingill, "Three-dimensional opto-electronic modules with electrical and optical interconnections and methods for making," U.S. Patent 6,690,845 B1 (2004).
12. T. Yoshimura, Y. Takahashi, J. Roman, M. T. McCormack, S. Beilin, W.-C. Wang, and M. Inao, "Optical coupling structures and the fabrication processes," U.S. Patent 6,684,007 B2 (2004).
13. T. Yoshimura, Y. Takahashi, K. Kuwabara, S. Beilin, M. Peters, W.-C. Wang, and M. Inao, "Optical reflective structures and method for making," U.S. Patent 6,706,546 B2 (2004).
14. T. Yoshimura, Y. Takahashi, J. Roman, M. T. McCormack, S. Beilin, W.-C. Wang, and M. Inao, "Single and multilayer waveguides and fabrication process," U.S. Patent 6,785,447 B2 (2004).
15. T. Yoshimura, T. Inoguchi, T. Yamamoto, S. Moriya, Y. Teramoto, Y. Arai, T. Namiki, and K. Asama, "Self-organized lightwave network based on

waveguide films for three-dimensional optical wiring within boxes," *J. Lightwave Technol.* **22**, 2091–2100 (2004).

16. T. Yoshimura, Y. Takahashi, M. Inao, M. Lee, W. Chou, S. Beilin, W.-C. Wang, J. Roman, and T. Massingill, "Multi-layer opto-electronic substrates with electrical and optical interconnections and methods for making," U.S. Patent 6,845,184 B1 (2005).
17. T. Yoshimura, M. Miyazaki, Y. Miyamoto, N. Shimoda, A. Hori, and K. Asama, "Three-dimensional optical circuits consisting of waveguide films and optical Z-connections," *J. Lightwave Technol.* **24**, 4345–4352 (2006).
18. T. Yoshimura, Y. Suzuki, N. Shimoda, T. Kofudo, K. Okada, Y. Arai, and K. Asama, "Three-dimensional chip-scale optical interconnects and switches with self-organized wiring based on device-embedded waveguide films and molecular nanotechnologies," *Proc. SPIE* **6126**, from Photonics West 2006, San Jose, California, 612609-1-15 (2006).
19. T. Yoshimura, *Optical Electronics: Self-Organized Integration and Applications*, Pan Stanford Publishing, Singapore (2012).
20. B. Chou, S. Beilin, H. Jiang, D. Kudzuma, M. Lee, M. McCormack, T. Massingill, M. Peters, J. Roman, Y. Takahashi, and W. V. Wang, "Multilayer high density flex technology," *Proc. 49th Electronic Components & Technology Conference (ECTC)*, 1181–1189 (1999).
21. T. Yoshimura, K. Kumai, T. Mikawa, O. Ibaragi, and O. Bonkohara, "Photolithographic packaging with selectively occupied repeated transfer (PL-Pack with SORT) for scalable film optical link multichip-module (S-FOLM)," *IEEE Trans. Electron. Packag. Manuf.* **25**, 19–25 (2002).
22. T. Yoshimura, J. Roman, W.-C. Wang, M. Inao, and M. T. McCormack, "Device transfer method," U.S. Patent 6,669,801 B2 (2003).

Chapter 4

Optical Waveguide Films with Vertical Mirrors and 3D Optical Circuits

The optical waveguide film with vertical mirrors is a basic element of the S-FOLM and is one of the core technologies in self-organized 3D integrated optical interconnects. Vertical mirrors in the optical waveguide film enable optical Z-connections in the 3D OE platforms consisting of stacked waveguide films. They are also necessary for couplings between optical waveguides and OE devices such as VCSELs and PDs in optical waveguide films.

There are a variety of fabrication processes of optical waveguide films, such as photolithography, laser ablation, and stamping. We developed a process called the built-in mask method, which enables duplication of optical waveguide films with vertical mirrors with simplified procedures.

In the present chapter, the built-in mask method is explained, and then, fabrication and characterization of optical waveguide films with vertical mirrors duplicated by the built-in mask method are reviewed. 3D optical circuits fabricated by stacking optical waveguide films are presented. OE board structures fabricated by stacking an optical waveguide film on an electrical substrate are also presented. In addition, fabrication of nanoscale waveguides required

Self-Organized 3D Integrated Optical Interconnects with All-Photolithographic Heterogeneous Integration
Tetsuzo Yoshimura

ISBN 978-981-4877-04-6 (Hardcover), 978-1-003-04305-8 (eBook)
www.jennystanford.com

in intra-LSI optical interconnects is demonstrated in a PRI sol–gel material.

4.1 Built-In Mask Method

Figures 4.1 and 4.2 respectively show schematic illustrations and flows for a duplication process of optical waveguide films with vertical mirrors using the built-in mask method [1, 2]. On a glass substrate, a reflective film, typically a metal film, is deposited with window patterns for waveguide cores to make a built-in mask. After a removable layer and an undercladding layer are successively formed on the mask, a waveguide core layer of a photo-definable material is coated. The core layer is selectively exposed to tilted ultraviolet (UV) light through a prism put on the back of the built-in mask. By this process, cured structures of beveled walls with designated angle slopes are constructed at waveguide core ends, providing waveguide core end facets for vertical mirrors. Vertical UV light exposure is carried out with a mask that covers the beveled wall

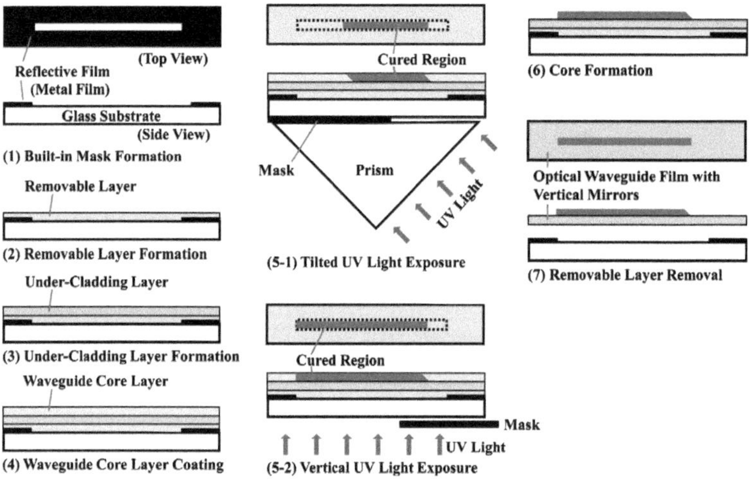

Figure 4.1 Duplication process of optical waveguide films with vertical mirrors: built-in mask method. Reprinted with permission from Yoshimura et al. [3].

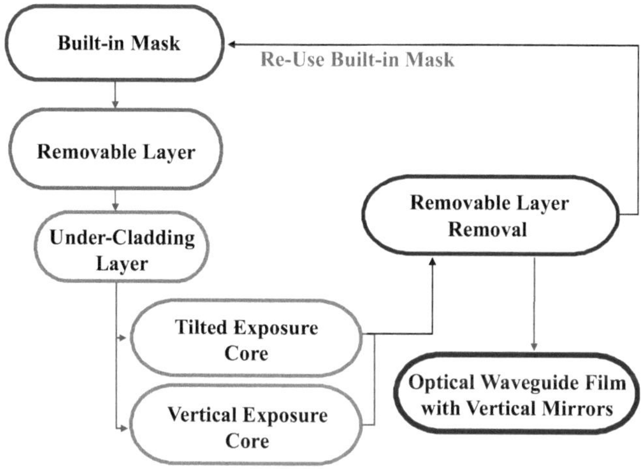

Figure 4.2 Process flow of the built-in mask method.

regions to construct planar waveguide cores. By developing, core formation is completed. Finally, the removable layer is removed to obtain an optical waveguide film with vertical mirrors. The built-in masks are reused for the next duplication of optical waveguide films.

4.2 Fabrication of Optical Waveguides and Vertical Mirrors

Fabrication of optical waveguides and vertical mirrors by utilizing the built-in mask method has been demonstrated [1–4].

Figure 4.3 shows CAD patterns for built-in mask formation and a whole view of a built-in mask. The mask size is $5 \times 5\ \text{cm}^2$. Various types of waveguide core patterns with widths ranging from 4 to 64 μm are included. For the reflective film of the built-in mask, Al films were used. The removable layer was 10 μm thick poly-vinyl-alcohol (PVA).

For the undercladding layer and waveguide cores, acrylic photo-definable materials (OPSTAR, JSR Corporation) were used [5]. For curing them, UV light from a high-pressure mercury lamp was used.

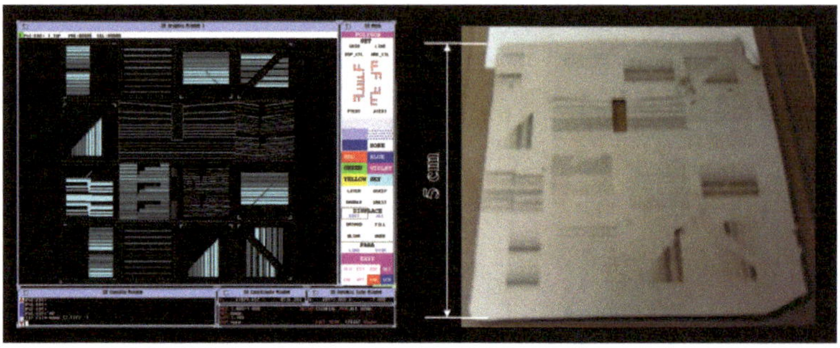

Figure 4.3 CAD patterns and a whole view of a built-in mask. Reprinted with permission from Yoshimura et al. [2].

The tilted and vertical UV light exposure was carried out by setting the built-in mask coated with a waveguide core layer on the apparatus shown in Fig. 4.4 to construct cores and vertical mirrors. The required energy density was ~1 J/cm^2. The refractive index of the undercladding layer was 1.56 and that of the core was 1.58. The removable layer was solved in water.

(a) Vertical Exposure (b) Tilted Exposure

Figure 4.4 Exposure apparatus for waveguide core and vertical mirror construction. The apparatus was made by Takashi Yamamoto.

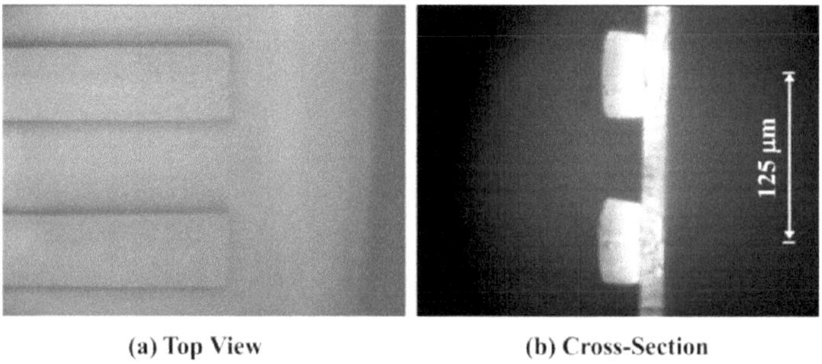

Figure 4.5 Top view and cross-section of cores in an optical waveguide film fabricated by vertical UV light exposure. Nominal core width is 64 μm. Reprinted with permission from Yoshimura et al. [2].

4.2.1 Waveguide Cores

Figure 4.5 shows the top view and cross-section of cores in an optical waveguide film obtained by vertical UV light exposure. The nominal core width and height are 64 μm and 32 μm, respectively. Sharp core patterns are formed, indicating that patterns of the built-in mask are appropriately duplicated into the core layer. Figure 4.6 shows near-field patterns (NFPs) of guided beams of wavelength 1.3 μm in a 64 μm × 32 μm core and a 32 μm × 32 μm core in optical waveguide films. Confinement of the guided beams into the core regions is

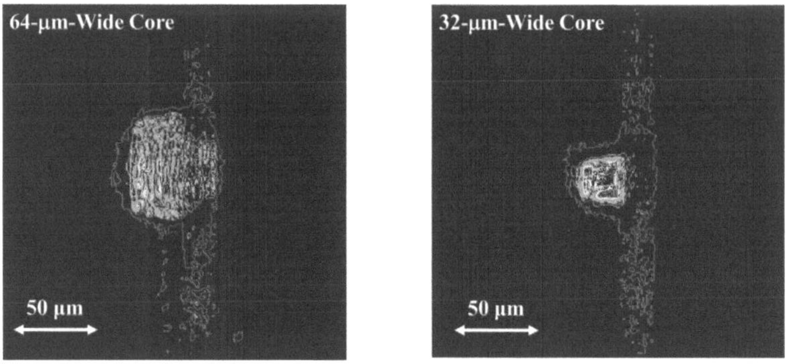

Figure 4.6 NFPs of guided beams of 1.3 μm in wavelength in a 64 μm × 32 μm core and a 32 μm × 32 μm core in optical waveguide films.

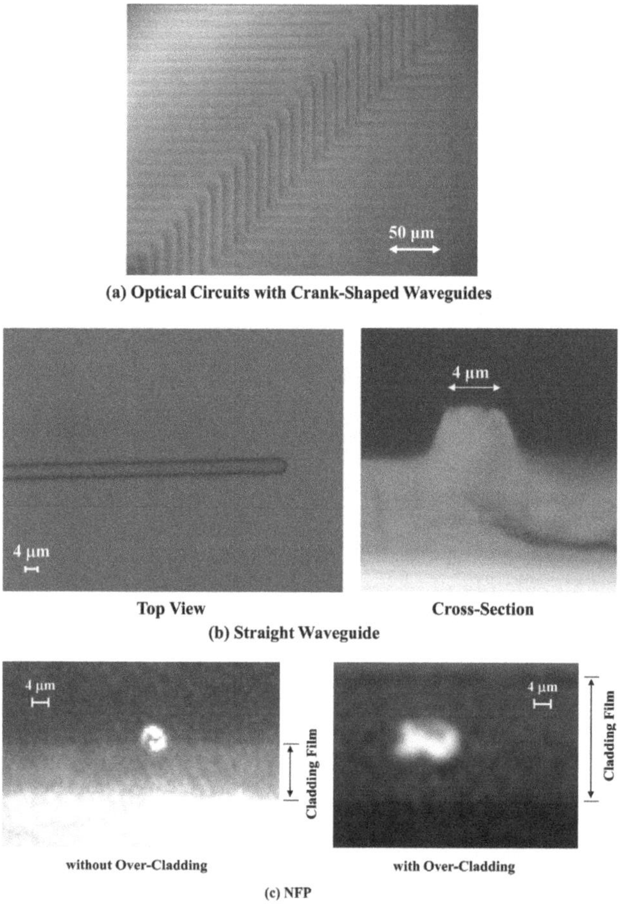

Figure 4.7 Optical waveguide films with a nominal core width of 4 μm. (a) Optical circuits consisting of crank-shaped waveguides, (b) straight waveguide, and (c) NFPs for optical waveguides without and with over-cladding. Reprinted with permission from Yoshimura et al. [6].

observed. The propagation loss measured by the cutback method is 0.8 dB/cm at 1.3 μm.

Figure 4.7 shows optical waveguide films with a nominal core width of 4 μm. As can be seen from Fig. 4.7a, 4 μm wide crank-shaped waveguides were duplicated by the built-in mask method. The top view and cross-section of a 4 μm wide straight waveguide are shown in Fig. 4.7b. Although slopes are found at core sides, a 4 μm wide core is obtained. As Fig. 4.7c shows, the width of the NFP is about 4 μm in the optical waveguide without overcladding, indicating strong confinement of the guided beams. In the optical waveguide with overcladding, the guided beam size increases due to evanescent waves, namely, the tunneling effect of photons. Air-cladding waveguides, which mean optical waveguides without overcladding, are preferable for intra-chip optical interconnects from a viewpoint of guided beam confinement. In the air-cladding waveguide shown in Fig. 4.7b, however, the bottom of the core contacts a cladding film. This causes reduction of the beam confinement strength for vertical directions. Ideal conditions for strong beam confinement can be achieved in all-air-cladding waveguides described in sections 4.6.2 and 5.5.2.

4.2.2 Vertical Mirrors

Figure 4.8 shows a side view of a duplicated optical waveguide film with a vertical mirror, and an NFP of a reflected beam of 1.3 μm in wavelength from the vertical mirror. A smooth mirror surface and a clear reflected beam are obtained. Although the UV light exposure was carried out with an incident angle of 45°, the mirror angle is deviated from 45°, making the mirror wall steeper. This is caused by the fact that we did not use a prism on the backside of the glass substrate during the tilted exposure. In other words, the refraction at the boundary between the air and the glass substrate changes the UV light direction.

Figure 4.9a shows exposure apparatus for the vertical mirror construction with a prism on the backside of the glass substrate. The top view and side view of an optical waveguide film with vertical mirrors fabricated by the apparatus are respectively shown in Figs. 4.9b and c. The far-field pattern (FFP) of a 1.3 μm reflected beam from the vertical mirror is shown in Fig. 4.9d. The nominal core width and height are 32 μm for lower three cores shown in the top view. Core end facets can be observed at the right edge of the

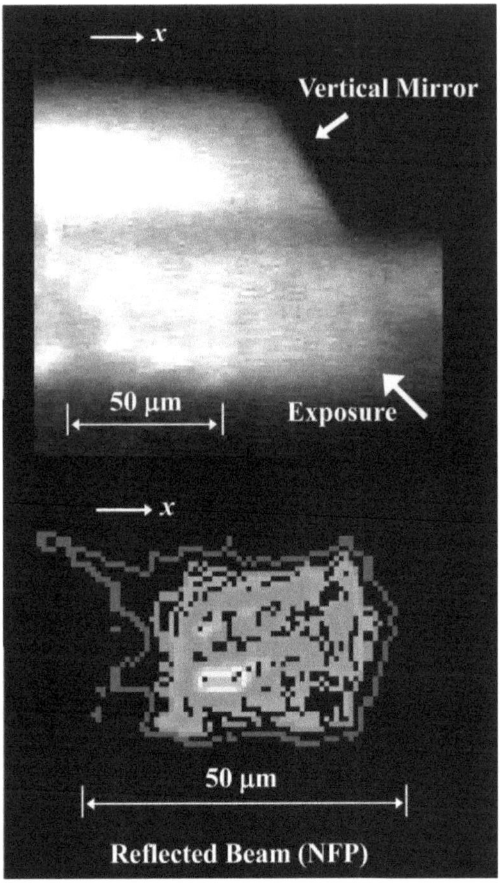

Figure 4.8 Side view of a duplicated optical waveguide film with a vertical mirror formed by tilted exposure with an incident angle of 45° without the prism on the backside of the glass substrate, and the NFP of a 1.3 μm reflected beam from the vertical mirror.

cores. Beveled sidewalls are also found in the cross marker. It can be seen in the side view that the core end facet has a flat surface with an exact 45° slope. The FFP for the reflected beams from the vertical mirror indicates that the beams are reflected in the surface-normal direction from the optical waveguide film, confirming that the mirror angle is exactly 45°. As Fig. 4.10 shows, a reflected beam from a vertical mirror of a 4 μm wide core is observed.

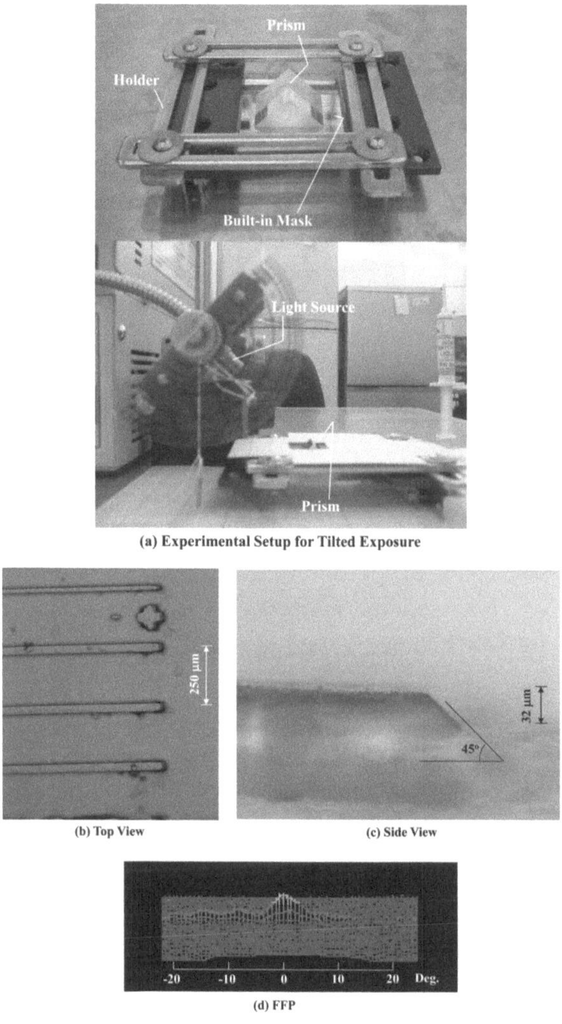

Figure 4.9 (a) Exposure apparatus for vertical mirror construction with a prism on the backside of the glass substrate. (b) Top view and (c) side view of an optical waveguide film with vertical mirrors fabricated by the built-in mask method. (d) FFP of a 1.3 μm reflected beam from the vertical mirror. Reprinted with permission from Yoshimura et al. [3].

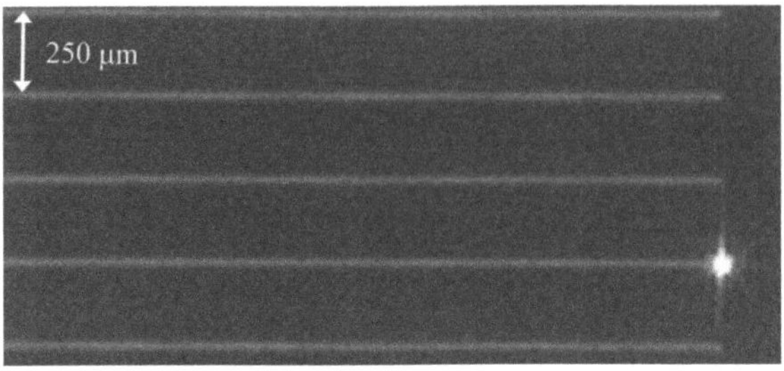

Figure 4.10 Reflected beam from a vertical mirror of a 4 μm wide core. Reprinted with permission from Yoshimura et al. [6].

4.3 Vertical Mirrors with Multi-Core-Layer Skirt-Type Structures

To increase the coupling efficiency between optical waveguides or between optical waveguides and OE devices, quality improvement of the vertical mirrors in the optical waveguides is important. To do this, we proposed and fabricated optical waveguide films with vertical mirrors with multi-core-layer skirt-type structures [6, 7].

4.3.1 Observation of Beam Leakage and Scattering at Vertical Mirrors

Figure 4.11 shows an observation of guided beam leakage and scattering at a vertical mirror in a conventional optical waveguide film. The observation was carried out by monitoring line growth in a PRI layer as follows. As Fig. 4.11a shows, the backside of the optical waveguide film was coated with a photopolymer to form a three-layer structure of PRI layer/cladding film/core. When a laser beam of 405 nm in wavelength was introduced into the core, as shown in Fig. 4.11b, four lines grown from the vertical mirror were observed in the PRI layer. One is in surface-normal direction, and the other three are in front, left-hand-side, and right-hand-side directions. The surface-normal line is the expected one grown by self-focusing of the laser beam reflected from the mirror. The other three lines are undesirable ones grown by self-focusing of the leaked and scattered guided beams.

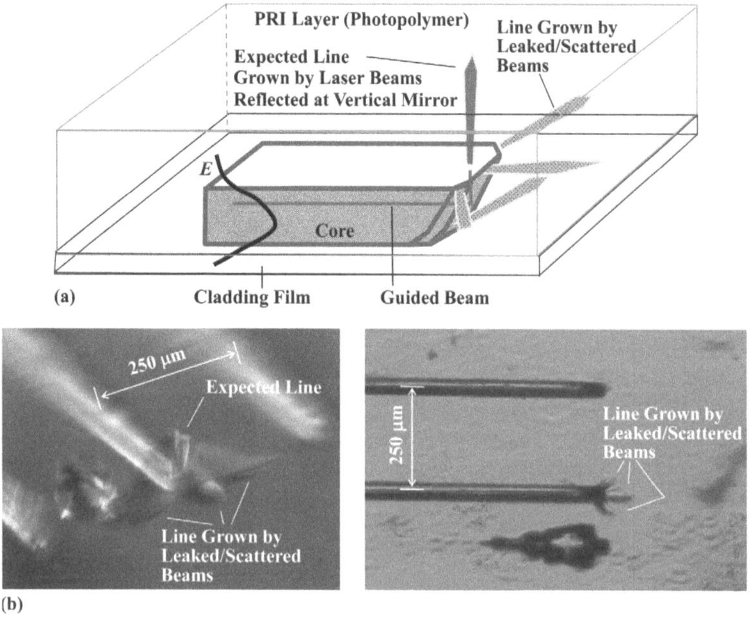

Figure 4.11 Observation of guided beam leakage and scattering at a vertical mirror in a conventional optical waveguide film. (a) Schematic illustration of core structures and line growth in a PRI layer. (b) Photographs of lines grown from the vertical mirror. Reprinted with permission from Yoshimura et al. [7].

As can be seen from Fig. 4.11a, the front line is attributed to the evanescent lightwave in the cladding film. That is to say, a portion of the guided beams penetrate into the cladding film by the tunneling effect and spread into the PRI layer. The beams that exist in the cladding film and the PRI layer cannot be reflected by the vertical mirror, passing through the mirror to grow the line in the front direction. Sometimes round-shaped core end corners appear due to photolithographic imperfection, resulting in guided beam reflection and refraction at the corners. The reflected and refracted beams from the round-shaped core end top and bottom corners also induce the front line. The left- and right-hand-side lines are attributed to laser beams reflected/refracted at the round-shaped core end side corners.

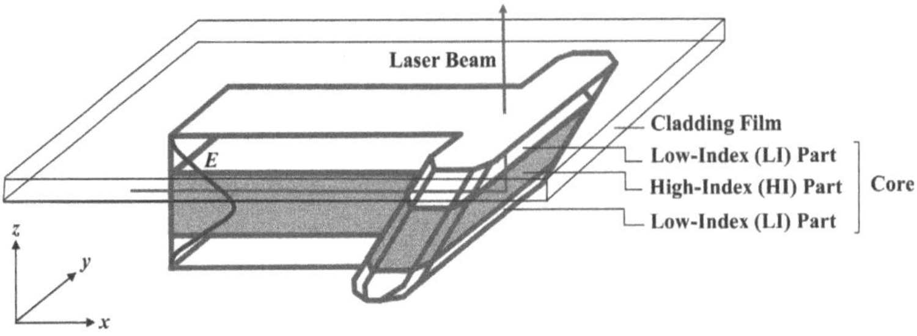

Figure 4.12 Structure of the proposed three-core-layer skirt-type vertical mirror. Reprinted with permission from Yoshimura et al. [7].

4.3.2 Three-Core-Layer Skirt-Type Vertical Mirrors

Figure 4.12 schematically shows a structure of the proposed three-core-layer skirt-type vertical mirror, where a core with a three-layered structure consisting of low-index (LI) part, high-index (HI) part, and LI part is constructed on a cladding film. In the conventional one-core-layer structure, as described in section 4.3.1, the tunneling of photons into the cladding film and the reflection/refraction at round-shaped core end corners give rise to beam leakage/scattering. In the three-core-layer structure, on the other hand, the LI part located between the HI part and the cladding film prevents guided beams from tunneling into the cladding film to reduce beam leakage/scattering. At the same time, the structure puts the round-shaped core end top and bottom corners away from the guided beam path, reducing beam leakage/scattering. Similarly, the skirt-type structure of core end facets puts the round-shaped core end side corners away from the guided beam path to reduce beam leakage/scattering.

4.3.3 Simulations of Beam Leakage/Scattering at Vertical Mirrors

Figure 4.13 shows results of the beam leakage/scattering effect calculated by the coupled simulation method with the beam propagation method (BPM) and the two-dimensional finite-difference time-domain (2D FDTD) method for 1.3 μm guided beams. The FDTD method is applied to calculations for the region, where a

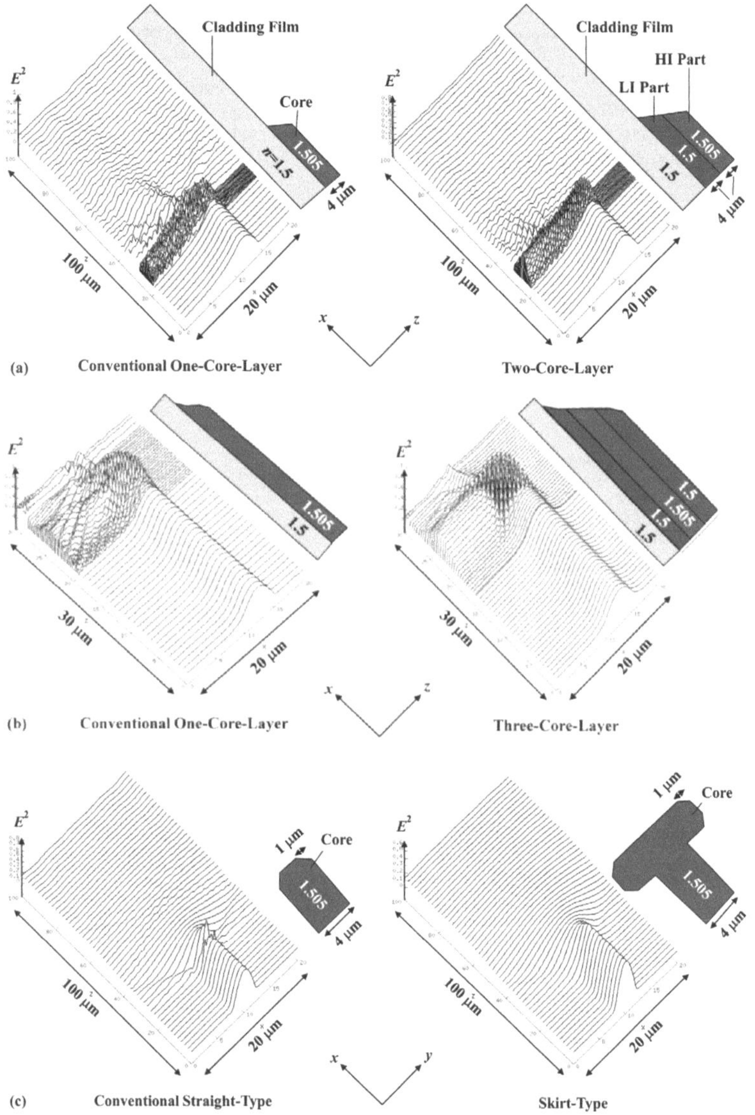

Figure 4.13 Simulation results of (a) beam leakage due to the tunneling, (b) beam scattering at the round-shaped core end top and bottom, and (c) beam scattering at round-shaped core end side corners. Reprinted with permission from Yoshimura et al. [7].

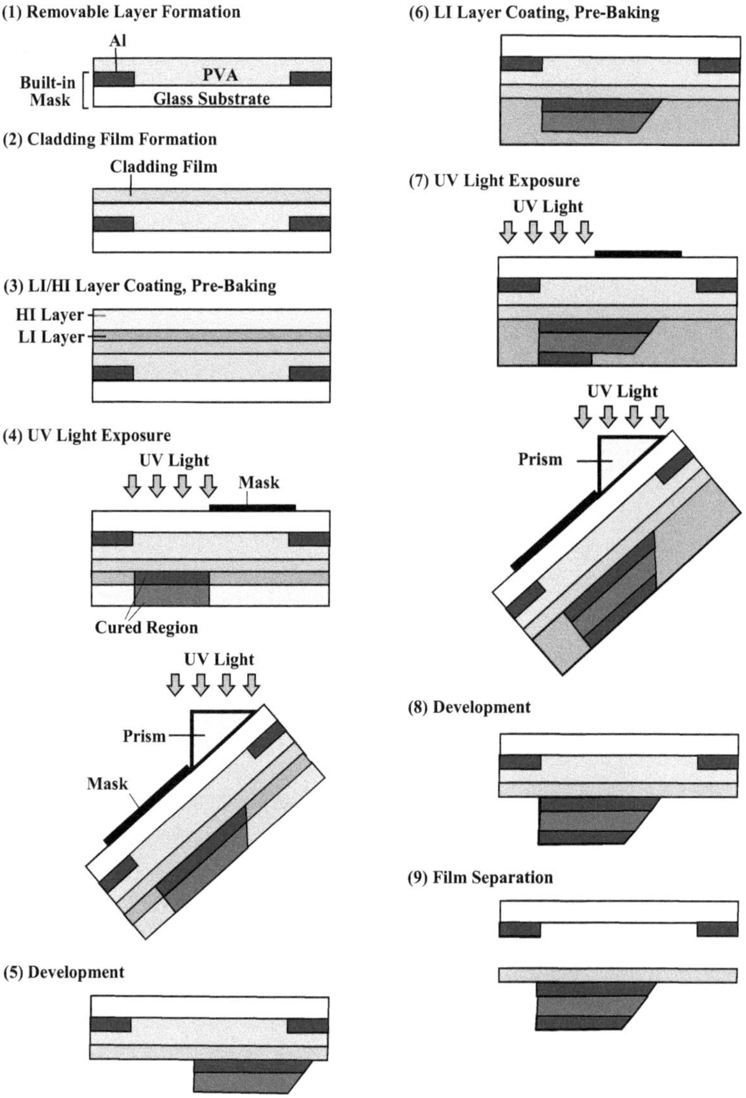

Figure 4.14 Fabrication process of optical waveguide films with three-core-layer vertical mirrors utilizing the built-in mask. Reprinted with permission from Yoshimura et al. [7].

mirror exists. Figure 4.13a shows the beam leakage due to the tunneling of the guided beams. In the conventional one-core-layer structure, the refractive index of the cladding film and that of the core is assumed 1.5 and 1.505, respectively. The core thickness is 4 μm. In the two-core-layer structure, the refractive index of the cladding film and that of the LI part is 1.5, and the refractive index of the HI part is 1.505. The thickness of the LI and HI parts is 4 μm. It is found that beam leakage is drastically reduced in the two-core-layer structure. The efficiency of the reflection from the vertical mirror increases to ~0.85 in the two-core-layer structure from ~0.4 in the conventional structure.

Figure 4.13b shows results of the beam scattering at the round-shaped core end top and bottom. In the conventional one-core-layer structure, disturbed reflected beams can be seen, while in the three-core-layer structure, smooth reflection of the propagating beams is observed at the vertical mirror part.

Figure 4.13c shows results of the beam scattering at round-shaped core end side corners. The refractive index of the cladding region and that of the core is assumed to be 1 and 1.505, respectively. The core width and skirt width are respectively 4 μm and 10 μm. Core corners are cut by 1 μm along directions parallel and perpendicular to the waveguide direction. In the conventional straight-type core structure, beam propagation in the left/right-hand-side directions appears, which is consistent with the observation shown in Fig. 4.11. In the skirt-type core structure, on the other hand, the beam propagation in the left/right-hand-side directions disappears.

These results indicate that the three-core-layer skirt-type structure is effective to suppress undesired beam leakage/scattering at vertical mirrors.

4.3.4 Fabrication of Multi-Core-Layer Skirt-Type Vertical Mirrors

Figure 4.14 shows a fabrication process of optical waveguide films with three-core-layer vertical mirrors. The process is basically the same as the built-in mask method shown in Fig. 4.1. After 10 μm thick PVA removable layer and a 20 μm thick cladding film are

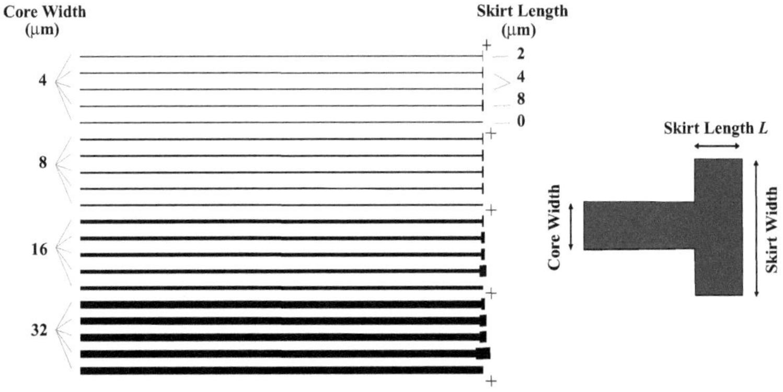

Figure 4.15 Photomask patterns for skirt-type core end facets. Reprinted with permission from Yoshimura et al. Reprinted with permission from Yoshimura et al. [7].

formed on a built-in mask, a 10 μm thick LI layer and a 32 μm thick HI layer are successively formed and pre-baked. The LI/HI layer is selectively exposed to UV lights vertically, constructing cured structures for planar waveguide cores. To construct core end facets for vertical mirrors, tilted UV light exposure is carried out through a prism by covering the planar waveguide core regions by a mask. By development, two-core-layer formation is completed. Next, an LI layer was formed and pre-baked again. By vertical and tilted UV light exposure followed by development, three-core-layer formation is completed. Finally, the removable layer is removed in water to obtain an optical waveguide film with three-core-layer vertical mirrors by separating the film and the built-in mask.

Figure 4.15 shows photomask patterns for skirt-type core end facets. Skirt width is four times the core width. The skirt length *L* varies as 1/2, 1/1, and 2 of the core width. Acrylic photo-definable materials with refractive index of 1.50 were used for the cladding film and the LI part, and those with refractive index of 1.52 for the HI part.

To examine the influence of *L* on reflected beams, an optical waveguide film with one-core-layer skirt-type vertical mirrors was fabricated. Figure 4.16a shows the top view and side view of the optical waveguide film. Nominal core width and thickness are 32 μm.

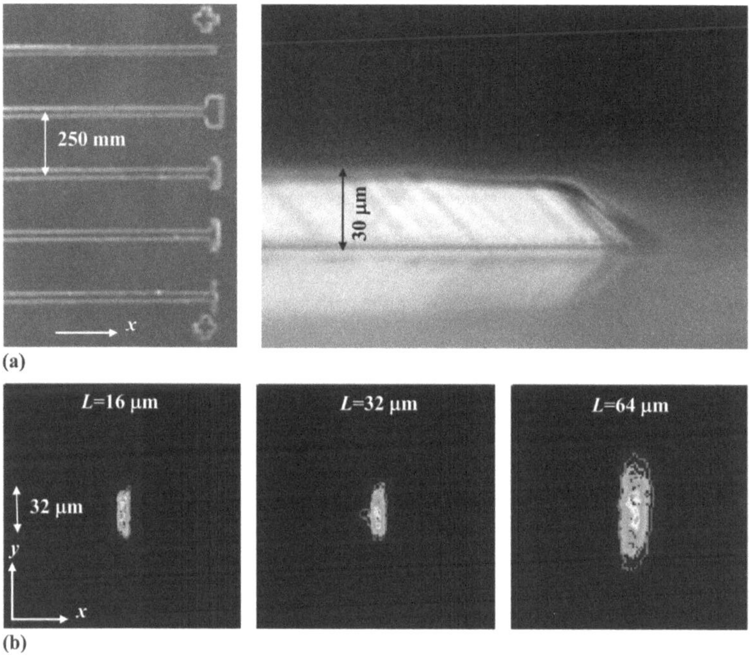

Figure 4.16 (a) Top and side views of an optical waveguide film with one-core-layer skirt-type vertical mirrors. (b) NFPs of 1.3 μm reflected beams from the vertical mirrors. Reprinted with permission from Yoshimura et al. [7].

It is found that vertical mirrors are constructed. NFPs of 1.3 μm reflected beams are shown in Fig. 4.16b for $L = 16$, 32, and 64 μm. Although with increasing L, the NFP size increases due to beam diameter expansion during propagation in a free space of the skirt region, drastic enlargement does not occur for $L < 32$ μm. Therefore, it is suggested that appropriate skirt length is around the core width.

Figure 4.17a shows cross-sections and NFPs of an optical waveguide film with two-core-layer structures of 32 μm and 4 μm widths. The wavelength of guided beams is 650 nm. It is found that two-core-layer structures are clearly formed, and guided beams are confined in the HI part. These results confirm that the two-core-layer structure can be constructed by the built-in mask method. Propagation loss of the optical waveguide was

Figure 4.17 (a) Cross-sections and NFPs of an optical waveguide film with two-core-layer structures. (b) Cross-section of an optical waveguide film with a three-core-layer structure. Reprinted with permission from Yoshimura et al. [7].

0.9 dB/cm at 1.3 μm. The relatively large propagation loss might be caused by air cladding—that is, it can be caused by scattering due to multi-mode characteristics of the optical waveguide with a high core/cladding index contrast. Reflected beams from the core end facet for the vertical mirror in the two-core-layer skirt-type structure were observed. Reflection loss of the vertical mirror was estimated to be ~2 dB. Although the structure is not perfect, an optical waveguide film with the three-core-layer vertical mirror was also constructed. The cross-section is shown in Fig. 4.17b.

To investigate the influence of round-shaped corners on light beam propagation, FFP measurements were carried out for the reflected light beams at vertical mirrors. The results are shown in Fig. 4.18. In a vertical mirror of three-core layer without skirt, the FFP is spread along the y direction—that is, a direction parallel to the film plane (see Fig. 4.12). This suggests that the guided beams are refracted to the left/right-hand-side directions by the round-shaped core end side corners, as can be seen in Fig. 4.11.

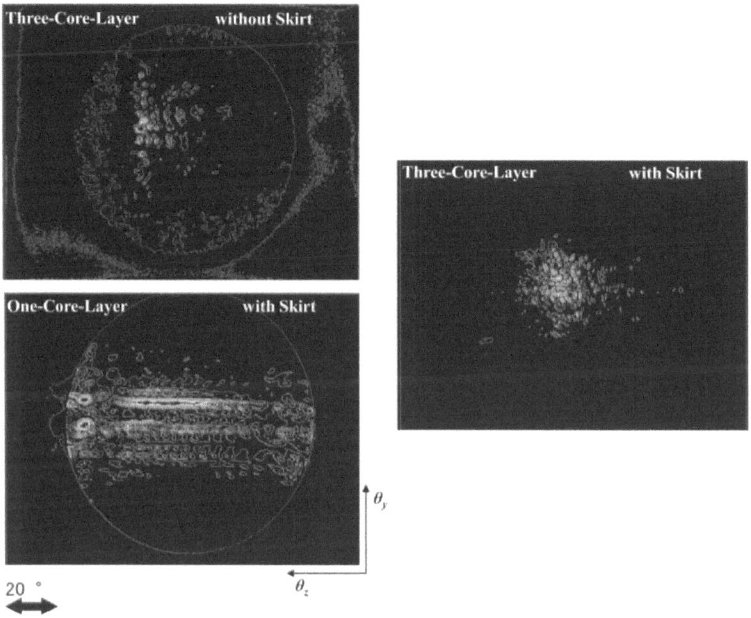

Figure 4.18 FFPs of 1.3 μm reflected beams from vertical mirrors.

In a vertical mirror of one-core layer with skirt, the FFP is spread along the *z* direction—that is, a direction perpendicular to the film plane. This suggests that the guided beams are reflected with a broad range of angles along the *z* direction by the round-shaped core end top and bottom corners. In a vertical mirror of three-core layer with skirt, the spread of the FFP is suppressed both along the *z* and *y* direction. This suggests that the influence of the round-shaped core end corners as well as tunneling is removed successfully.

In addition, the three-core-layer skirt-type structure is effective in widening the alignment tolerances between core end facets and metal (or dielectric) films for mirrors (or filters) because the structure can avoid direct deposition of the metal (or dielectric) films on the top surface of the HI part, allowing us to deposit the metal (or dielectric) films on areas wider than the beveled end facet surfaces.

4.4 3D Optical Circuits

By stacking optical waveguide films duplicated by the built-in mask method, we demonstrated a possibility to construct 3D optical circuits utilizing optical Z-connections [3].

4.4.1 Structures

Figure 4.19 shows two types of 3D optical circuit structures investigated in present experiments. Type I structure is "stacked waveguide films with vertical mirrors." Two optical waveguide films, Film A and Film B, are stacked by contacting their undercladding layers. A probe beam is introduced into *Input* in Film A. The beam is transferred into Film B through *Optical Z-Connection* to reach *Output* in Film B.

Type II structure is "optical waveguide films with vertical waveguides." A vertical waveguide is constructed above a vertical mirror aperture in an optical waveguide film. When a write beam is introduced into *Input*, it is reflected into the PRI layer to grow a vertical waveguide at *Optical Z-Connection*. The vertical waveguide

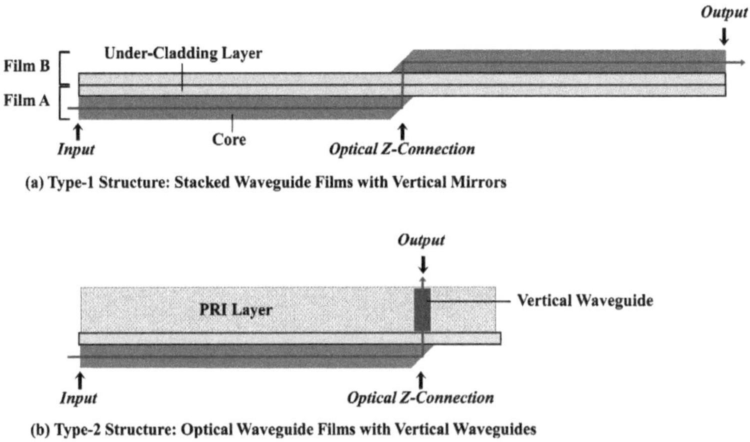

Figure 4.19 Schematic illustrations of 3D optical circuits. Reprinted with permission from Yoshimura et al. [3].

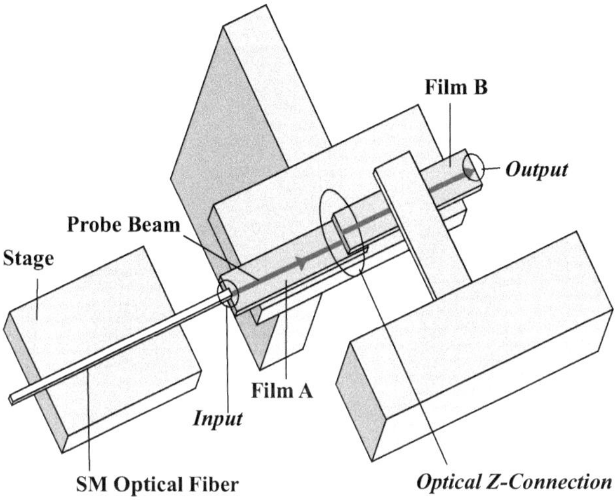

Figure 4.20 Fabrication/measurement setup for the type I structure of 3D optical circuits consisting of stacked waveguide films. Reprinted with permission from Yoshimura et al. [3].

acts as an optical via in the PRI layer, and confines a reflected probe beam to guide it efficiently to *Output* at the PRI layer surface.

4.4.2 Type I: Stacked Waveguide Films with Vertical Mirrors

4.4.2.1 Demonstration of 3D optical wiring

In Fig. 4.20 a fabrication/measurement setup for the type I structure of 3D optical circuits consisting of stacked waveguide films is schematically illustrated. Film B put on a stage is stacked on Film A by aligning the positions of the vertical mirrors in Film A and Film B to construct *Optical Z-Connection*. A single-mode (SM) optical fiber for 1.3 μm in wavelength with a core diameter of ∼9 μm is connected to *Input* of a core in Film A by butt joint. A probe beam is introduced into *Input* and propagates to *Output* in Film B through *Optical Z-Connection*.

Figure 4.21 shows propagation of a 650 nm probe beam before and after alignments between Film A and Film B. Before alignments, as Fig. 4.21a shows, since the vertical mirror in Film B is not located

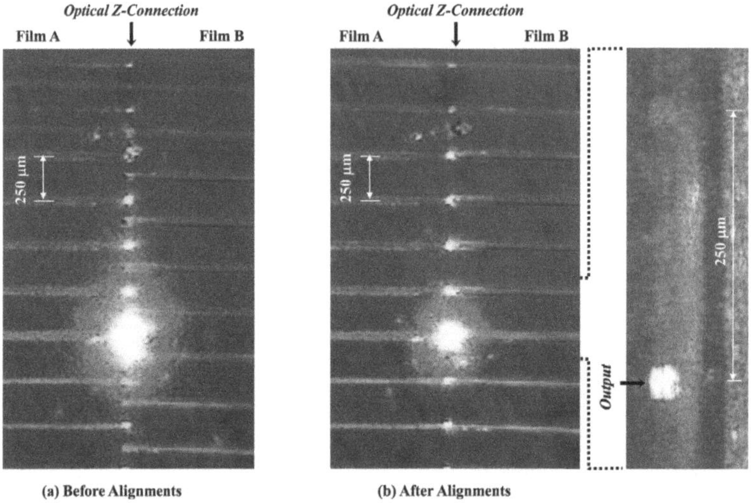

Figure 4.21 Propagation of a 650 nm probe beam in the 3D optical circuit consisting of stacked waveguide films. Reprinted with permission from Yoshimura et al. [3].

above the vertical mirror in Film A, the probe beam reflected by the vertical mirror in Film A propagates through undercladding layers to reach a camera set above the vertical mirror. After alignments, the two vertical mirrors are correctly aligned to form *Optical Z-Connection*, as can be seen in Fig. 4.22. The reflected probe beam is transmitted into Film B through *Optical Z-Connection* to reach *Output* (Fig. 4.21b). A ~30× 30 μm^2 NFP is observed at *Output*, demonstrating the 3D optical wiring operation in the 3D optical circuit.

4.4.2.2 Loss measurements

The cutback method revealed that the propagation loss of the optical waveguide film is 0.8 dB/cm for 1.3 μm probe beams. Reflection loss of the vertical mirror was measured using the setup shown in Fig. 4.23. The loss was estimated to be 2 dB by collecting the reflected 1.3 μm probe beams by a multimode (MM) optical fiber with a core diameter of ~50 μm.

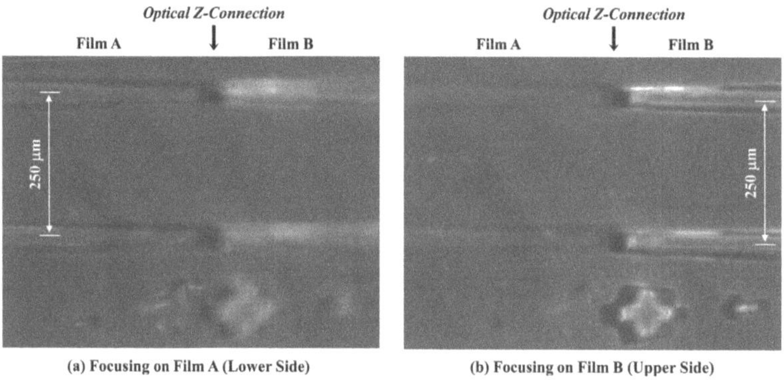

Figure 4.22 Photographs of the *Optical Z-Connection* part. Reprinted with permission from Yoshimura et al. [3].

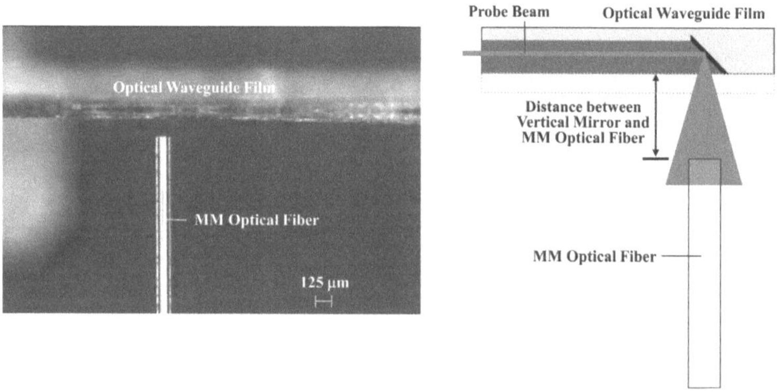

Figure 4.23 Setup for measurements of coupling efficiency from an optical waveguide film to an MM optical fiber.

Figure 4.24 shows dependence of the coupled 1.3 μm probe beam power from the optical waveguide film to the MM optical fiber on the distance between the vertical mirror and the MM optical fiber. For an air-cladding-type waveguide, where only one side of the core contacts a cladding layer, the coupled power decreases with the distance rapidly. For an overcladding-type waveguide, where three sides of the core are covered by a cladding region, the rate of the coupled power decrease is low compared to the case of the air-cladding-type waveguide.

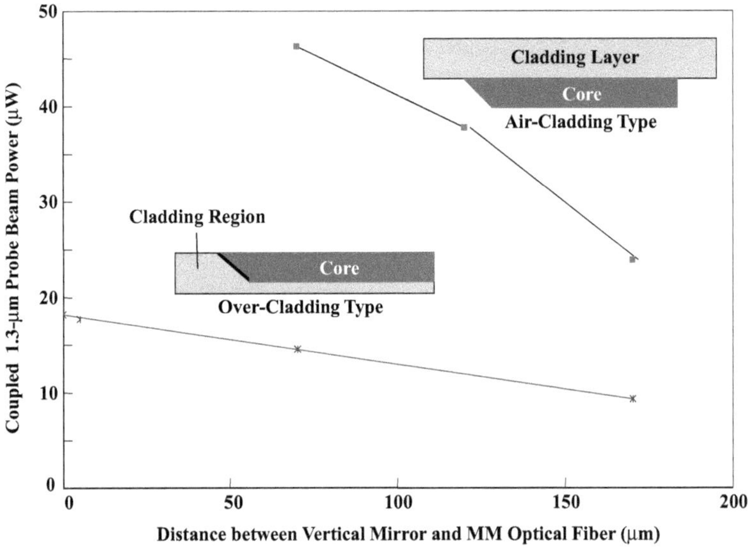

Figure 4.24 Dependence of the coupled 1.3 μm probe beam power from the optical waveguide film to the MM optical fiber on the distance between the vertical mirror and the MM optical fiber.

These results can be explained by models schematically illustrated in Fig. 4.25. The air-cladding-type waveguide can guide light beams in higher modes compared to the overcladding-type waveguide because the refractive index difference between the core and the cladding region is larger in the former than in the latter. This implies that the air-cladding-type waveguide exhibits large numerical aperture (NA) compared to the overcladding-type waveguide, allowing light beams to be emitted from the edge with large angles.

NFPs and FFPs of probe beams reflected from vertical mirrors confirm the above-mentioned explanation as shown in Fig. 4.26. It is found from the NFPs that the air-cladding type waveguide exhibits more multimode-like characteristics compared to the overcladding-type waveguide. The FFPs reveal that in the air-cladding-type waveguide higher modes are excited and a large beam spread is induced along the *y* direction, in which direction the light beams are strongly confined by the air cladding located in both sides. The larger

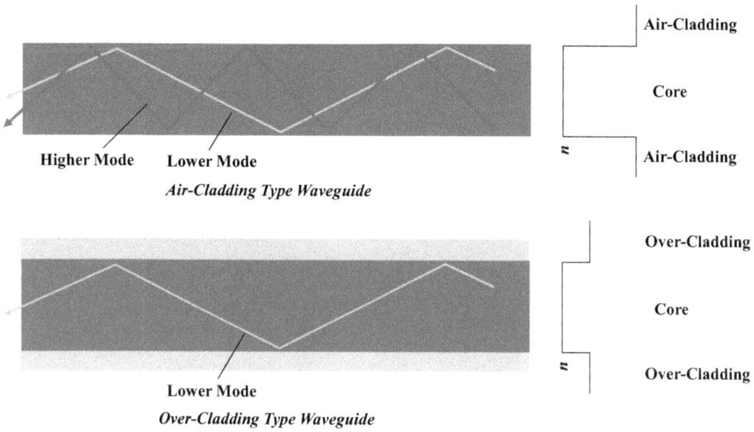

Figure 4.25 Schematic illustration of light beam confinement into an optical waveguide.

the emitting angle becomes, the higher the rate of the coupled power decrease with the distance between the vertical mirror and the MM optical fiber becomes. This causes the difference in the rate of the coupled power decrease shown in Fig. 4.24.

For the beam spread along the z direction, no difference is observed between both types of optical waveguides since in this direction the light beams are confined similarly by an air-cladding region and a cladding region in both types.

The measured reflection loss of 2 dB for the vertical mirrors is relatively large. This might be attributed to the NA mismatch between the optical waveguide and the MM optical fiber. In the present experiments, in which air-cladding-type waveguides are used, since the optical waveguide involves air-cladding regions, the NA of the optical waveguide is larger than that of the MM optical fiber. The NA mismatch results in overestimation of the loss. The reflection loss might be minimized by use of an optical fiber with large core-cladding refractive index differences or a coupling lens.

4.4.2.3 Loss at *Optical Z-Connection*

In the present experimental setup, the total propagation length from *Input* to *Output* is 5 cm, giving rise to a loss of 8 dB for the

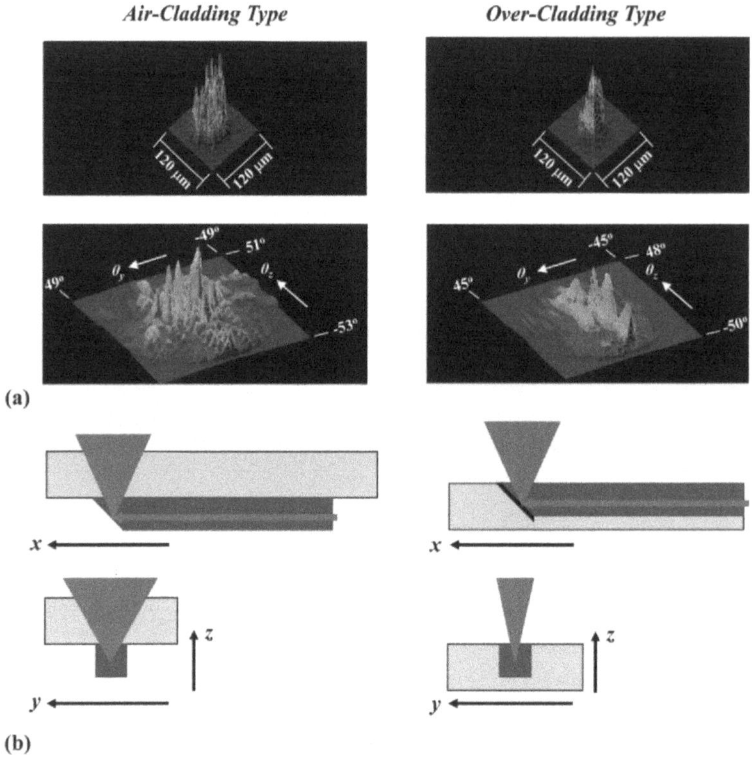

Figure 4.26 (a) NFPs and FFPs of reflected probe beams from vertical mirrors, and (b) models for degree of spread of the reflected probe beams.

propagation loss and the reflection loss of two vertical mirrors. The insertion loss of the 3D optical circuit was measured to be ~22 dB. These results suggest that the loss at *Optical Z-Connection* is ~14 dB. The main origin of the loss might be leakage of probe beams reflected from the vertical mirror with large diverging angles from the surface-normal direction.

For easy handling of film samples, we used 70 μm thick undercladding layers, resulting in a long free-space path length of ~170 μm for the probe beams reflected by the vertical mirror in Film A at *Optical Z-Connection*. So, only a limited portion of the reflected probe beams can be received by the vertical mirror in Film B. Furthermore, in the air-cladding waveguide, light beams in higher modes propagate in the core with a critical angle of ~40° to the core sidewall. As a result, the emitting angle range of the reflected probe beams from the vertical mirror in Film A is widely spread, as shown in Fig. 4.26, increasing the loss at *Optical Z-Connection*. These conditions cause considerable probe beams passing through Film B even after the film alignment as can be seen in Fig. 4.21b.

Decreases in the undercladding layer thickness might reduce the loss at the *Optical Z-connection*. A slight curvature implementation to the vertical mirror surfaces might also reduce the loss. Another promising challenge to reduce the loss at the *Optical Z-Connection* is optical via formation discussed in the following section.

4.4.3 Type II: Waveguide Films with Vertical Waveguides

For loss reduction at *Optical Z-Connection*, implementation of vertical waveguides on optical waveguide films is promising. A fabrication/measurement setup for the type II structure of 3D optical circuits is schematically illustrated in Fig. 4.27. The back of an undercladding layer of an optical waveguide film is coated with a ~500 μm thick PRI layer. For the PRI material, the monomer/monomer-type photopolymer described in Chapter 7 is used.

A write beam of 405 nm in wavelength is introduced from an SM optical fiber into *Input* of the optical waveguide film with power

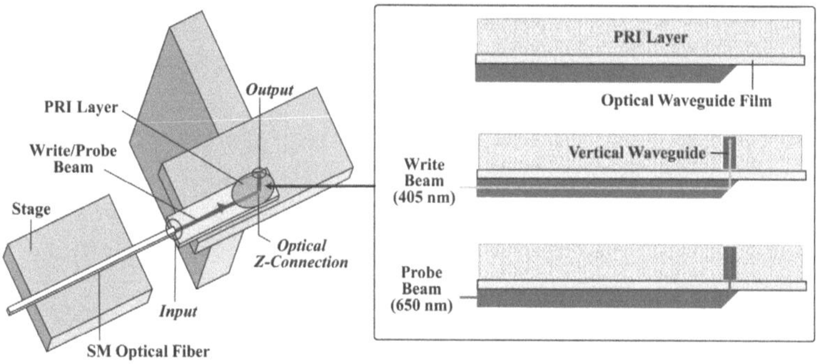

Figure 4.27 Fabrication/measurement setup for the type II structure of 3D optical circuits consisting of optical waveguide films with vertical waveguides. Reprinted with permission from Yoshimura et al. [3].

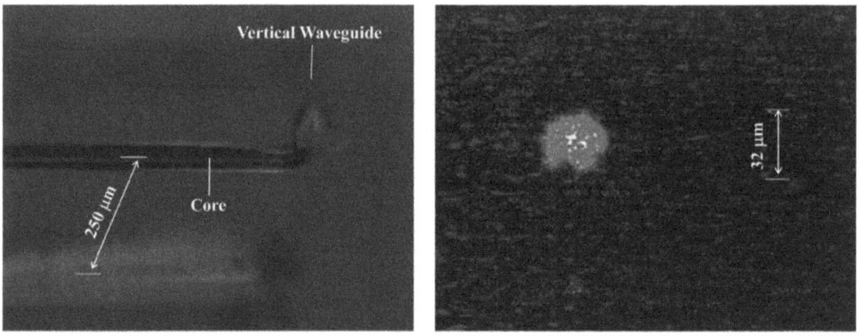

(a) Vertical Waveguide
(View from above with an angle of ~30°)

(b) Near Field Pattern

Figure 4.28 (a) Optical waveguide grown vertically on the undercladding layer above the vertical mirror aperture. (b) NFP of 650 nm probe beam guided in the vertical waveguide at the PRI layer surface. Reprinted with permission from Yoshimura et al. [3].

of ~100 μW or less. The propagated write beam is reflected to the surface-normal direction by a vertical mirror in the waveguide. The write beam passes through the undercladding layer and is injected into the PRI layer. Since the refractive index of the PRI material increases upon write beam exposure, the write beam is focused along the center of the propagation axis. The self-focusing effect grows a vertical waveguide in the PRI layer. Then, a 650 nm probe beam is introduced into *Input*. The beam is guided in the vertical waveguide at *Optical Z-Connection* to reach *Output* at the PRI layer surface.

In Fig. 4.28a, an optical waveguide grown vertically on the undercladding layer is shown. The vertical waveguide is constructed in the PRI layer above the vertical mirror aperture. An NFP of a guided probe beam at *Output* is shown in Fig. 4.28b. An NFP of $\sim 30 \times 30\ \mu m^2$, which is approximately the same as the core size of the optical waveguide film, is observed, demonstrating the optical via of the vertical waveguide for 3D optical wiring.

It should be noted that the vertical waveguide is formed by write beams through the undercladding layer, which is a free space with no beam-confining structures. In other words, optical vias of vertical waveguides can be formed even when some spacers are inserted between vertical mirrors and PRI layers.

As a future challenge, implementation of SOLNET vertical waveguides that are inserted between two vertical mirrors by combining type I and II structures is interesting. By connecting two optical waveguide films with optical vias of the SOLNET vertical waveguides, light beams can be guided from one vertical mirror to the other with strong confinement, reducing the leakage of the light beams at *Optical Z-Connection* even if there are misalignments between the vertical mirrors. The details are described in Chapter 7.

4.5 Optical Waveguide Films Stacked on Electrical Boards

By stacking optical waveguide films on electrical substrates, various types of OE board structures can be constructed. In the present section, examples of optical waveguide films and stacked structures of optical waveguide films and a PCBs are presented [8–10].

4.5.1 Process Flow

Figure 4.29 shows a process flow example for optical waveguide film preparation and film stacking on PCB. The film material is the fluorinated polyimide (Hitachi Chemical Co.). Because the fluorinated polyimide is not photosensitive, the core and vertical mirror formation is performed by conventional processes like reactive ion etching (RIE) and laser ablation. The refractive indexes of the cladding region and cores are ~1.52 and ~1.54, respectively.

On an Al substrate, a 15 μm thick undercladding layer and a 30 μm thick core layer are successively formed. MM optical waveguide core arrays of 250 μm pitch are formed by RIE with oxygen or oxygen/argon mixture or UV (THG)-YAG laser ablation. The typical core size is 30 μm × 30 μm. An overcladding layer is coated on it. The waveguide film size is 6 inch × 6 inch. The maximum film thickness is about 60 μm.

Beveled walls with an angle of 45° for vertical mirrors in optical waveguides are formed by 45° trench cuts using oblique RIE or

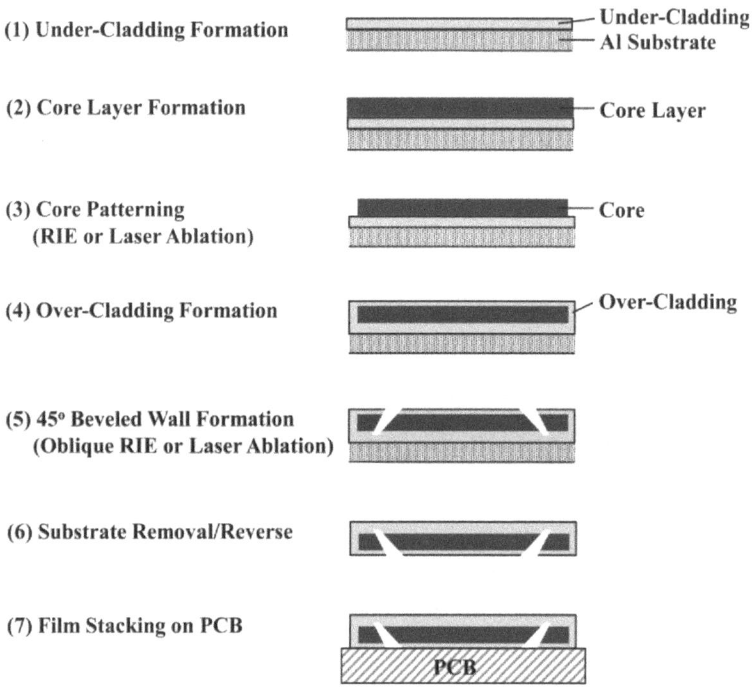

Figure 4.29 Process flow example for optical waveguide film preparation and film stacking on PCB.

laser ablation. Depending on applications, we should make outward mirrors or inward mirrors in the optical waveguides. For example, for the structure shown in Fig. 3.6a, inward mirrors are necessary. In the Film/Z-connection approach, it does not matter which type of mirrors are formed at this step, because we can reverse the waveguide film upside down in the lamination step mentioned below. This is one of the merits of adopting the Film/Z-connection. In the present trial, since it is easier to make outward mirrors than inward mirrors, we made outward mirrors.

Then the Al substrate is removed to obtain a free-standing waveguide film. The film is reversed upside down and stacked on a PCB with a lamination process. Thus, we obtain an optical waveguide film with inward vertical mirrors on the PCB.

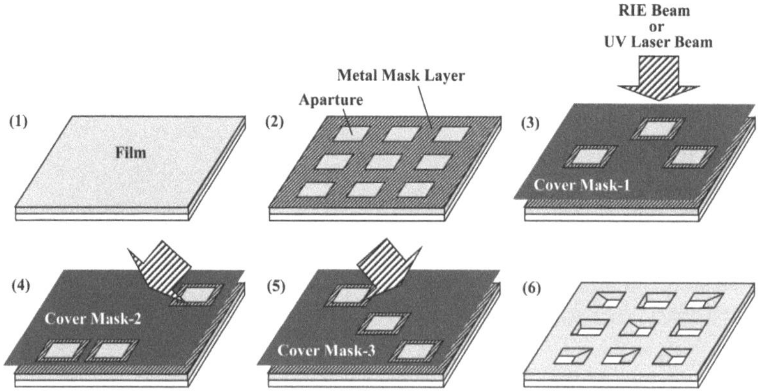

Figure 4.30 Process flow of MNA and MNE. Reprinted with permission from Yoshimura et al. [10].

In S-FOLMs, vertical mirrors of various directions are needed with high density corresponding to LSI pad pitches and 2D optical fiber array pitches. Although this can be achieved by repeating metal mask layer patterning steps for beveled wall formation, the many-step photolithographic processes raise the fabrication cost. To solve the problem, we proposed the following process schematically shown in Fig. 4.30.

(1) A film is prepared.
(2) On the film, a metal mask layer is formed with apertures at all the sites, where walls will be made, regardless of the wall directions.
(3) A cover mask 1 having apertures at the sites, where vertical walls should be formed, is put on the film. The film is exposed vertically to the RIE beam or laser beam to form vertical walls.
(4) A cover mask 2 having apertures at the sites, where beveled walls of direction A should be formed, is put on the film. Oblique exposure of the RIE beam or laser beam along direction A is performed.
(5) A cover mask 3 having apertures at the sites, where beveled walls of direction B should be formed, is put on the film. Oblique

exposure of the RIE beam or laser beam along direction B is performed.
(6) Thus, vertical walls and beveled walls with two different directions are fabricated.
(7) If necessary, more steps for other directions are added.

We call this process as *moving neon etching* (MNE) or *moving neon ablation* (MNA). In the MNE, RIE is used, and in the MNA, laser ablation. These processes enable wall fabrication with arbitrary directions/locations by only one step for a fine mask layer deposition, which is performed by photolithography, with the assistance of cover masks.

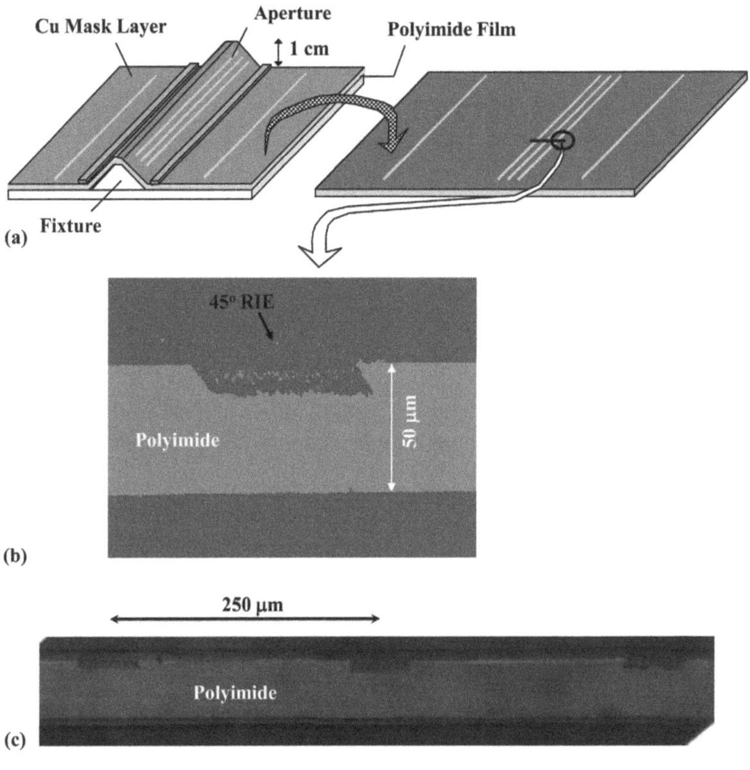

Figure 4.31 Wall formation of three different angles with 250 μm pitch in a polyimide film by MNE. Reprinted with permission from Yoshimura et al. [10].

In Fig. 4.31, an MNE example is shown. On a 50 μm thick Kapton polyimide film surface, a 200 nm thick Cu mask was deposited with line-shaped apertures as shown in Fig. 4.31a. The pitch of the apertures is 250 μm, and the width ranges from 30 μm to 100 μm. The film was set on a stainless fixture with a 45° slope to apply oblique exposure of RIE beams that was demonstrated by Chen et al. [11]. RIE was carried out in pure oxygen gas of 2 Pa with 450 W rf power. The etching rate was ~5 μm/h. A beveled wall structure was obtained as shown in Fig. 4.31b. Three-step RIE for oblique angles of −45°, 0°, and +45° was carried out by changing cover masks step by step. In Fig. 4.31c, 250 μm pitch walls of three different directions can be seen. The result indicates feasibility of the MNE for forming vertical mirrors in optical waveguides with arbitrary angles at arbitrary locations with one fine-mask-formation step to realize S-FOLM-compatible vertical mirror density.

4.5.2 Waveguide-Film Stacking on PCBs

Figure 4.32 shows schematic illustrations for two types of waveguide routing configurations, type I and type II, in OE transceivers of S-FOLMs with 2D connections to 2D optical fiber arrays. In this figure, a 4 × 4 array case is drawn. For the chip-mounting parts, corresponding to pad positions of the chips, surface-normal couplers like vertical mirrors and wavelength filters are distributed for optical couplings between optical waveguides and OE devices embedded just under the pad locations (OE devices are not drawn in the figures). For the 2D-connection parts in type I, surface-normal couplers are distributed for optical couplings between optical waveguides in S-FOLMs and 2D optical fiber arrays. Optical waveguides are routed by corner-turning mirrors and S-bending waveguides from the chip-mounting parts to the 2D-connection parts. In type II, optical waveguides from the chip-mounting parts are led into a number of film parts, which are twisted and stacked to construct 2D arrayed configurations for the 2D connection parts.

Figure 4.33a shows a photomask for optical waveguide films, which contain patterns both for type I and type II routing configurations. The mask size is 6′ × 6′, and the array size is

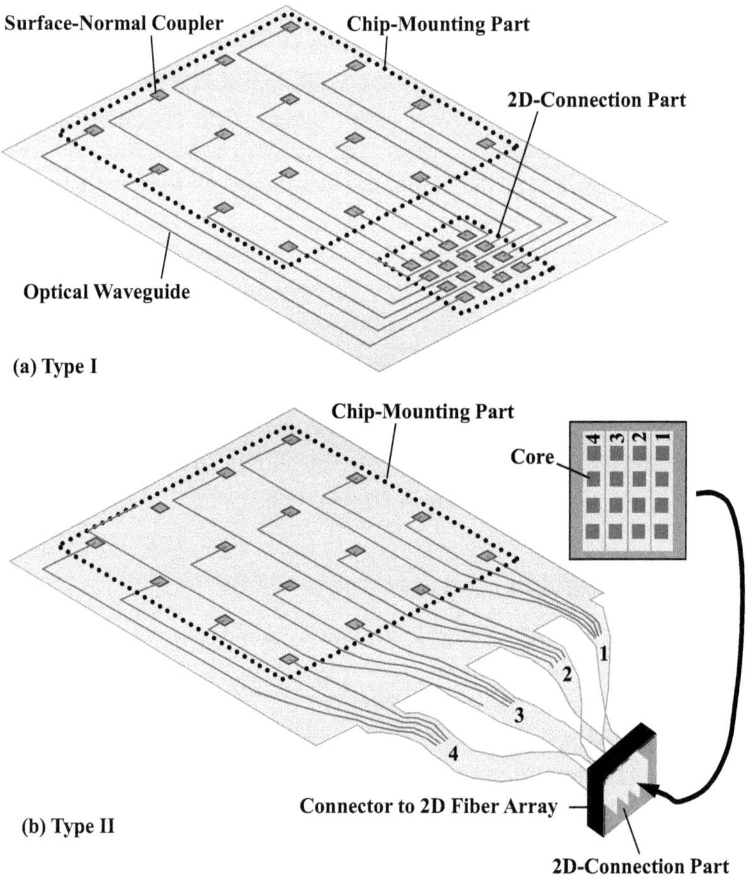

Figure 4.32 Waveguide routing configurations, type I and type II, in OE transceivers of S-FOLMs with 2D connections. Reprinted with permission from Yoshimura et al. [9].

16 × 16. In the chip-mounting parts, the surface-normal couplers are distributed with a pitch of 1.2 mm. A built-in mask fabricated using the photomask, and a sample for the optical waveguide film duplicated by the built-in mask are shown in Figs. 4.33b and c, respectively. Figure 4.34a shows a photograph of waveguide routing configurations, and Fig. 4.34b shows a high-magnification photograph of the 2D-connection part for type I.

(a) Photomask for Optical Waveguide Films

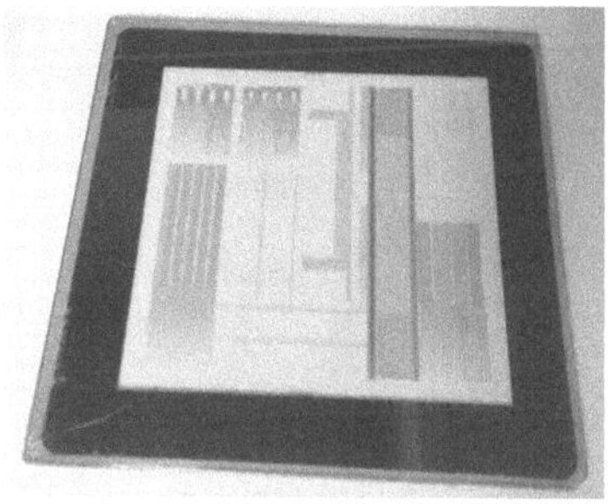

(b) Built-In Mask

(c) Optical Waveguide Film

Figure 4.33 (a) 6′ × 6′ photomask for optical waveguide films, which contain patterns for both type I and type II routing configurations. (b) Built-in mask fabricated using the photomask. (c) A sample of the optical waveguide film duplicated by the built-in mask.

Figure 4.35a shows a cross-section of a fluorinated polyimide waveguide film. The core patterns were formed by RIE under oxygen gas pressure of 4 Pa. A slight undercut is observed. RIE gas pressure reduction will suppress the undercut. Figure 4.35b shows a side view of a vertical mirror formed by UV (THG)-YAG laser ablation in the waveguide film. Figures 4.35c and d show propagation and reflection from the vertical mirror of 640 nm guided beams.

Figure 4.36 shows photographs of a 6′ × 6′ optical waveguide film made of fluorinated polyimide. Figure 4.37a shows samples of an optical waveguide film attached onto a flexible PCB. Coexistence of optical and electrical interconnect lines is observed in the

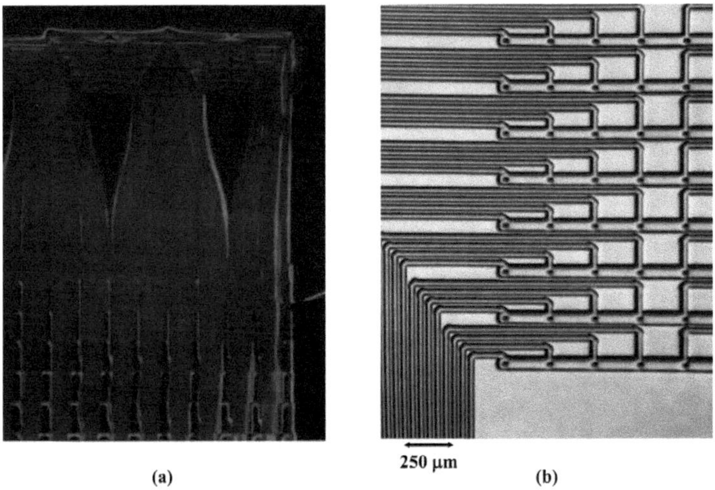

Figure 4.34 (a) Photograph of an optical waveguide film with 2D connection parts, and (b) a high-magnification photograph of a part of the 2D-connection part for type I. Reprinted with permission from Yoshimura et al. [9].

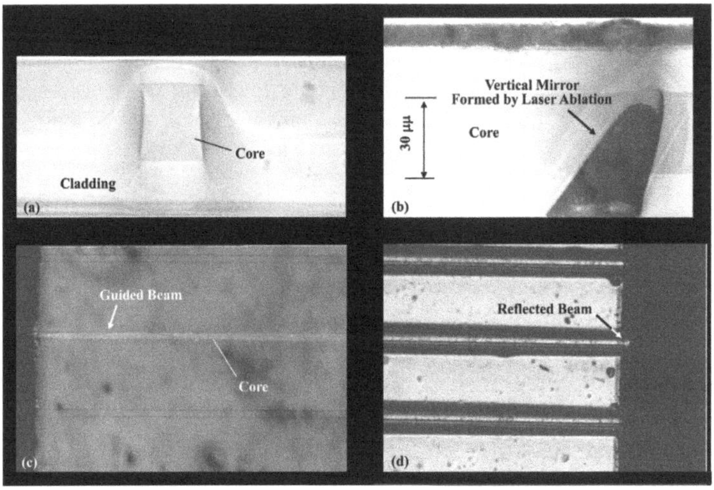

Figure 4.35 (a) Cross-section of a fluorinated polyimide waveguide film, (b) side view of a vertical mirror, (c) beam propagation in the waveguide, and (d) beam reflection from the vertical mirror. Reprinted with permission from Yoshimura et al. [9].

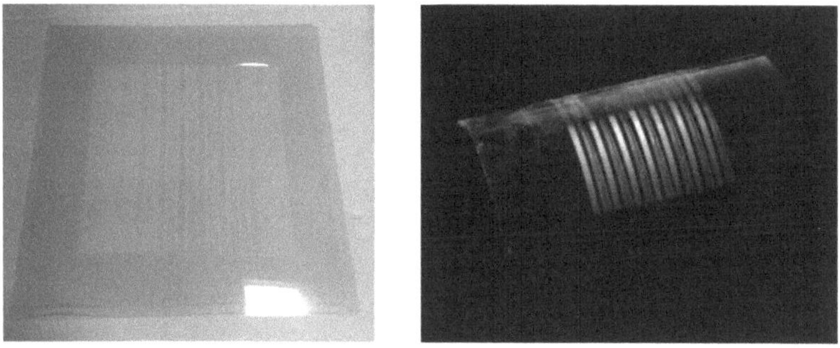

6' by 6' Optical Waveguide Film

Figure 4.36 Photographs of a 6' × 6' optical waveguide film made of fluorinated polyimide. Reprinted with permission from Yoshimura et al. [9].

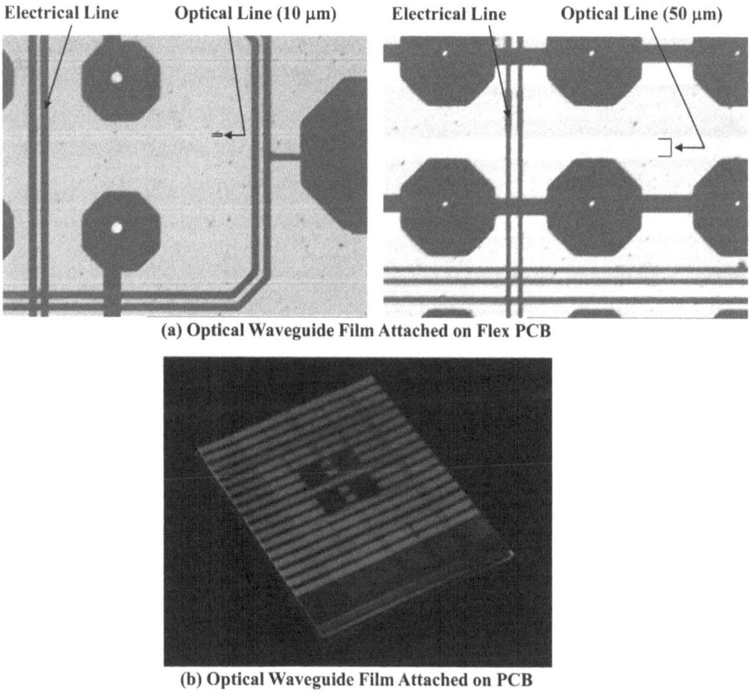

Figure 4.37 Samples of an optical waveguide film attached (a) on a flexible PCB and (b) on a PCB.

substrate. Figure 4.37b shows a sample of an optical waveguide film attached on a PCB. The optical waveguide film and PCB are laminated without peeling. These results indicate the feasibility of the Film/Z-connection approach extending to real OE substrate construction.

4.6 Nanoscale Waveguides Made of PRI Sol–Gel Thin Films

Nanoscale optical circuits are required in intra-LSI optical interconnects as well as light modulators, optical switches, and wavelength filters. Si waveguides with SiO_2 cladding [12] are usually used for the nanoscale optical circuits because they have large refractive index contrast between the core and cladding. Although the fabrication processes of Si waveguides are well established in the LSI industry, there are several problems in them such as complicated fabrication processes and narrow bandgap characteristics that prevent short-wavelength lights from being transmitted in waveguide cores.

The PRI sol–gel material developed by Nissan Chemical Industries Ltd. is suitable for nanoscale optical circuits due to its large-refractive-index-increase capability, superior processability, and wide bandgap characteristics. Thin films of the PRI sol–gel material are deposited by spin-coating. The refractive index of the PRI sol–gel thin films increases from ~1.65 to ~1.85 at $\lambda = 633$ nm by UV light exposure and baking, giving rise to a refractive index contrast of ~13%. Although the contrast is smaller than ~60% in Si waveguides, the contrast of ~13% is sufficient to produce nanoscale waveguide structures as described below. Note that the refractive index contrast of ~1% in typical polymer waveguides is insufficient for nanoscale waveguide structures.

The large refractive index contrast available in the PRI sol–gel thin films enables us to fabricate high-index-contrast (HIC) waveguides without etching/developing and vacuum processes, simplifying the fabrication processes of nanoscale waveguides. The wide bandgap characteristics of PRI sol–gel materials enable us to use short-wavelength lightwaves that are favorable for miniaturizing optical circuits. That is, the dimension of optical circuits can be reduced in proportion to λ, where λ is the lightwave wavelength, achieving volume reduction by λ^3. For example, the volume required to guide 600 nm light beams is one order of magnitude smaller than that required to guide 1300 nm light beams. Furthermore, PRI sol–gel thin films realize optical circuits consisting of HIC waveguides in ultrathin films of a 100 nm-order thickness. This is a favorable property for stacking optical circuits on LSIs, and for constructing 3D optical circuits with stacked films. Therefore, PRI sol–gel thin films are considered promising for nanoscale optical circuits in 3D OE platforms.

In the present section, experimental demonstrations for nanoscale waveguides in PRI sol–gel thin films are reviewed [13].

4.6.1 Linear, Bending, and Branching Waveguides

4.6.1.1 Fabrication processes

Various kinds of nanoscale waveguides can be fabricated in PRI sol–gel thin films by the process shown in Fig. 4.38. On a Si substrate with a 2 μm thick SiO_2, a layer of the sol–gel material, which is a silicon oxide-based material incorporating titanium oxide, is formed by spin-coating, and by pre-baking the layer becomes a PRI sol–gel thin film. The thin film is selectively exposed to UV lights from a high-pressure mercury lamp through a photomask, which has 1 μm wide apertures for the waveguide core patterns, with light intensity of ~5 mW/cm^2 and energy density of 9000 mJ/cm^2. UV light exposure

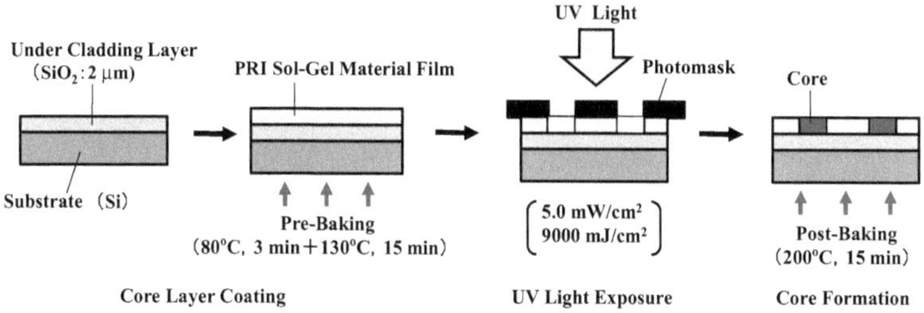

Figure 4.38 Fabrication process of optical waveguides using the PRI sol–gel material. Reprinted with permission from Ono et al. [13].

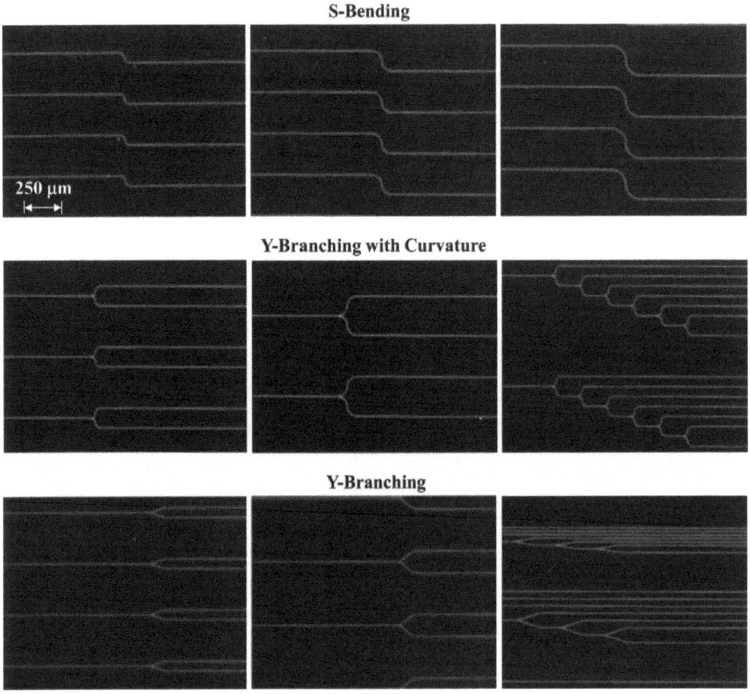

Figure 4.39 Nanoscale optical circuits fabricated in the PRI sol–gel thin films by the process shown in Fig. 4.38. Reprinted with permission from Ono et al. [13].

has a role in activating the hydrolytic condensation of titanium oxide. During post-baking at 200°C for 15 min, hydrolytic condensation proceeds more rapidly in UV-exposed areas than in unexposed areas, yielding higher refractive index in the UV-exposed areas than in the surrounding areas to produce cores. Examples of nanoscale optical circuits in PRI sol–gel thin films are shown in Fig. 4.39. The core thickness is ~230 nm and the core width is ~1000 nm.

The difference in the refractive index between the cores and cladding regions (Δn), which is proportional to the difference in the amount of activation of hydrolytic condensation, is controllable with UV light dose. Initial refractive index, which corresponds to the refractive index of the cladding region, is 1.65. Δn increases with UV light dose, and is saturated to be ~0.2 when the UV light dose reaches 9000 mJ/cm^2. Δn is basically not affected by the pre-/post-baking temperature below 200°C. However, high-temperature baking (ex. over 300°C) induces hydrolytic condensation even in unexposed areas, and hence, Δn tends to reduce with increasing baking temperature.

In intra-chip optical interconnects, 850 nm is popular as the signal light wavelength. However, from a viewpoint of detector sensitivity, short-wavelength light is preferable [14]. Short-wavelength light is also preferable to miniaturize optical circuits, increasing wiring density. The measurements of optical circuits reported in the present section were carried out using a He-Ne laser of 633 nm.

4.6.1.2 Linear waveguides

Figure 4.40 shows output beams from a core edge of a linear waveguide. Propagation loss at 633 nm was found from the cutback method to be 1.86 dB/cm for TE mode (polarization//x axis) and 1.89 dB/cm for TM mode (polarization//y axis), indicating small polarization dependences.

Guided beam confinement was investigated by FFPs of output beams from the core edge. Spread angles of the output beams are estimated by the width of the FFPs at light intensity of $1/e$ of the

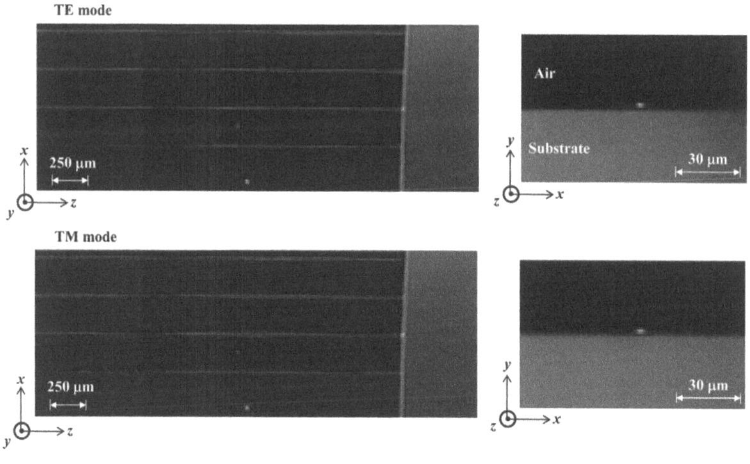

Figure 4.40 Linear waveguides and output beams from the core edges. Reprinted with permission from Ono et al. [13].

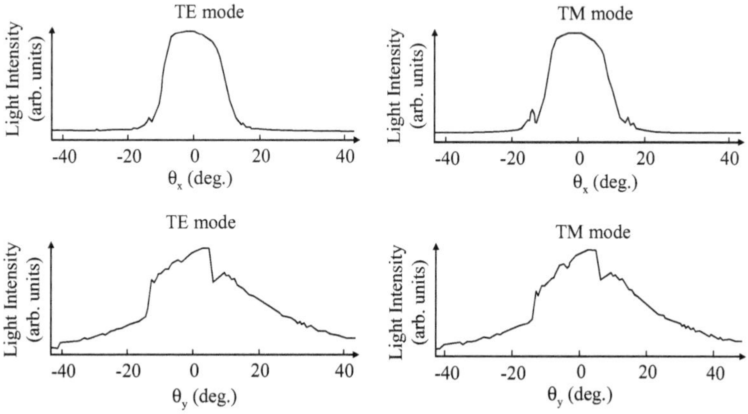

Figure 4.41 FFPs of output beams from the core edge of the linear waveguide. Reprinted with permission from Ono et al. [13].

maximum height. As can be seen in Fig. 4.41, the spread angles along core width direction (//x axis) and along core thickness direction (//y axis) are respectively 20° and 40° for both TE mode and TM mode.

Spot radius w of guided beams can be calculated from the spread angles of output beams using the following equation:

$$w = \frac{180 \times \lambda}{2\theta \times \pi^2 n} \tag{4.1}$$

Here, 2θ [°] is the spread angle, λ [μm] is the guided beam wavelength, and n is the refractive index of free space, where emitted beams propagate. In the present case, n is 1. Spot radii of guided beams along x axis and along y axis were calculated to be 600 nm and 300 nm, respectively, for both TE and TM modes. These results indicate that PRI sol–gel thin films realize guided beam confinement within a cross-section area less than 1.0 μm^2 with small polarization dependence.

Light intensity profiles in optical waveguides made of PRI sol–gel thin films were simulated by BPM. The model is shown in Fig. 4.42. The core thickness and width are respectively 230 nm and 1000 nm. The refractive indexes of the core and cladding are respectively

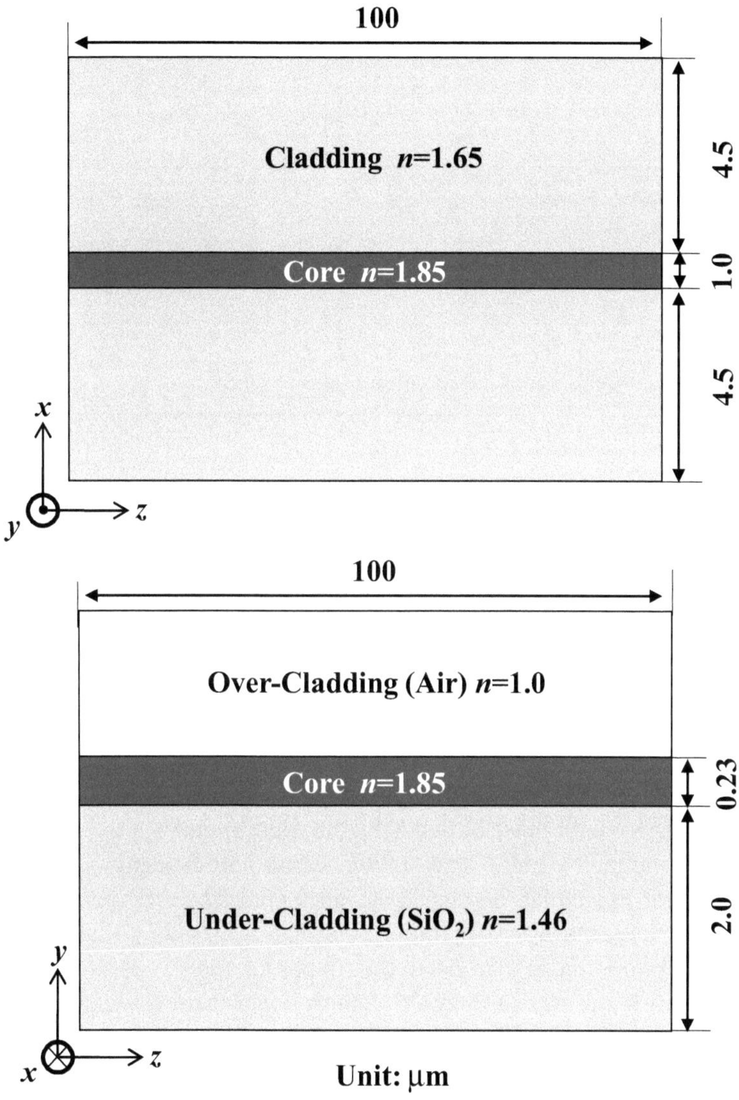

Figure 4.42 Model for BPM simulations of light intensity profiles in optical waveguides made of PRI sol–gel thin films. Reprinted with permission from Ono et al. [13].

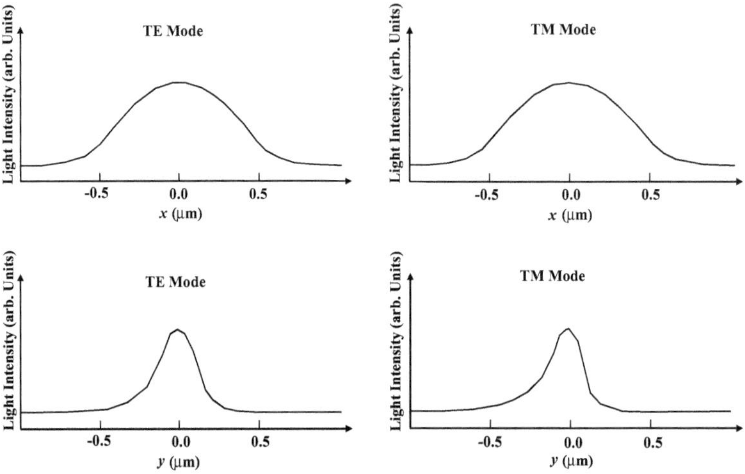

Figure 4.43 Results of the BPM simulations of light intensity profiles in the optical waveguides made of PRI sol–gel thin films. Reprinted with permission from Ono et al. [13].

1.85 and 1.65. The thickness of the SiO_2 undercladding is 2 μm, and the refractive index is 1.46. The wavelength of guided beams is 633 nm. The results are shown in Fig. 4.43. Spot radii of guided beams along *x* axis and along *y* axis are respectively 500 nm and 200 nm for both TE and TM modes. The spot radii obtained by the simulations fairly agree with those obtained by the experiments.

It should be noted that guided beam confinement is slightly stronger in simulations than in experiments. The difference might be attributed to the fact that Δn is slightly smaller in fabricated optical waveguides than Δn in the simulation model, resulting in an increase in the evanescent wave penetration into the cladding region to increase the spot sizes of guided beams.

Si waveguides enable strong guided beam confinement in 0.5 μm wide cores due to its high core refractive index of ~3.5. However, the high refractive index causes a decrease in the propagation speed of optical signals in the Si waveguides. PRI sol–gel material waveguides, on the other hand, exhibit faster propagation speed of optical signals by a factor of ~2 compared with Si waveguides because the core refractive index is ~1.85. Considering the tradeoff relationship between guided beam confinement and propagation speed, it is expected that PRI sol–gel material waveguides can preferably be applied to global wiring in LSIs and MCMs.

It is expected that further reduction of core widths is achieved in PRI sol–gel thin films by optimizing the fabrication processes to increase Δn, which enables core widths less than 1000 nm.

4.6.1.3 Bending and branching waveguides

S-bending waveguides and 633 nm guided beams in them are shown in Fig. 4.44 for TE modes. The core thickness and width of optical waveguides are respectively ~230 nm and ~1000 nm. The beams

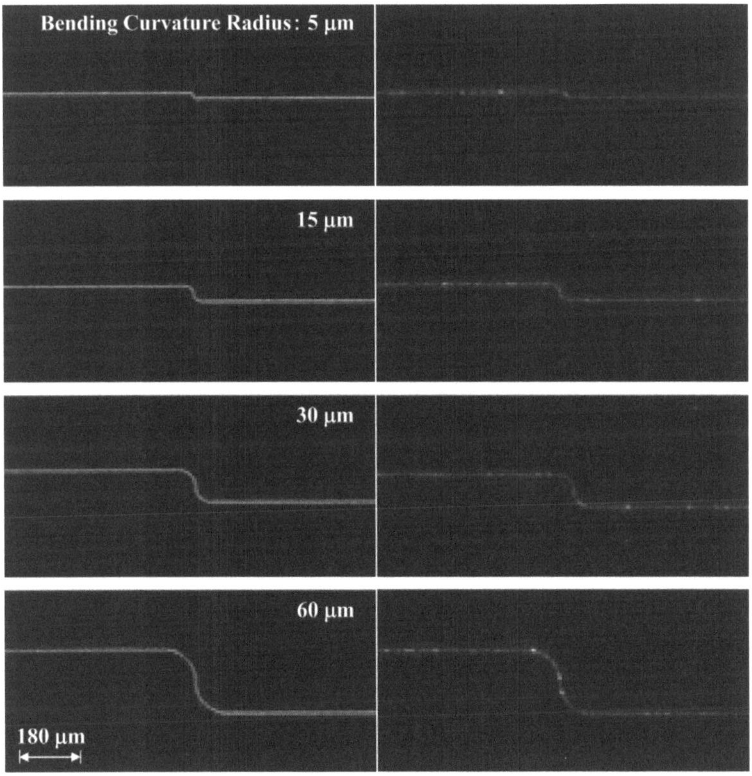

Figure 4.44 S-bending waveguides and 633 nm guided beams in them for TE mode. Reprinted with permission from Ono et al. [13].

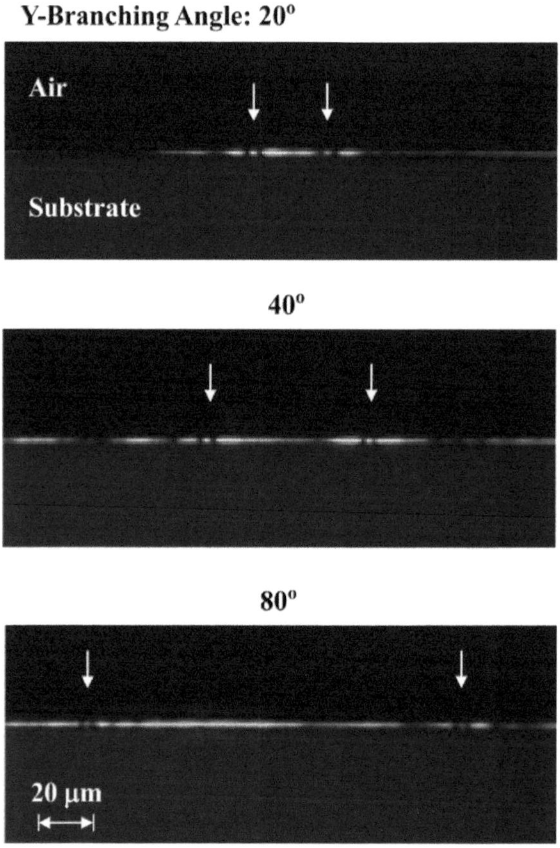

Figure 4.45 Output beams from core edges of Y-branching waveguides for TE mode. Reprinted with permission from Ono et al. [13].

propagate in the S-bending waveguides with bending curvature radii of 5, 15, 30, and 60 μm. TM mode beam propagation is similarly observed in these waveguides.

Output beams from core edges of Y-branching waveguides are shown in Fig. 4.45 for TE modes. Input beams are divided into two branches with branching angles of 20°, 40°, and 80°. For TM mode, same kinds of divided beams were observed.

The S-bending loss is evaluated with following process (Fig. 4.46).

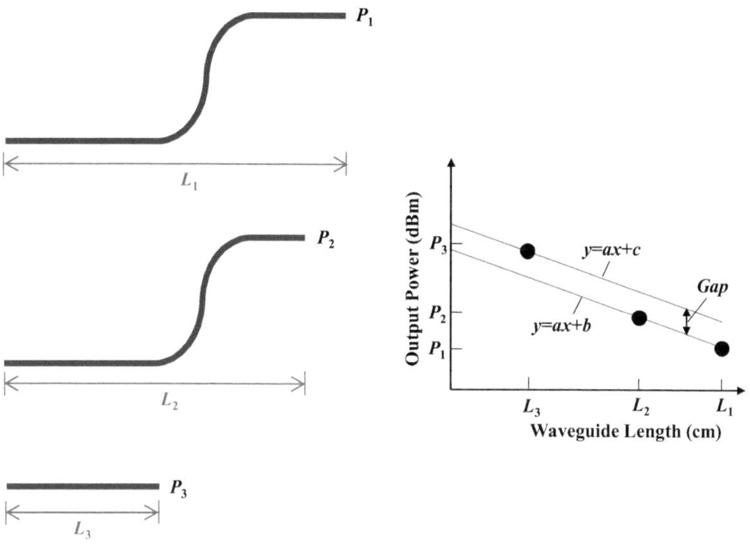

Figure 4.46 Evaluation procedure for bending loss. Reprinted with permission from Ono et al. [13].

(1) Output power P_1 at waveguide length L_1 is measured.
(2) After cutting the waveguide, output power P_2 at waveguide length L_2 is measured.
(3) After the optical waveguide is cut beyond the S-bend, output power P_3 at waveguide length L_3 is measured.
(4) The averaged output power in the unit of dBm at each measurement point is plotted against waveguide length.
(5) (L_1, P_1) and (L_2, P_2) are connected with a straight line of $y = ax + b$, where, slope a gives propagation loss of the linear waveguide. Another straight line of slope a, namely, $y = ax + c$, is traced on (L_3, P_3) to represent power versus waveguide length characteristics before reaching the S-bend part.
(6) Bending loss is obtained from the gap at the S-bend part, $c - b$.

The Y-branching loss is also measured by the same method. The bending and Y-branching losses are expressed as follows. The bending loss for one bend is a half of the S-bending loss.

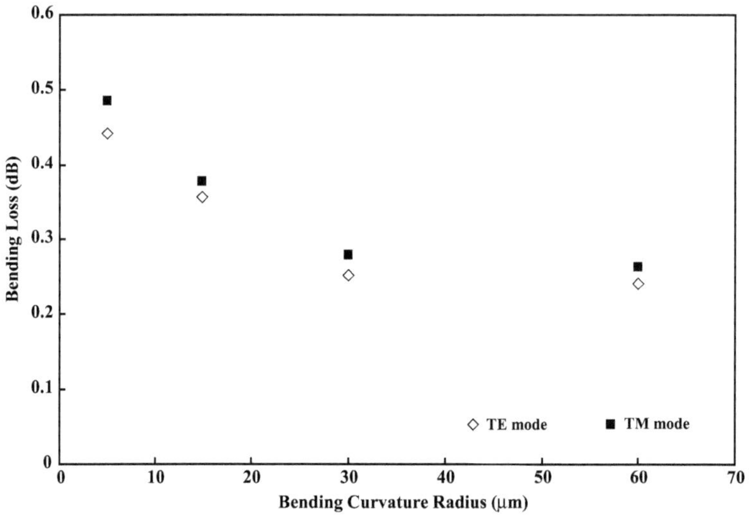

Figure 4.47 Dependence of bending loss on the bending curvature radius. Reprinted with permission from Ono et al. [13].

$$\left.\begin{array}{l}\text{S-bending loss} = c - b\ [dB] \\ \text{Bending loss} = (c - b)/2\ [dB] \\ \text{Y-branching loss} = c - b - 3\ [dB]\end{array}\right\} \quad (4.2)$$

As can be seen in Fig. 4.47, the bending loss is around 0.25 dB for the bending curvature radius from 60 μm to 30 μm. After that, the bending loss increases with a decrease in the bending curvature radius. These results suggest that the PRI sol–gel thin film can produce miniaturized optical circuits with bending curvature radii of several tens μm if bending loss of ~0.3 dB is permitted. For Y-branching, as can be seen in Fig. 4.48, by increasing the branching angle from 20° to 80°, the Y-branching loss increases from ~0.1 dB to ~1.35 dB. If Y-branching loss of ~0.3 dB is permitted, the PRI sol–gel thin film can be used to fabricate optical circuits with a branching angle of several tens of degrees.

Fabrication process optimization will reduce bending and branching losses as well as propagation loss. In fine patterns having

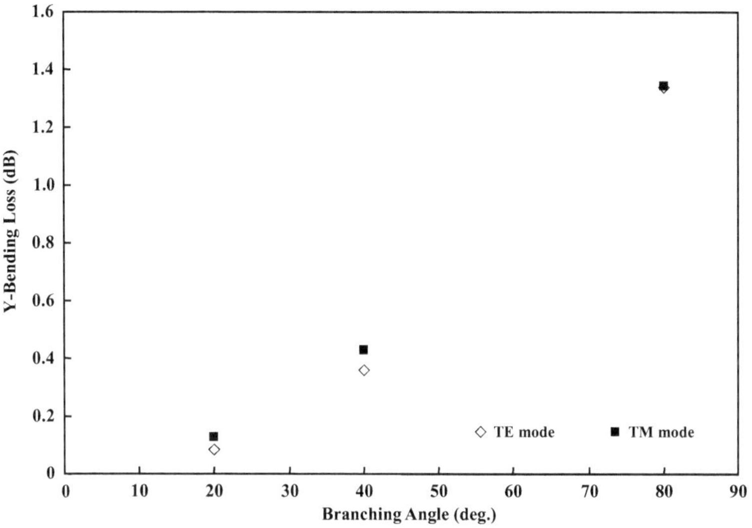

Figure 4.48 Dependence of Y-branching loss on the branching angle. Reprinted with permission from Ono et al. [13].

sizes less than 1000 nm, electron beam lithography might be required for accurate pattern formation.

4.6.2 Vertical Mirrors and All-Air-Cladding Waveguides

Vertical mirrors in the nanoscale waveguide were constructed by tilted exposure using UV light with an incident angle of 45°. In this preliminary experiment, the prism for tilted exposure was not used. So, the slope of the beveled wall in the vertical mirror deviated from 45°. The expected slope was estimated to be 65° due to refraction at the air/sol–gel material boundary. FFP measurements revealed that the reflected beams from the vertical mirror propagate along an axis tilted from the surface-normal direction by 40°. This confirms that the fabricated vertical mirror has a slope of 65°.

The PRI sol–gel material becomes insolvable in Al etchant by exposing the material to UV/blue light followed by post-baking. By using this property, fine 3D structures including all-air-cladding waveguides can be fabricated.

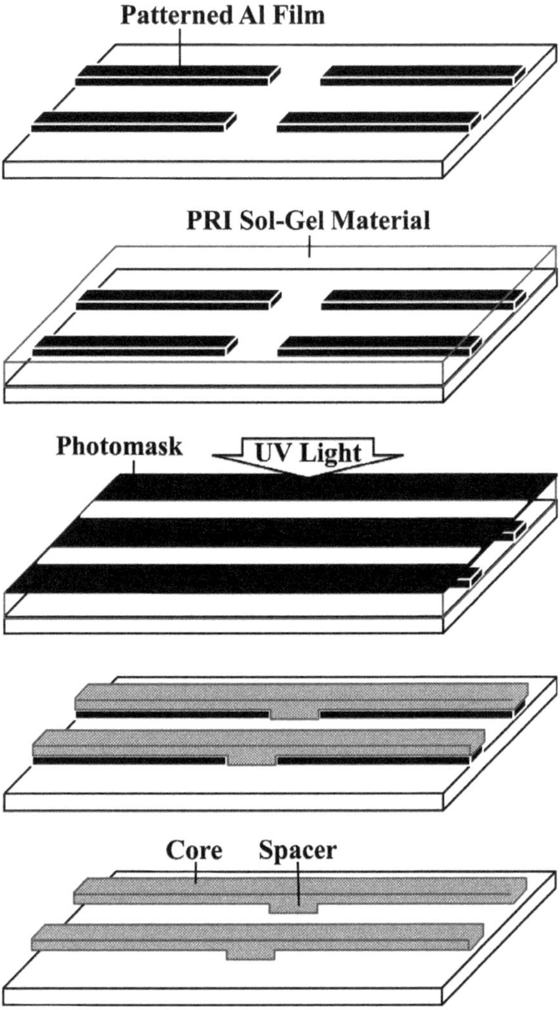

Figure 4.49 Process for fabrication of a 3D waveguide structure.

Figure 4.49 shows an example of the process for fabrication of fine 3D structures. After patterned Al films are deposited on a substrate, a PRI sol–gel thin film is formed on it using spin-coating followed by pre-baking. Through a photomask, the thin film is exposed to UV light with a pattern of waveguide cores and is post-baked. By putting the sample in Al etchant, the unexposed

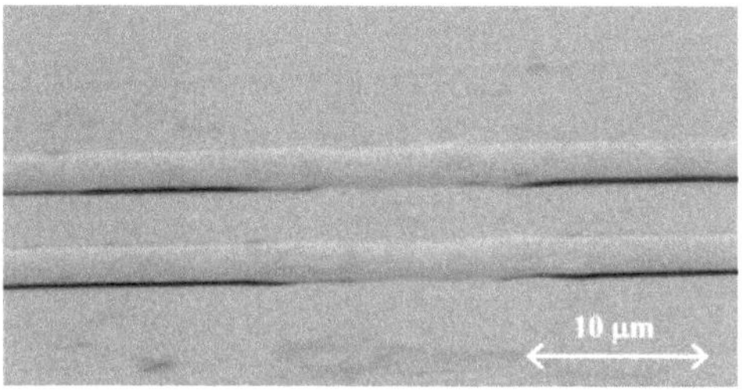

Figure 4.50 SEM image of an all-air-cladding waveguide structure with spacers.

regions of the PRI sol–gel thin film and the Al film are solved. Thus, fine 3D patterns for all-air-cladding waveguides are constructed. Figure 4.50 shows a scanning electron microscope (SEM) image of a 3D structure for all-air-cladding waveguides with spacers.

References

1. T. Inoguchi, S. Moriya, T. Yamamoto, Y. Arai, T. Yoshimura, and K. Asama, "Film waveguides with micro mirrors for three-dimensional optical circuits fabricated by the built-in mask method," *J. Electronics Jisso* **8**, 237–242 (2005) [in Japanese].
2. T. Yoshimura, T. Inoguchi, T. Yamamoto, S. Moriya, Y. Teramoto, Y. Arai, T. Namiki, and K. Asama, "Self-organized lightwave network based on waveguide films for three-dimensional optical wiring within boxes," *J. Lightwave Technol.* **22**, 2091–2100 (2004).
3. T. Yoshimura, M. Miyazaki, Y. Miyamoto, N. Shimoda, A. Hori, and K. Asama, "Three-dimensional optical circuits consisting of waveguide films and optical Z-connections," *J. Lightwave Technol.* **24**, 4345–4352 (2006).
4. T. Yoshimura, T. Kofudo, T. Kashiwazako, K. Naito, K. Ogushi, and Y. Kitabayashi, "A material-saving optical waveguide fabrication process with selective transfers of cores," *Opt. Eng.* **47**, 014601-1-7 (2008).
5. F. Huang, H. Takase, Y. Eriyama, and T. Ukachi, "Optical fiber interconnection by using self-written waveguides," *Proc. 9th Int. Conf. Rad. Tech. Asia*, Japan, 637–639 (2003).
6. T. Yoshimura, K. Ogushi, Y. Kitabayashi, K Naito, Y. Miyamoto, and M. Miyazaki, "Optical waveguide films with two-layer skirt-type core end

facets for beam leakage reduction at 45° mirrors," *Proc. SPIE* **6899**, from Photonics West 2008, San Jose, California, 689913-1-11 (2008).

7. T. Yoshimura, K. Ogushi, Y. Kitabayashi, Y. Miyamoto, M. Miyazaki, "Optical waveguide films with two-layer skirt-type core structures for beam leakage/scattering reduction at tapered mirrors," *IEEE Photonics Technol. Lett.* **19**, 1427–1429 (2007).
8. T. Yoshimura, J. Roman, Y. Takahashi, M. Lee, B. Chou, S. I. Beilin, W. V. Wang, and M. Inao, "Proposal of optoelectronic substrate with Film/Z-connection based on OE-film," *Proc. 3rd IEMT/IMC Symposium*, 140–145 (1999).
9. T. Yoshimura, J. Roman, Y. Takahashi, M. Lee, B. Chou, S. I. Beilin, W. V. Wang, and M. Inao, "Optoelectronic scalable substrates based on Film/Z-connection and its application to film optical link module (FOLM)," *Proc. SPIE* **3952**, from Photonics West 2000, San Jose, California, 202–213 (2000).
10. T. Yoshimura, J. Roman, P. Reilly, Y. Takahashi, M. Lee, B. Chou, S. I. Beilin, W. V. Wang, and M. Inao, "Proposal of film optical link module (FOLM) and beveled cut process for the high density waveguide couplers," *Proc. 1999 IEEE LEOS Annual Meeting Conference*, San Francisco, California, 27–28 (1999).
11. Y. Lin, L. Lin, C. Choi, B. Bihara, and R. Chen, "Optoelectronic integration of polymer waveguide array and metal-semiconductor-metal photodetector through micromirror couplers," *IEEE Photonics Technol. Lett.* **13**, 355–357 (2001).
12. E. Cassan, S. Laval, S. Lardenois, and A. Koster, "On-chip optical interconnects with compact and low-loss light distribution in silicon-on-insulator rib waveguides," *IEEE J. Sel. Top. Quantum Electron.* **9**, 460–464 (2003).
13. S. Ono, T. Yoshimura, T. Sato, and J. Oshima, "Fabrication and evaluation of nano-scale optical circuits using sol–gel materials with photo-induced refractive index variation characteristics," *J. Lightwave Technol.* **27**, 1229–1235 (2009).
14. A. Bhatnagar, S. Latif, C. Debaes, D. A. B. Miller, "Pump-probe measurements of CMOS detector rise time in the blue," *J. Lightwave Technol.* **22**, 2213–2217 (2004).

Chapter 5

Resource-Saving All-Photolithographic Heterogeneous Integration: PL-Pack with SORT

As described in Chapter 2, the S-FOLM fabricated by heterogeneous integration is expected to solve the *optics excess* problems in optical interconnects within boxes of computers.

Currently the bulk-chip packaging based on flip-chip bonding is a major technology [1, 2] for OE module fabrication. However, it is not applicable to S-FOLM fabrication because thick bulk chips cannot be buried in films, and furthermore, it is material/space-consuming and requires large efforts to assemble chips with precise alignments. Meanwhile, as mentioned in section 2.4, heterogeneous integration utilizing the ELO [3–7] enables us to embed small/thin OE dies in films to fabricate OE-Films that can easily be stacked to build 3D-stacked structures. It also enables reductions in material/space and OE device alignment efforts.

In the process utilizing the ELO, usually, small/thin OE dies are placed on substrates one by one by the pick-and-place method, which is time-consuming. The essential issue in the heterogeneous integration utilizing the ELO is how to distribute small/thin OE dies on substrates with minimized step counts.

Self-Organized 3D Integrated Optical Interconnects with All-Photolithographic Heterogeneous Integration
Tetsuzo Yoshimura

ISBN 978-981-4877-04-6 (Hardcover), 978-1-003-04305-8 (eBook)
www.jennystanford.com

In the present chapter, a resource-saving all-photolithographic heterogeneous integration process, "Photolithographic Packaging with Selectively Occupied Repeated Transfer (PL-Pack with SORT)," is reviewed [8–14]. PL-Pack with SORT is performed by multi-step device transfers. In each transfer step, a large number of small/thin OE dies are placed on a substrate at one time with designated distributions to reduce the step counts. PL-Pack with SORT uses photolithography-based processes similar to those for LSI fabrication to avoid individual chip assembling, and achieves cost/resource saving.

First, after advantages of PL-Pack with SORT over conventional packaging technology are explained, process designs based on PL-Pack with SORT are presented for S-FOLM fabrication. Next, the material consumption and module costs are estimated. Process step counts for SORT and the conventional pick-and-place method are compared. Then, advantages of PL-Pack with SORT against flip-chip-bonding-based packaging are discussed from the viewpoint of cost/resource saving. Finally, experimental demonstrations of SORT for transfers of waveguide lenses, waveguide cores, and transfers of two kinds of model devices are presented.

5.1 Advantages of PL-Pack with SORT over Conventional Packaging

In Fig. 5.1, the S-FOLM (heterogeneous integration in waveguide) fabricated by PL-Pack with SORT is compared with the conventional OE module fabricated by flip-chip-bonding-based packaging. VCSELs and PDs are assembled on a substrate, where optical waveguides and electrodes/pads/lines/vias are distributed.

In flip-chip-bonding-based packaging, a VCSEL chip, a PD chip, and a D/A IC chip are mounted on a substrate, providing a conventional OE module.

In PL-Pack with SORT, first, a small/thin VCSEL die array, a small/thin PD die array, and a small/thin D/A IC die array are grown on a VCSEL growth wafer, a PD growth wafer, and a D/A growth wafer, respectively. Next, VCSEL dies and PD dies

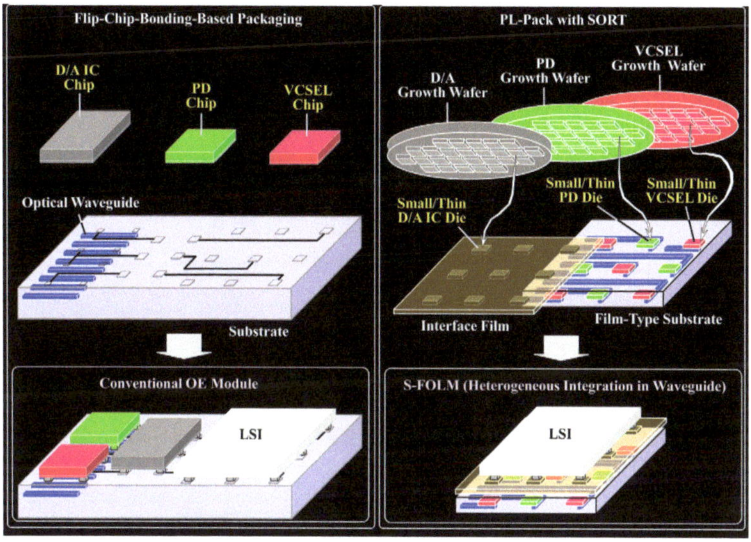

Figure 5.1 (a) Conventional OE module fabricated by flip-chip-bonding-based packaging, and (b) S-FOLM fabricated by PL-Pack with SORT.

in the arrays are selectively transferred to appropriate sites in a film-type substrate, and embedded with optical waveguides and electrodes/pads/lines/vias. Small/thin D/A IC dies are embedded in an interface film. Finally, the interface film is stacked on the film-type substrate to complete S-FOLM (heterogeneous integration in waveguide). The S-FOLM fabricated by PL-Pack with SORT has the following advantages over conventional OE modules fabricated by the flip-chip-bonding-based packaging.

- Space saving (by embedding and stacking)
- Noise reduction (by replacing long metal lines with short vias)
- Cost/resource saving (to be described in section 5.3.1)
- Thermal stress reduction (to be described in section 5.3.2)

PL-Pack with SORT can remove the conventional flip-chip-bonding-based batch-type packaging, which is called *jisso gijutsu* in Japanese, to realize the heterogeneous integration achieved only by semiconductor device fabrication processes based on photolithography. These features are effective to reduce *optics excess*.

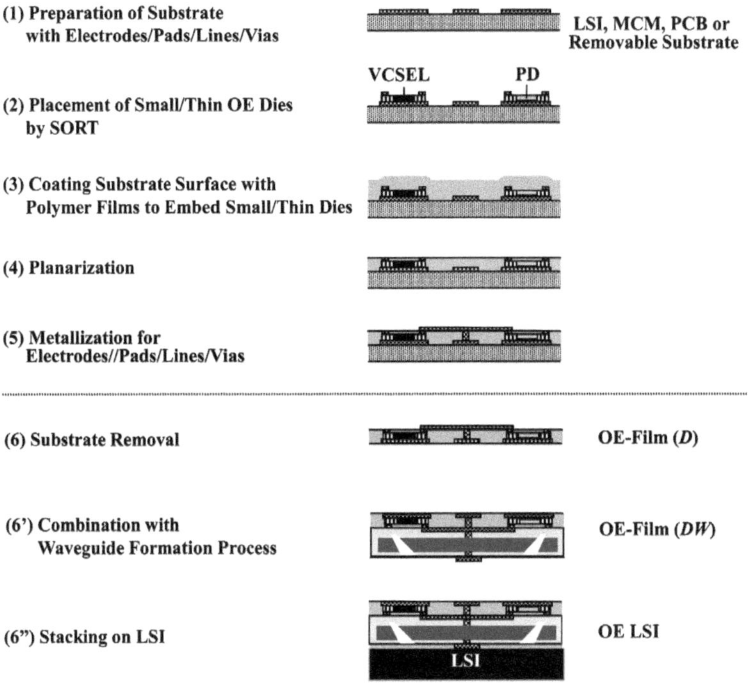

Figure 5.2 Whole process flow of the PL-Pack with SORT. Reprinted with permission from Yoshimura et al. [8].

5.2 PL-Pack with SORT

5.2.1 Whole Process Flow of PL-Pack with SORT

The whole process flow of PL-Pack with SORT is shown in Fig. 5.2. A substrate (LSI, MCM, PCB, removable substrate, etc.) with electrodes/pads/lines/vias is prepared. Small/thin OE dies, which are produced by ELO, polishing, or something like that, are placed on pads of the substrate by SORT described in section 5.2.2. After coating the substrate surface with polymer films to embed small/thin OE dies, the surface is planarized. Then, metallization for electrodes/pads/lines/vias are carried out.

By substrate removal, a small/thin-OE-die-embedded OE-Film, which corresponds to OE-Film (*D*) described in section 3.1, is obtained. By combining process steps (1)–(5) with waveguide formation processes, OE-Film (*DW*) described in section 3.1 is

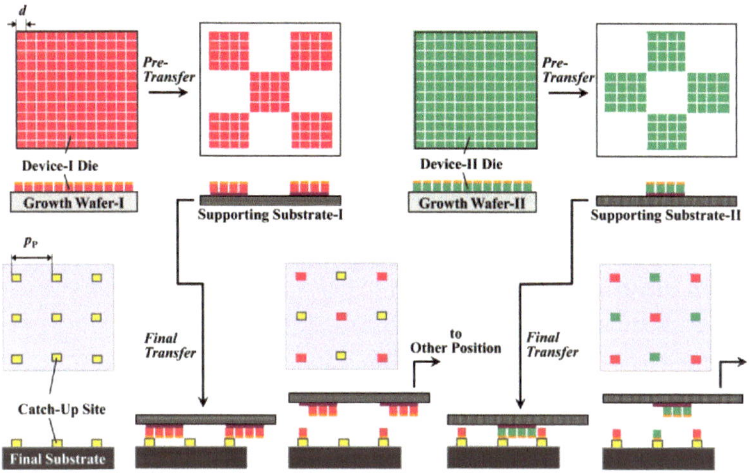

Figure 5.3 Process flow of the SORT for integration of two kinds of devices. Reprinted with permission from Yoshimura et al. [10].

obtained. By stacking OE-Film (*DW*) on an LSI, an OE LSI is constructed.

Here, the critical part is step (2), that is, the process for small/thin OE die placement on a substrate by SORT, whose details are described in section 5.2.2.

5.2.2 Process Flow of SORT

Figure 5.3 shows a process flow of SORT. For simplicity, two kinds of devices are considered although, in general, many kinds of devices are integrated. A thin film for Device-I die array and a thin film for Device-II die array are respectively grown on Growth Wafer-I and Growth Wafer-II. The small/thin dies are obtained by dividing the thin films with boundary grooves. The die size is $d \times d$. All the Device-I dies are picked up on a pick-up substrate at one time and are selectively pre-transferred from the pick-up substrate to Supporting Substrate-I. The dies left on the pick-up substrate are provided to another supporting substrate. Similarly, the Device-II dies are selectively pre-transferred to Supporting Substrate-II, and the dies left on the pick-up substrate are provided to another supporting substrate. The distribution patterns of the small/thin dies on Supporting Substrate-I and Supporting Substrate-II are respectively designed to match the pad distribution patterns for Device-I and Device-II on final substrates. For the final transfer, Supporting Substrate-I is attached to one of the final substrates, where catch-up sites with a pitch of p_p are formed, to place Device-I dies at sites for Device-I on the final substrate selectively. Then, Supporting Substrate-II is attached to the final substrate to place Device-II dies at sites for Device-II. Supporting Substrates-I and II are further attached to other final substrates or other places of the final substrate to provide Device-I dies and Device-II dies selectively. By repeating the procedure, all the dies on the supporting substrates are transferred to final substrates.

For the case shown in Fig. 5.3, there are $4 \times 4 = 16$ dies for one catch-up site. This implies that one supporting substrate can provide dies to 16 final substrates. Because two supporting substrates are available for Device-I and II, respectively, 32 final substrates, on which Device-I dies and Device-II dies are distributed with a desired configuration, can be obtained from Growth Wafer-I and Growth Wafer-II.

It should be noted that, in the SORT process, the die placement is performed by aligners similar to those used in the photolithography for the LSI fabrication. This enables precise alignments of a number of small/thin OE dies at one time, reducing the alignment efforts in the assembling processes. For fabrication of OE LSIs and 3D OE platforms such as multilayer OE boards and 3D-stacked OE MCMs, Device-I could be light modulators or VCSELs, and Device-II could be PDs. For fabrication of 3D MOSSs, Device-I and Device-II could be optical switches, tunable wavelength filters, waveguide lenses, and so on.

An example of the precise SORT flow for two kinds of devices is illustrated in Figs. 5.4a–e.

(a) A Device-I die array is grown on Growth Wafer-I, with a pitch of d, which is assumed to equal the die size by neglecting the boundary groove width. By ELO, the die array is transferred to a pick-up substrate, Sub-I.1. Similarly, a Device-II die array grown on Growth Wafer-II is transferred to a pick-up substrate, Sub-II.1.

(b) For pre-transfer, Sub-I.1 is attached to a supporting substrate, Sub-I.2(1), where adhesive layers are formed. Here, the adhesive layer pattern is designed so that the adhesive regions match Device-I sites on the final substrate. Then, Device-I dies are selectively placed on Sub-I.2(1). The Device-I dies left on Sub-I.1 are placed on Sub-I.2(2).

(c) With similar procedures, Device-II dies are selectively pre-transferred on Sub-II.2(1) and Sub-II.2(2).

(d) Final substrates, Substrate(1), (2), . . ., (32) are prepared. The substrates have catch-up sites with a pitch of p_P. For final transfer, by attaching Sub-I.2(1) to Substrate(1), Device-I dies are selectively placed at sites for Device-I on Substrate(1). By attaching Sub-I.2(1) to Substrate(2), Device-I dies are selectively placed at sites for Device-I on Substrate(2). The same procedure is repeated to provide the dies to other final substrates. At last, Sub-I.2(1) can make $4 \times 4 = 16$ final substrates of Substrate(1), (2), . . ., and (16) with Device-I dies. Similarly, Sub-I.2(2) makes 16 more final substrates of Substrate(17), (18), . . ., and (32) with Device-I dies. Thus, 32 final substrates with Device-I dies are made.

(e) On the Substrate(1) with Device-I dies, Sub-II.2(1) is attached to put Device-II dies selectively at sites for Device-II on Substrate(1). By repeating the procedure, finally, 32 final substrates, on which Device-I dies and Device-II dies are distributed with a designated configuration, are obtained.

In SORT the marker alignment procedure, which is used in the photolithography for LSI fabrication, can be applied to die transfer by using aligners. This means that SORT can be regarded as an all-photolithographic process, which eliminates *jisso gijutsu*, namely, the conventional flip-chip-bonding-based batch-type packaging.

In the process flow model shown in Fig. 5.4, the supporting substrates and the final substrates have the same size. However, the model can work in the case that final substrates are larger than the supporting substrates just by replacing "Substrate(1)" and "Substrate(2)" with "Region(1)" and "Region(2)." In other words, dies are transferred from a supporting substrate to a plurality of regions in the same final substrate.

SORT can be performed more stably by using an adhesive, whose adhesion strength is controlled with "adhesion strength hierarchy" or "dynamic adhesion strength control." In the former, the adhesion

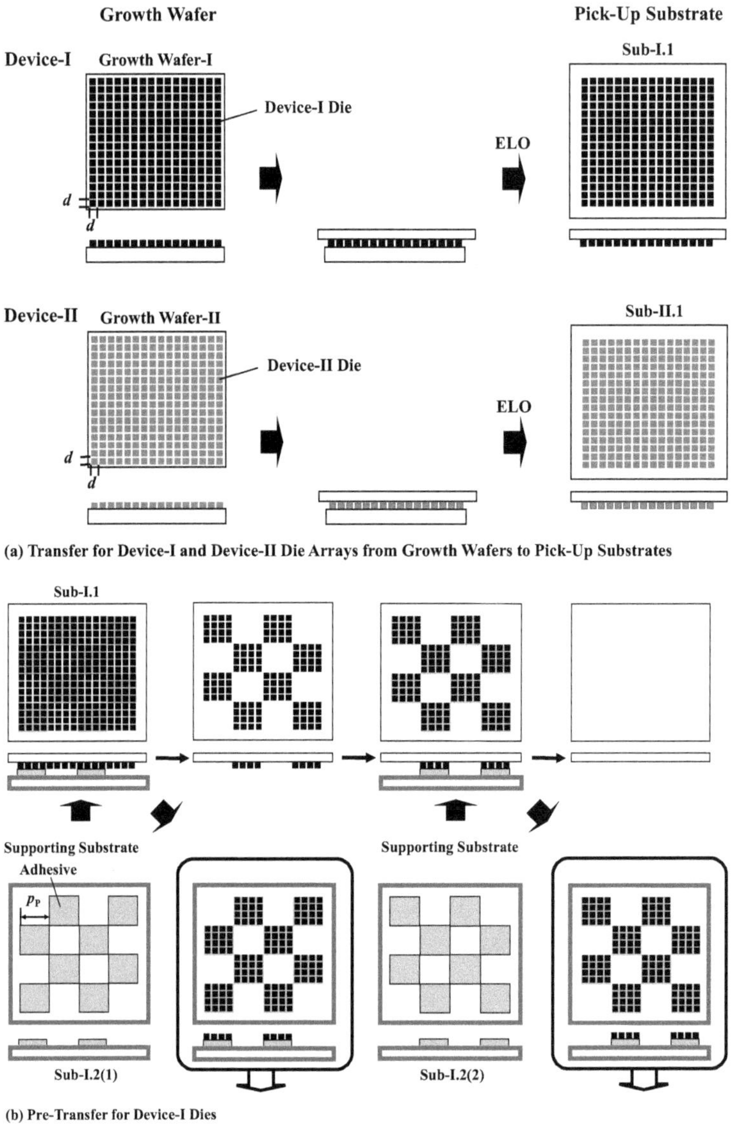

Figure 5.4 Example of the SORT process for two kinds of devices. (a) Transfer for Device-I die array and Device-II die array from growth wafers to pick-up substrates by the ELO, (b) pre-transfer for Device-I dies, (c) pre-transfer for Device-II dies, (d) final-transfer for Device-I dies, and (e) final-transfer for Device-II dies. Reprinted with permission from Yoshimura et al. [9].

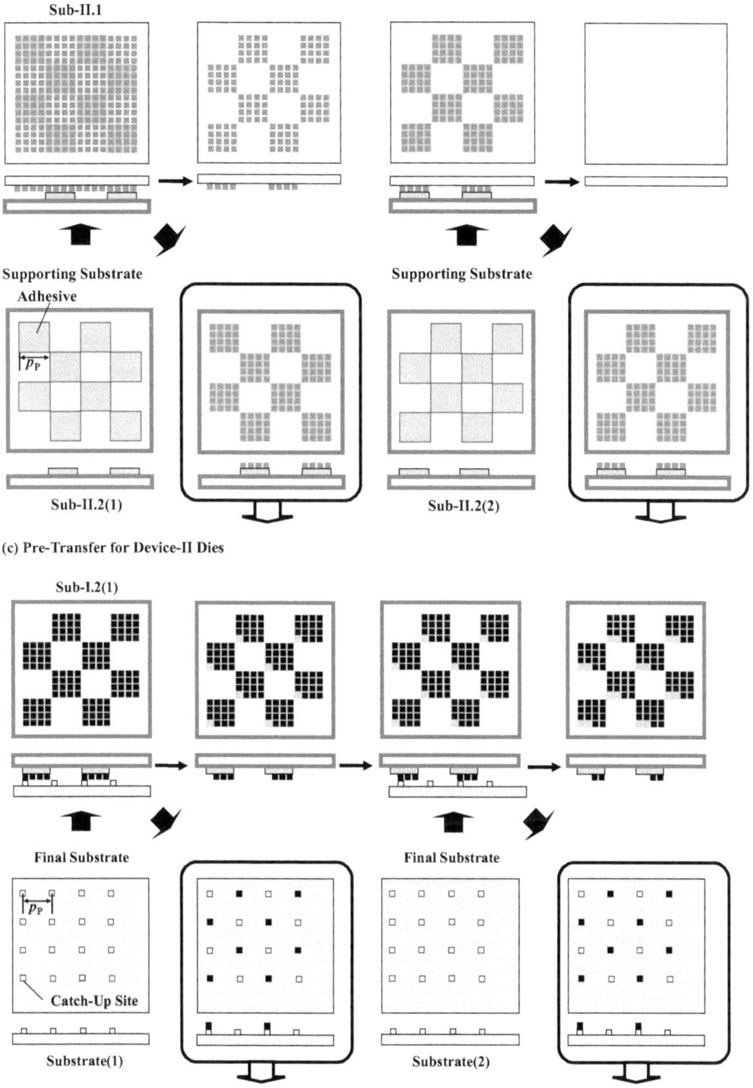

(c) Pre-Transfer for Device-II Dies

(d) Final-Transfer for Device-I Dies

Figure 5.4 (*Continued*)

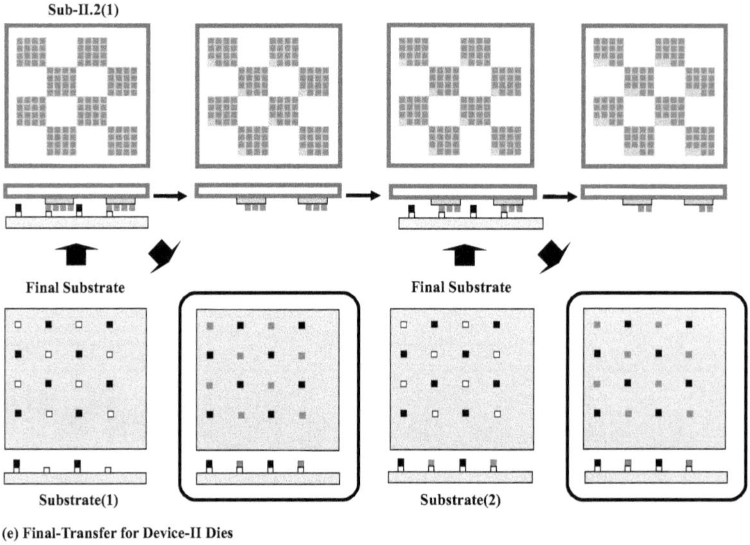

Figure 5.4 (*Continued*)

strength is increased with step progress, that is, in the order of pick-up substrates, supporting substrates, and final substrates. In the latter, adhesion strength of the adhesive is controlled during the process by, for example, UV light exposure using UV-curable materials, which corresponds to light-assisted SORT described in section 5.6.1.

The SORT process described above is one of the examples. There are a lot of process variations. The essential point is that a number of dies are distributed on final substrates with designated configurations with minimum steps by successive device transfers using aligners and supporting substrates, where patterned selective adhesion regions are formed.

5.3 Impacts of PL-Pack with SORT

Structures built by PL-Pack with SORT and the conventional flip-chip-bonding-based packaging are compared in Fig. 5.5 for devices made of III–V materials like VCSELs and PDs. In flip-chip-bonding-based packaging, a bulk chip of III–V materials is put on a substrate.

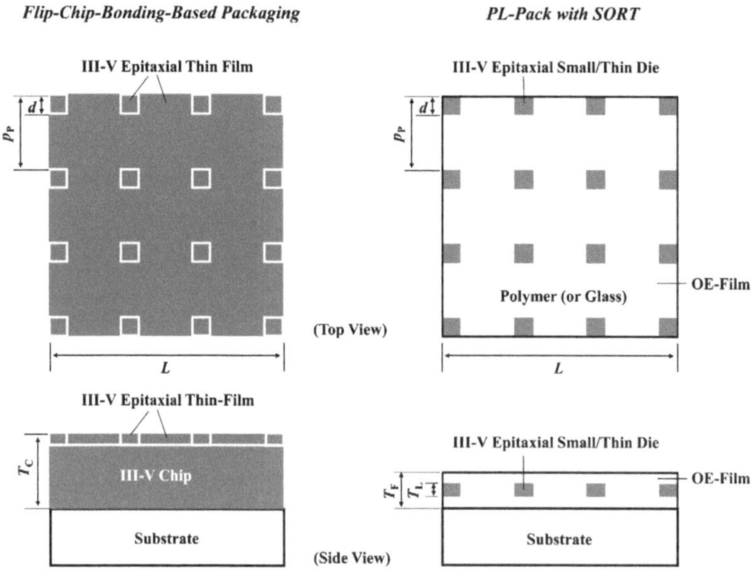

Figure 5.5 Structures built by the PL-Pack with SORT and the conventional flip-chip-bonding-based packaging. Reprinted with permission from Yoshimura et al. [10].

In PL-Pack with SORT, an OE-Film, in which III–V epitaxial small/thin dies with a length/width of d are embedded with a pitch of p_P, is put on a substrate. The structural difference brings remarkable advantages to the PL-Pack with SORT, as described in sections 5.3.1–5.3.4

5.3.1 Material Consumption and Costs

One of the advantages of PL-Pack with SORT is material-saving characteristics. In flip-chip-bonding-based packaging, as shown in Fig. 5.5, the whole area is occupied by expensive III–V epitaxial materials. In PL-Pack with SORT, on the other hand, epitaxial materials exist only at necessary sites.

The effects of III–V epitaxial material saving can be measured by the material saving factor (MSF) of PL-Pack with SORT defined by

$$MSF = [1/R_{\text{SORT}}]/[1/R_{\text{FC Bonding}}] = (p_P/d)^2. \qquad (5.1)$$

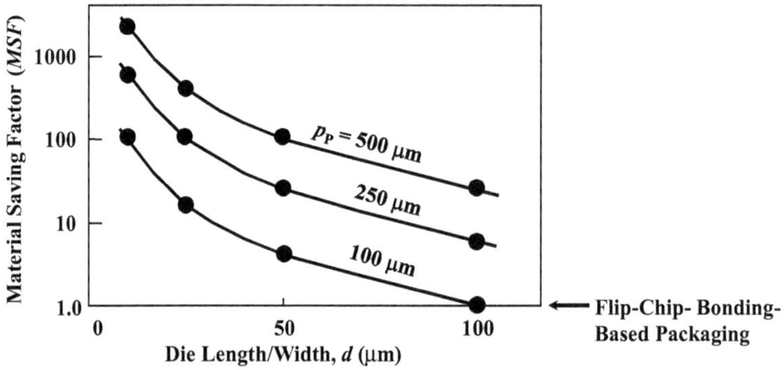

Figure 5.6 Dependence of *MSF* on die length/width d for various die pitches p_P. Reprinted with permission from Yoshimura et al. [10].

Here, R_{SORT} and $R_{FC\ Bonding}$ represent III–V epitaxial material consumption in PL-Pack with SORT and in flip-chip-bonding-based packaging, respectively. The die length/width d and pitch p_P in the OE-Film in Fig. 5.5 respectively correspond to the die pitch on the growth wafer and the catch-up site pitch in Fig. 5.4. Figure 5.6 shows *MSF* as a function of d for various p_P. *MSF* for flip-chip-bonding-based packaging is 1. For PL-Pack with SORT, when p_P is 100 μm, *MSF* equals 1 for d of 100 μm. This implies that the whole area of the OE-Film is occupied by III–V epitaxial materials because p_P is equal to d. *MSF* increases with an increase in p_P and with a decrease in d, reaching more than 100 ~ 1000 typically.

The resource-saving characteristics of PL-Pack with SORT results in cost reduction of OE systems. Stirk et al. reported "cost versus device pitch on growth wafers" characteristics [15]. Based on their results, the impact of PL-Pack with SORT on the cost for an OE module with a VCSEL array can be estimated as shown in Fig. 5.7. As the 0th approximation, costs are assumed to be the same in PL-Pack with SORT and flip-chip-bonding-based packaging except for the III–V epitaxial material cost. Due to the epitaxial-material saving effect in PL-Pack with SORT, the OE module cost is reduced with a decrease in d. For example, in the case that p_P and d are respectively 200 μm and 50 μm, cost reduction of ~1/6 is expected. By further decrease in d, cost reduction of <1/10 is expected.

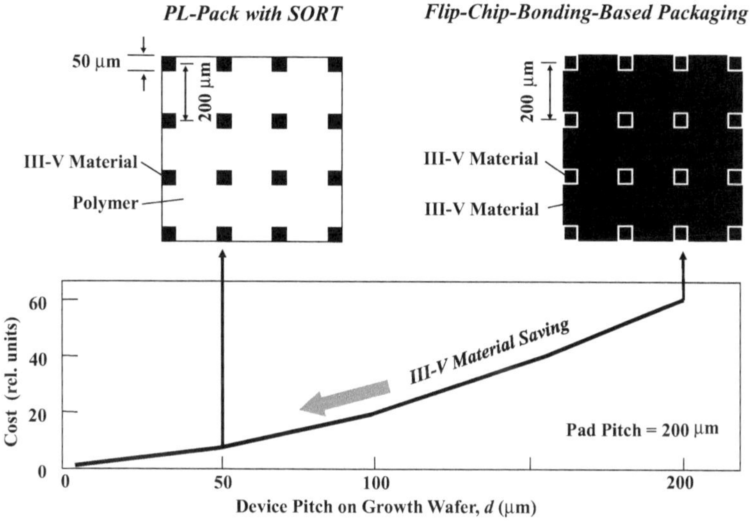

Figure 5.7 "Cost versus device pitch on growth wafers" characteristics for an OE module with a VCSEL array.

Another cost reduction factor in PL-Pack with SORT is process simplicity. As shown in Fig. 5.1, heterogeneous integration can be performed by delivering small/thin OE dies from a set of growth wafers, on which die arrays consisting of a single kind of devices are grown. For example, in an OE module for WDM optical interconnects with 10 wavelengths, 10 kinds of laser diodes (LDs) or VCSELs are involved. The heterogeneous integration can be achieved by delivering 10 kinds of OE devices from 10 growth wafers for individual wavelengths to a final substrate. Each growth wafer has same-wavelength devices made by a simple growth method. Sophisticated epitaxial growth methods for integrating OE devices of different wavelengths into one chip is not necessary.

5.3.2 Mechanical Properties

PL-Pack with SORT has the potential to reduce thermal stress. As an example, consider the case that a two-dimensional VCSEL array of L^2 μm^2, in which VCSELs are distributed with a pitch of p_P μm, is mounted on a plastic substrate as depicted in Fig. 5.5.

For flip-chip-bonding-based packaging, an L^2 μm^2 bulk III–V chip with a thickness of T_C μm is mounted on a plastic substrate. For PL-Pack with SORT, an L^2 μm^2 OE-Film with a thickness of T_L μm, where T_F μm thick III–V epitaxial small/thin dies with an area of d^2 μm^2 are embedded with a pitch of p_P μm, is mounted on a plastic substrate.

In flip-chip-bonding-based packaging, due to the large difference in CTE between the III–V material and the plastics as well as due to the large thickness of the III–V chip with a large Young's modulus, a large thermally induced stress between the chip and the substrate is caused. In PL-Pack with SORT, the thickness of the III–V material dies is reduced by a factor of T_F/T_C from that of the III–V chip, and the area by a factor of d^2/L^2, indicating that the III–V material volume is reduced by a factor of $(T_F/T_C)(d^2/L^2)$. For the case that $T_F = 15$ μm, $T_C = 500$ μm, $d = 20$ μm, and $L = 1$ cm, roughly speaking, the power inducing the stress is reduced by seven orders of magnitude in PL-Pack with SORT.

Figure 5.8 shows the material volume fraction (*MVF*) of the III–V material in an OE-Film. The *MVF* is defined by

$$MVF = [d^2 T_F]/[p_P^2 T_L]. \tag{5.2}$$

MVF for flip-chip-bonding-based packaging is 1. For PL-Pack with SORT, *MVF* decreases with a decrease in d. In the case that $d = 20$ μm, $T_F = 15$ μm, $p_P = 200$ μm, and $T_L = 46$ μm, *MVF* is calculated to be 0.003, suggesting that the CTE of the OE-Film is close to that of the host matrix polymer with a low Young's modulus. Thus, stress between the OE-Film and the plastic substrate is drastically reduced compared to that between the III–V chip and the plastic substrate.

Furthermore, basically, no large thermal expansion differences arise between OE-Films because the CTE of the films is almost determined by the host matrix polymer. Then, it is expected that thermally induced displacement at optical Z-connections might be neglected in S-FOLMs, where stacked films are pinned by electrical pad connections.

The arguments mentioned above are applicable to other devices such as PDs, light modulators, optical switches, microre-

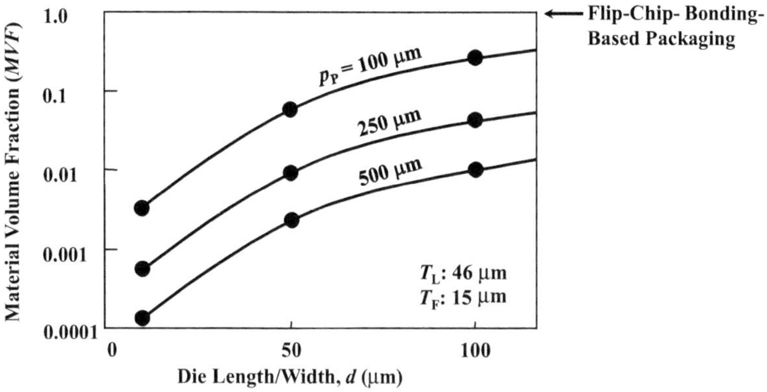

Figure 5.8 Dependence of *MVF* on die length/width d for various die pitches p_P. T_L and T_F are respectively 46 μm and 15 μm. Reprinted with permission from Yoshimura et al. [10].

flectors/lenses/prisms, photonic crystals, ICs, dielectric films like lithium niobate (LN) and PLZT, organic compounds, glass, and so on.

5.3.3 Transfer Step Count

The main concern in the SORT process is the transfer step counts required to place small/thin dies on the final substrates. The die transfer step counts can be calculated using a model shown in Fig. 5.9a. Two kinds of devices, Device-I and Device-II, are considered, namely, the number of device types, M, is 2. The Device-I dies and Device-II dies are assumed to be placed alternately on a final substrate.

Device-I dies are on a pick-up substrate with a pitch of d. The effective area of the substrate occupied by the dies is $L \times L$. The Device-I dies are transferred to M supporting substrates by pre-transfers. Similarly, the Device-II dies on another pick-up substrate are transferred to M more supporting substrates. Thus, M^2 steps are involved in the pre-transfer for M kinds of devices. By final transfers, individual Device-I dies are placed on $M\ (p_P/d)^2$ final substrates, where catch-up pads are formed with a pitch of p_P. Device-II dies are placed on $M\ (p_P/d)^2$ final substrates, on which the Device-I dies have already been transferred. Consequently, $M\ (p_P/d)^2$ final substrates,

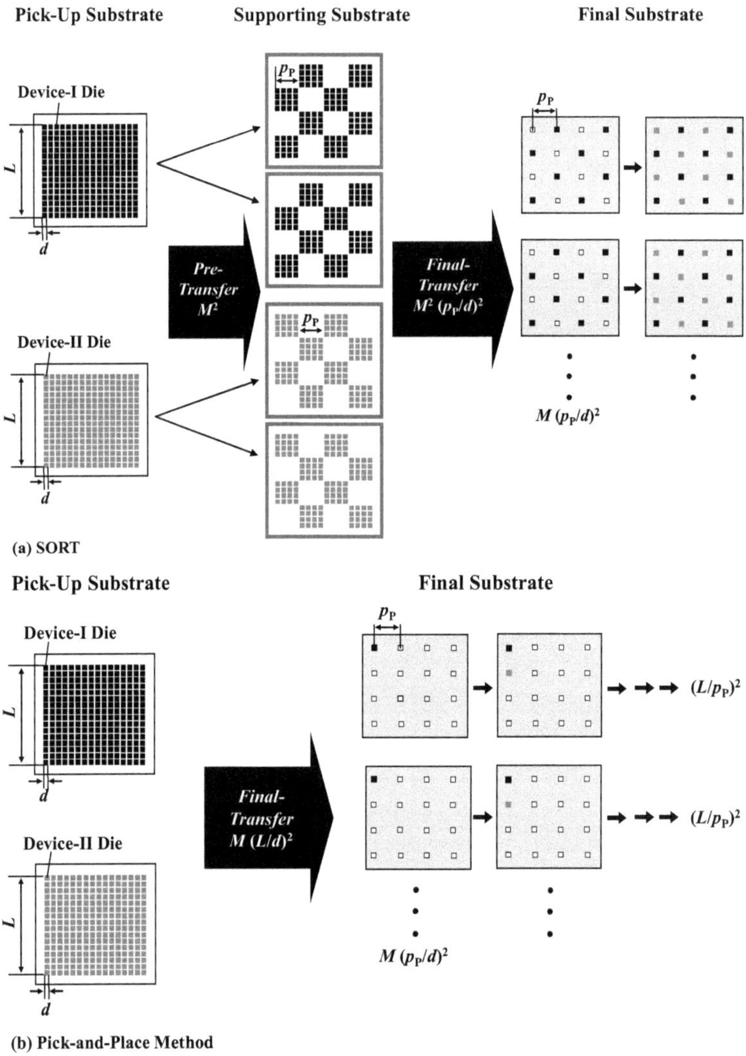

Figure 5.9 Models for calculations of die transfer step counts for (a) the SORT and (b) the pick-and-place method. Reprinted with permission from Yoshimura et al. [9].

on which Device-I dies and Device-II dies are placed alternately, are obtained. The step count involved in the final transfer for M kinds of devices is $M^2(p_P/d)^2$. Therefore, the total step count necessary for SORT is

$$\text{Step Count (SORT)} = M^2(1 + (p_P/d)^2). \quad (5.3)$$

In the present example, $M = 2$, so, M^2 is 4. When $p_P = 200$ μm and $d = 50$ μm, 68 steps are required to make 32 final substrates.

For comparison, die transfer step counts in the conventional pick-and-place method were calculated using the model shown in Fig. 5.9b. Device-I dies and Device-II dies are transferred from pick-up substrates to $M\ (p_P/d)^2$ final substrates one by one. In order to complete one final substrate, $(L/p_P)^2$ steps are necessary. Therefore, the total step count in the pick-and-place method is

$$\text{Step Count (Pick and Place)} = M(p_P/d)^2(L/p_P)^2 = M(L/d)^2. \quad (5.4)$$

When $p_P = 200$ μm, $d = 50$ μm, and $L = 1$ cm, $M\ (L/d)^2 = 80{,}000$. Namely, 80,000 steps are required to make 32 final substrates, revealing that the step count in SORT is smaller by three orders of magnitude than that in the pick-and-place method.

In Fig. 5.10, die transfer step counts are compared for the SORT and the pick-and-place method in the case that $p_P = 250$ μm and $d = 25$ μm. The horizontal axis represents the number of device types, M. For example, in the case that 10 kinds of VCSELs of $\lambda_1 - \lambda_{10}$ in wavelength are placed in final substrates for 10-λ-WDM communications, $M = 10$. In the pick-and-place method, the step count increases with an increase in the effective area on the pick-up substrates because the number of dies involved is proportionally increased with the effective area. In SORT, on the other hand, the step count is constant regardless of the effective area due to its "at-one-time transfer" characteristics. In a range of $L = 1{\sim}5$ cm, and $M = 2 - 100$, the step count for SORT is smaller by a factor of $1/10 - 1/10,000$ than that for the pick-and-place method.

5.3.4 Small/Thin-Die Placement Density

While in SORT, as can be seen in Fig. 5.3, there are no restrictions on the die placement density, slope limits arise in the pick-and-

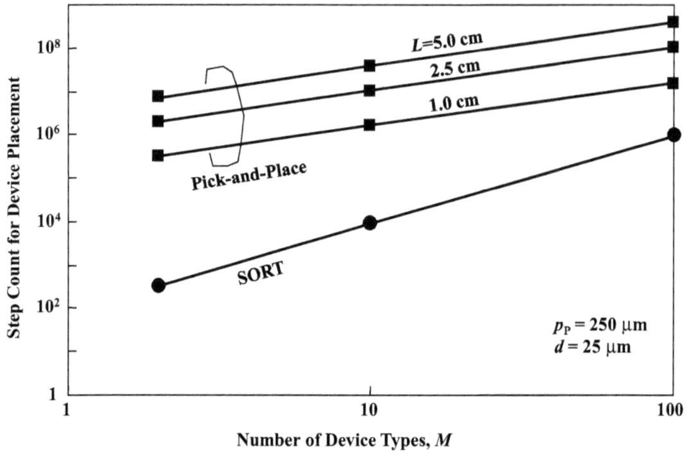

Figure 5.10 Die transfer step counts for the SORT and the pick-and-place method. p_P and d are respectively 250 μm and 25 μm. Reprinted with permission from Yoshimura et al. [9].

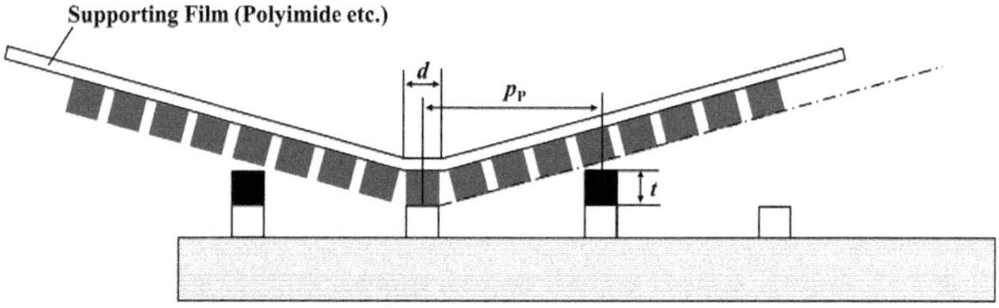

Figure 5.11 Model for estimation of the slope limit in the pick-and-place method. Reprinted with permission from Yoshimura et al. [9].

place method. The limit can be estimated using the model shown in Fig. 5.11. Dies to be placed are on a polyimide film with a pitch of d. By deforming the film, the dies are transferred one by one to pads with a pitch of p_P on a final substrate. In some conditions, dies that have already been placed on the neighboring pads prevent the dies from being placed on unoccupied pads. The slope limit condition is expressed as

$$\text{Slope} = t/(p_P - d) < \tan\theta_d, \tag{5.5}$$

where t is the device thickness and $\tan\theta_d$ the deformation limit of the polyimide film.

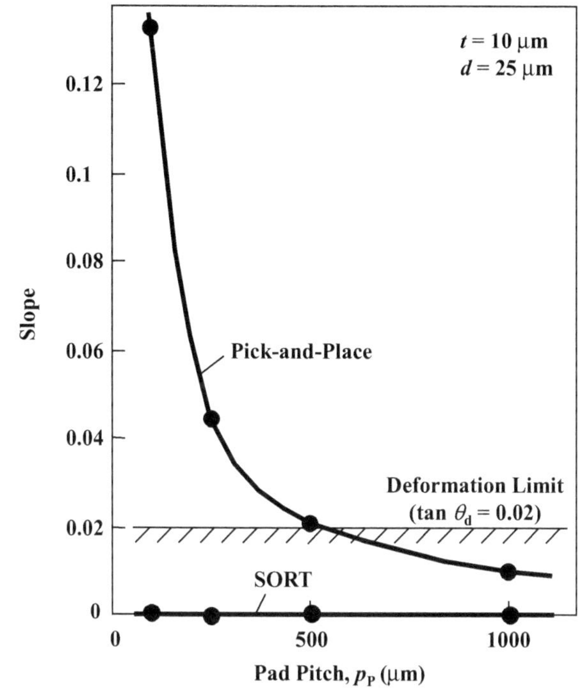

Figure 5.12 Dependence of slopes necessary for the die placement on the pad pitches. Reprinted with permission from Yoshimura et al. [9].

In Fig. 5.12, slopes necessary for the die placement are plotted as a function of pad pitches on the final substrate for $t = 10\,\mu\text{m}$ and $d = 25\,\mu\text{m}$. In the pick-and-place method, the required slope rapidly increases with decreasing pad pitch. If, for example, $\tan\theta_d = 0.02$ is assumed, a slope limit arises around a pad pitch of 500 μm, ruling out high-density die placements. In SORT, since there is no such slope limit, high-density die placements are possible.

5.4 SORT of Polymer Waveguide Lenses

As an experimental demonstration of SORT, SORT of polymer waveguide lenses was attempted [11]. The process is shown in Fig. 5.13. On a glass substrate Al patterns of 50 nm thickness with

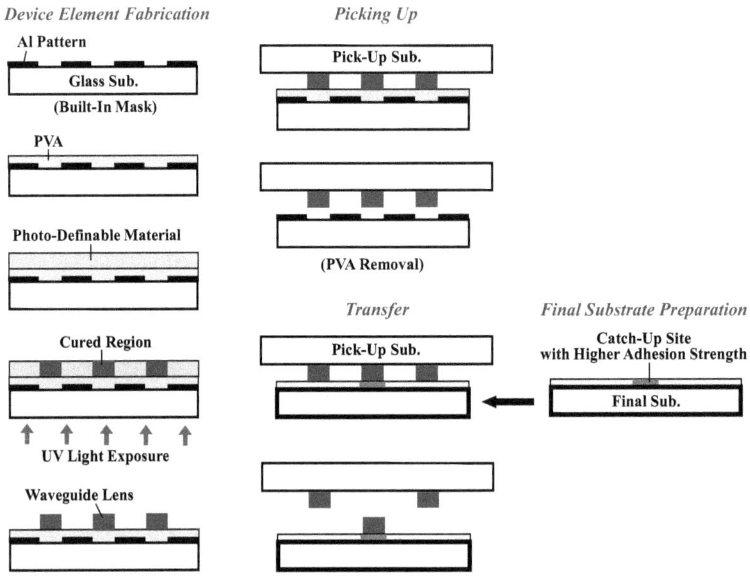

Figure 5.13 Process flow of SORT for polymer waveguide lenses fabricated by the built-in mask method. Reprinted with permission from Yoshimura et al. [11].

apertures of waveguide lens shapes were formed to make a built-in mask. The surface was spin-coated with a 10 μm thick PVA film. After drying the PVA in ambient atmosphere at room temperature for 1 hour, a 40 μm thick acrylic photo-definable material film was deposited on it by spin-coating. The film was exposed to 400 mJ/cm^2 UV light from a high-pressure mercury lamp through the built-in mask to form a polymer waveguide lens array by developing. The lens diameter is 300 μm, and the pitch in the array is 450 μm. For picking up, a pick-up substrate, which is a glass substrate covered with an adhesive film (GelPak film-x8 provided by Gel-Pak), was attached to the top of the lens array. By solving the PVA film in water, the polymer waveguide lens array was transferred onto the pick-up substrate.

A final substrate was prepared by exposing a glass substrate covered with a UV-curable-material-coated film (UV tape UE-111AJ provided by Nitto Denko) to 180 mJ/cm^2 UV light through a photomask. The photomask was designed so that the UV light did not

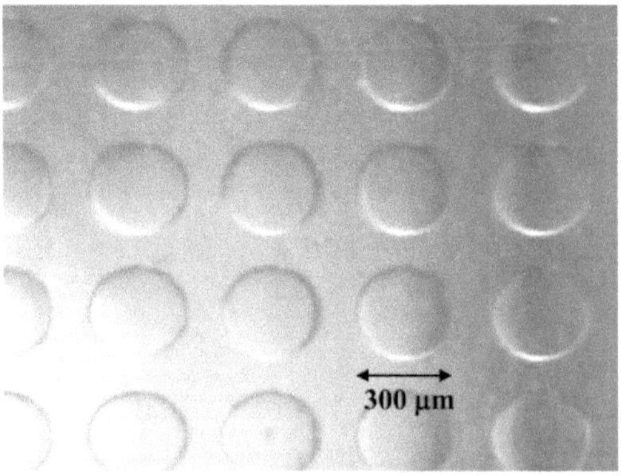

(a) Transfer onto Pick-Up Substrate

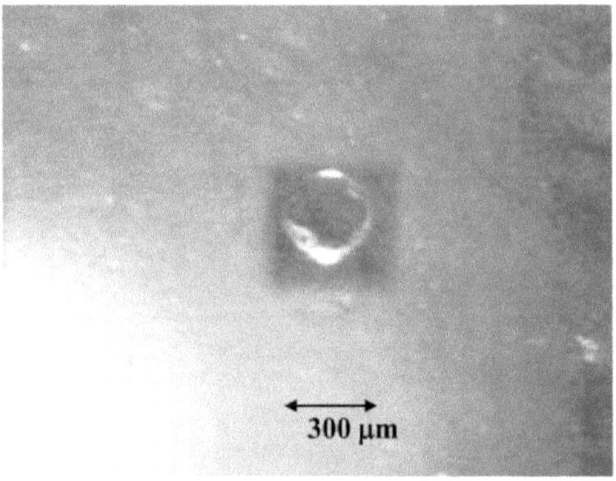

(b) Transfer onto Final Substrate

Figure 5.14 Microscopic photographs for (a) a polymer waveguide lens array picked-up on a pick-up substrate, and (b) a polymer waveguide lens transferred onto a final substrate. The dark square observed under the polymer waveguide lens on the final substrate is photo-mask pattern for catch-up site formation. Reprinted with permission from Yoshimura et al. [11].

reach the catch-up site areas. The adhesion strength of the exposed regions decreased by about two orders of magnitude compared with unexposed regions, providing a final substrate with catch-up sites. The pick-up substrate was attached to the final substrate, and then was detached from it to transfer the polymer waveguide lenses onto the catch-up sites selectively.

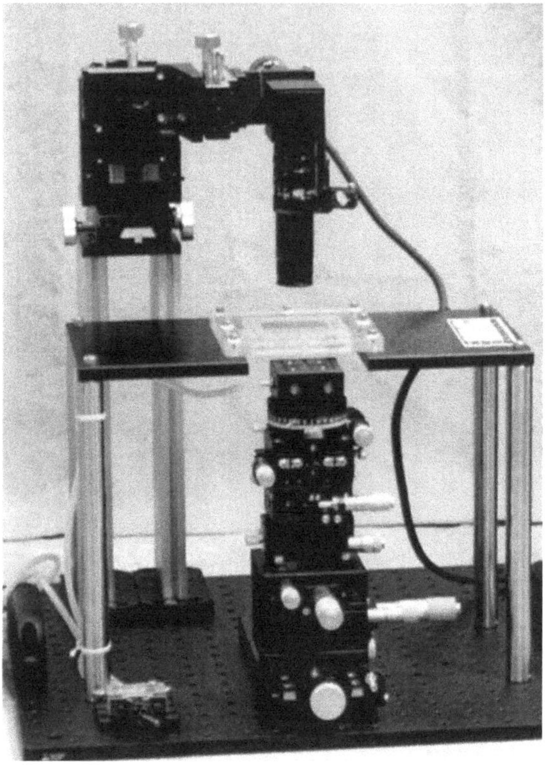

Figure 5.15 SORT equipment. Reprinted with permission from Yoshimura et al. [12].

The photographs in Figs. 5.14a and b respectively represent a polymer waveguide lens array transferred onto a pick-up substrate at one time and a polymer waveguide lens transferred selectively onto a catch-up site of a final substrate. This demonstrates the feasibility of the SORT process.

For the device element transfer the SORT equipment shown in Fig. 5.15 was used. The equipment was proposed by the author and was assembled in Suruga Seiki Co., Ltd., using components for fiber optics with positional accuracy in the order of 0.1–1 μm. Alignments between two substrates were achieved by an optical stage, on which a lower substrate was set, and a substrate holder, below which an upper substrate was set. One problem of the equipment is that it has a one-eye alignment system, which reduces the angular alignment accuracy. The problem can be solved by implementing a two-eye alignment system into the equipment. The improved SORT equipment with a two-eye alignment system is commercially available from Suruga Seiki as "Wafer Bonding Unit."

5.5 SORT of Waveguide Cores

In order to reduce the material consumption of optical waveguides, and to realize all-air-cladding waveguides for strong guided beam confinement, we attempted SORT of waveguide cores made of photo-definable polymers onto a substrate to fabricate optical waveguides [12].

5.5.1 SORT Process for Optical Waveguides

Figure 5.16 shows a process flow of SORT for optical waveguides. Waveguide cores formed on a glass substrate by the built-in mask method are transferred onto a pick-up substrate using the same process as the polymer waveguide lens formation shown in Fig. 5.13. On the pick-up substrate, waveguide cores are placed in an array with high density. On final substrates, catch-up sites are formed. By attaching the pick-up substrate to the final substrate and detaching it, the cores located at areas of the catch-up sites are selectively transferred onto the final substrate to construct optical waveguides. In the case shown in Fig. 5.16, since the core pitch on the final substrate is six times larger than that on the pick-up substrate, one pick-up substrate can provide cores to six final substrates. This implies that waveguide core materials are saved by six times.

In usual optical waveguides for optical interconnects, devices such as vertical mirrors and wavelength filters fabricated by coating the core end facets with metal/dielectric films coexist. To make these devices, complicated deposition/photolithography processes are involved. SORT enables us to deliver waveguide cores with vertical mirrors/wavelength filters from one growth substrate to a lot of final substrates, drastically saving the material consumption and process efforts in optical circuit fabrication.

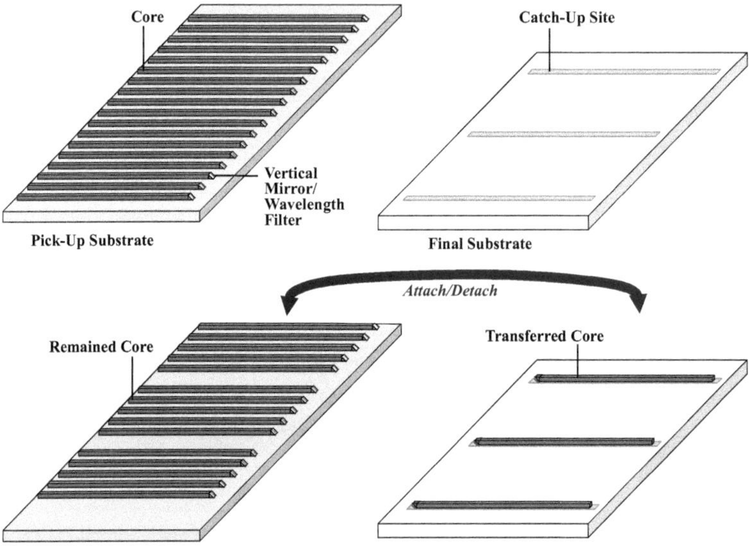

Figure 5.16 Process flow of SORT for optical waveguides. Cores in a high-density array placed on a pick-up substrate are selectively transferred onto a final substrate with catch-up sites. Reprinted with permission from Yoshimura et al. [12].

SORT is applicable to fabrication of all-air-cladding waveguides depicted in Fig. 5.17. Because for intra-chip optical interconnects fine-pattern optical circuits with light-beam-confinement sizes of less than 1 μm are necessary, refractive index contrast between cores and cladding regions should be high. For this purpose, all-air-cladding waveguides, where all sides of cores are surrounded by air, are desirable. By transferring cores on spacers of final substrates, all-air-cladding waveguides are obtained. Furthermore, by building spacers/cores up in multilayer configurations, 3D optical wiring consisting of all-air-cladding waveguides is constructed.

The all-air-cladding structure is a kind of goal to be reached in the future. In all-air-cladding waveguides, however, the losses caused by spacer edge scattering will be a concern. Usage of low-index materials for spacers and high-index materials for cores, as well as narrowing the spacers, will contribute to scattering loss reduction.

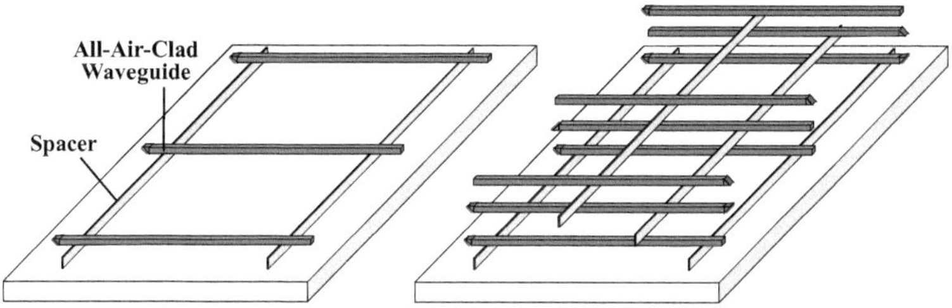

Figure 5.17 All-air-cladding waveguides, which can be fabricated by SORT, for fine-pattern optical interconnects. Reprinted with permission from Yoshimura et al. [12].

5.5.2 Experimental Demonstration of SORT for Optical Waveguides

Core transfers onto a glass substrate with catch-up sites were attempted. A linear array of 3 cm long cores with $32 \times 32\,\mu m^2$ cross sections was formed with a 250 μm pitch on a built-in mask using a photo-definable material with a refractive index of 1.58. The linear array was transferred to a pick-up substrate from the built-in mask. Meanwhile, catch-up sites were formed with 500 μm pitch on a final substrate using UV-curable-material-coated films. In the case that one kind of device elements are arranged in a linear array, a direct transfer from the pick-up substrate to the final substrate are possible without the supporting substrate. By attaching/detaching the two substrates, cores were transferred to the final substrate selectively.

Figures 5.18a and b respectively show the cores on the pick-up substrate and the final substrate after selective transfer. It can be seen that a core is transferred onto the catch-up site of the final substrate, while a core is missed on the pick-up substrate. As Fig. 5.18c shows, a guided beam of 650 nm in wavelength is clearly observed at the core edge. These results confirm that SORT can be applied to the transfer and integration of microscale optical circuits, providing the proof-of-concept of the material-saving waveguide fabrication process.

Figure 5.19 shows an all-air-cladding waveguide made of fluorinated photo-definable material developed by DIC. A core formed by the built-in mask method was transferred on a GelPak

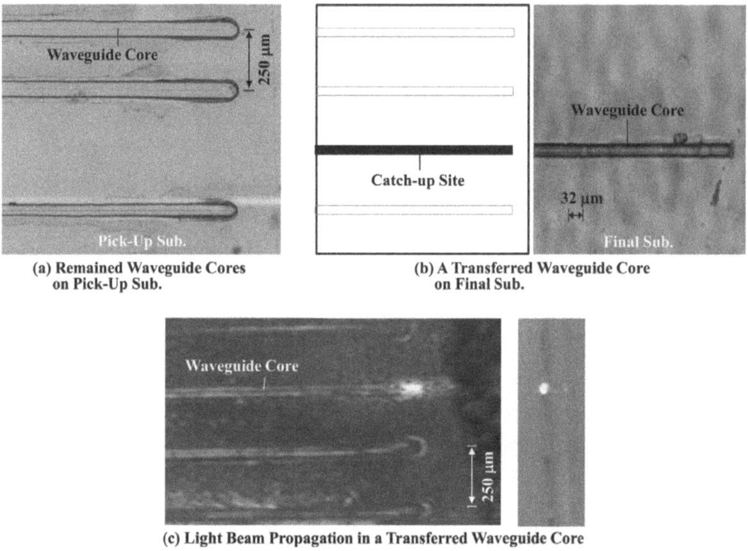

Figure 5.18 SORT of Waveguide cores. (a) Remained cores on the pick-up substrate and (b) a transferred core on the final substrate. (c) Guided beams in an optical waveguide consisting of a transferred core. Reprinted with permission from Yoshimura et al. [12].

Figure 5.19 All-air-cladding waveguide on a GelPaK film.

film. The film has a rough surface, which can be regarded as a surface with natural spacers to produce all-air-cladding waveguides. A 650 nm guided beam and a 650 nm beam emitted from the core edge are observed. The cutback method revealed that the propagation loss is ~0.5 dB/cm.

5.6 Light-Assisted SORT

5.6.1 LA-SORT

Figure 5.20 shows a process flow of light-assisted SORT (LA-SORT) [12]. Catch-up sites on a supporting substrate are prepared by exposing a UV-curable material layer on the substrate to patterned UV light with a distribution that the catch-up site areas are dark. Consequently, unexposed regions have stronger adhesion strengths compared to exposed regions, forming catch-up sites.

For catch-up sites of final substrates, any materials that induce adhesion strength contrasts can be used. One candidate is a UV-curable material similar to the catch-up site material for the supporting substrate. In this case, however, there is a concern about reliability, that is, adhesion strength between dies and catch-up sites decreases gradually with time. To avoid this problem, photo-definable materials might be promising for catch-up sites in the case that the devices such as waveguide cores to be transferred are made of a photo-definable material. The devices preserve strong adhesion strength in the catch-up sites made of another photo-definable material after curing. Pads made of solder, diffusable metal, and

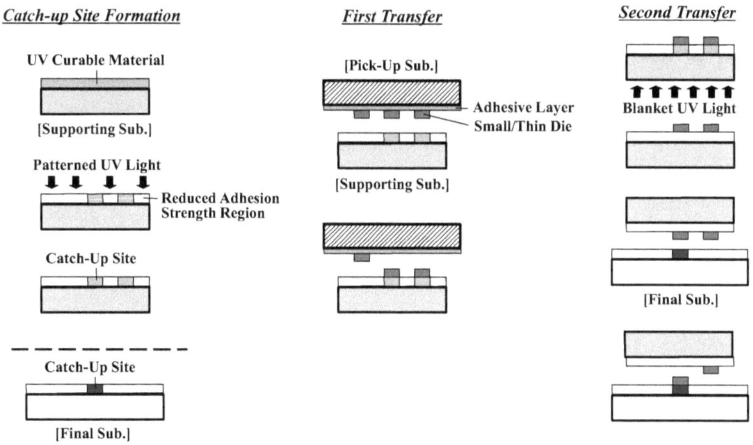

Figure 5.20 Process flow of LA-SORT. Reprinted with permission from Yoshimura et al. [12].

polymers are also applicable to catch-up sites with the assistance of thermal, supersonic, or optical excitation. When Au and InSn are used for diffusable metal, bonding of Au-InSn catch-up sites with Au pads of small/thin dies can occur by alloying InSn with Au [5]. When polyimide is used for pads, polyimide bonding can be performed.

For the first transfer, small/thin dies on a pick-up substrate are attached to the supporting substrate to be transferred onto the catch-up sites. For the second transfer, the catch-up sites are exposed to blanket UV lights from the back of the substrate, which reduces the adhesion strength of all the catch-up sites. Then, the dies on the supporting substrate are attached to the final substrate to distribute the dies on the final substrate.

5.6.2 Experimental Demonstration of LA-SORT

Using Device-I and Device-II dies made of polymers, we carried out LA-SORT, shown in Fig. 5.20. To form catch-up sites of supporting and final substrates, UV-curable-material-coated films were used.

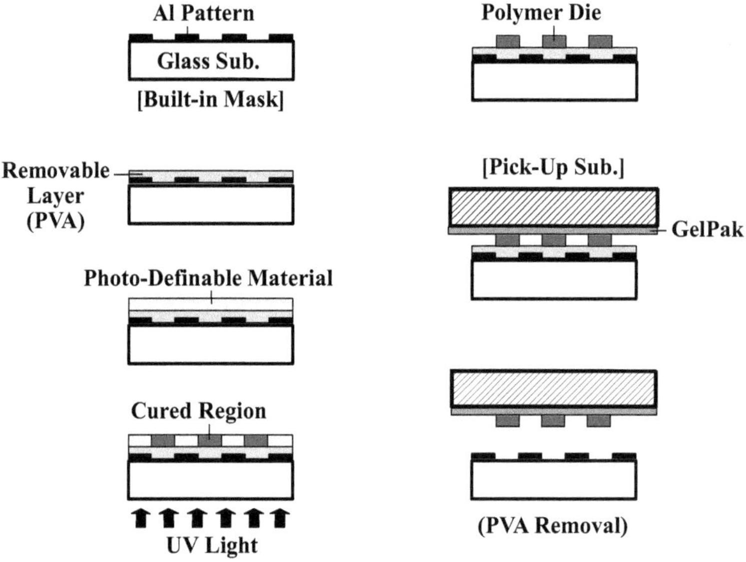

Figure 5.21 Fabrication process of polymer dies using the built-in mask method. Reprinted with permission from Yoshimura et al. [12].

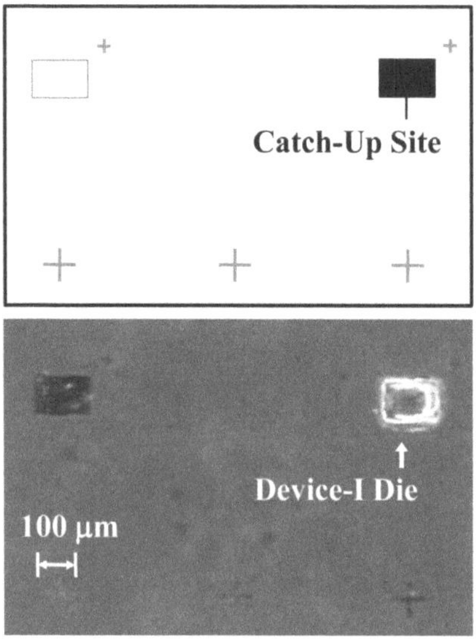

(a) First Transfer of Device-I Die onto Supporting Sub.

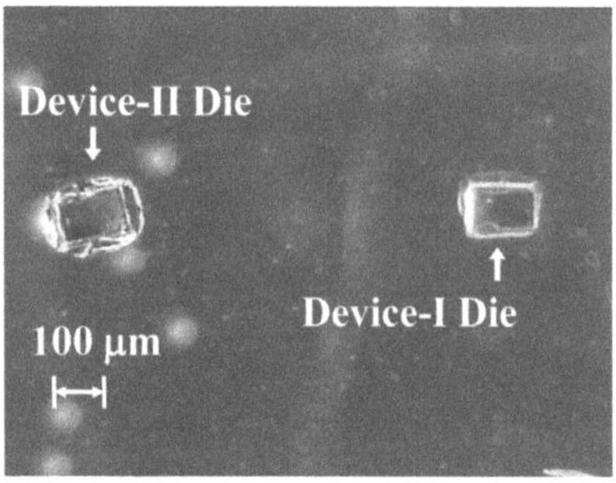

(b) Second Transfer of Device-I and II Dies onto Final Sub.

Figure 5.22 (a) Selective transfer of Device-I die onto a supporting substrate by the first transfer, and (b) heterogeneous transfer of Device-I and II dies onto a final substrate by the second transfer. Reprinted with permission from Yoshimura et al. [12].

The UV light exposure energy for the catch-up site formation was 270 mJ/cm^2. For die transfers, the SORT equipment shown in Fig. 5.15 was used.

A fabrication process of polymer dies is shown in Fig. 5.21. The process is similar to that shown in Fig. 5.13. An array of rectangular polymer dies (Device-I dies) with a die size of $100 \times 130\ \mu m^2$ and a pitch of 1000 μm was fabricated by the built-in mask method, and it was transferred to a pick-up substrate with GelPak film-x8. An array of Device-II dies was obtained by depositing Au thin films on the array of Device-I dies by vacuum evaporation.

In the first transfer, after the dies on the pick-up substrate were aligned to the corresponding catch-up sites on the supporting substrate, the pick-up substrate was attached to the supporting substrate, and then detached from it, resulting in a selective transfer of the dies onto the supporting substrate, as shown in Fig. 5.22a. In the second transfer, after the UV-curable-material-coated film with transferred dies on the supporting substrate was exposed to blanket UV light from the back, the supporting substrate was attached to the final substrate and detached from it, resulting in a selective transfer of the dies onto the final substrate. Figure 5.22b shows an example of heterogeneous integration of Device-I die and Device-II die on the final substrate. They were successively transferred on to the final substrate using the process shown in Fig. 5.20. It

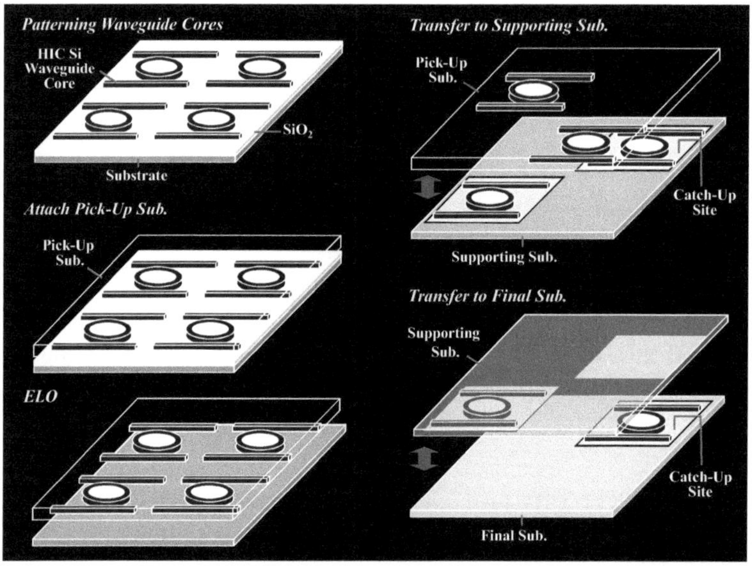

Figure 5.23 Nanoscale integration: SORT of HIC waveguides. Reprinted with permission from Yoshimura et al. [16].

should be noted that cross-shaped markers consisting of lines 10 μm wide and 100 μm long were transferred on substrates together with the dies.

These results indicate the feasibility of the LA-SORT process for heterogeneous integration in micrometer-scale resolutions. The procedure described above might be applicable to dies made of semiconductor, glass, and dielectric materials with process condition tuning.

5.7 SORT for Nanoscale Heterogeneous Integration

Figure 5.23 shows a concept of nanoscale integration for HIC waveguides [12]. HIC Si waveguide cores are picked up onto a pick-up substrate by the ELO, and are transferred to a final substrate via a supporting substrate using SORT. In Fig. 5.24, photonic crystals (PhCs) and HIC waveguides, which respectively correspond to Device-I dies and Device-II dies in Fig. 5.3, are integrated in an optical waveguide film. To realize such nanoscale transfers,

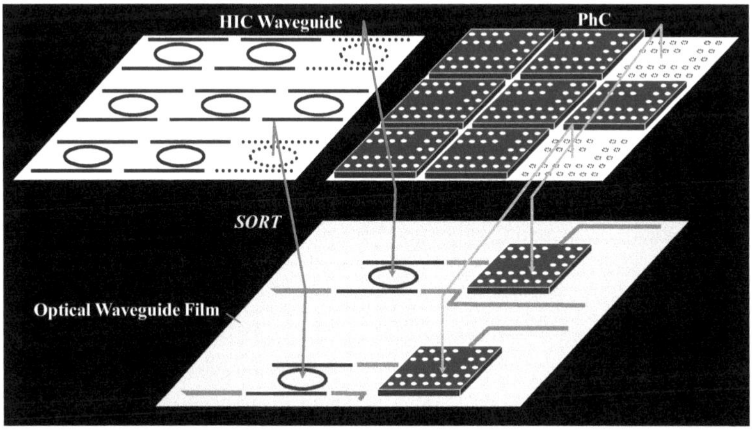

Figure 5.24 Nanoscale heterogeneous integration: SORT of photonic crystals/HIC waveguides. Reprinted with permission from Yoshimura et al. [16].

improvement of alignment accuracy of the SORT equipment to submicrometer ranges is necessary.

Nanoscale optical circuits are expensive due to the ultrafine fabrication processes using submicrometer photolithography. Meanwhile, in optical interconnects, nanoscale optical circuits are placed with pitches of $\sim$100 μm corresponding to I/O pads of chips. SORT enables us to deliver nanoscale optical circuits from one growth substrate to a lot of final substrates, say, 100 substrates, saving materials and process steps for nanoscale optical circuit fabrication. Therefore, SORT greatly contributes to reduction in material consumption and process effort in optical interconnects.

References

1. M. E. Robertsson, O.-J. Hagel, G. Gustafsson, A. Dabec, M. Popall, L. Cergel, P. Wennekers, P. Kiely, M. Lebby, and T. Lindahl, "O/e MCM packaging with new, patternable dielectric and optical materials," *Proc. 48th Electronic Components & Technology Conference (ECTC)*, 1423–1421 (1998).
2. S. Koike, F. Shimokawa, T. Matsuura, and H. Takahara, "New optical and electrical packaging techniques using optical waveguides for optoelectronic multichip modules," *IEEE Trans. Compon. Packag. Manuf. Technol., Part B* **19**, 124–130 (1996).
3. E. Yablonovitch, T. Gmitter, J. P. Harbison, and R. Bhat, "Extreme selectivity in the lift-off of epitaxial GaAs films," *Appl. Phys. Lett.* **51**, 2222–2224 (1987).
4. W. Chang, G. Pike, C. Kao, and E. Yablonobitch, *MRS 1997 Fall Meeting, Abstracts*, H10.1, 210 (1997).
5. R. Pu, C. Duan, and C. Wilmsen, "Hybrid integration of VCSEL's to CMOS integrated circuits," *IEEE J. Sel. Top. Quantum Electron.* **5**, 201–208 (1999).
6. J. J. Chang, S. Jung, M. Vrazel, K. Jung, M. Lee, M. A. Brooke, N. M. Jokerst, and S. Wills, "A hybrid optically interconnected microprocessor: an InP I-MSM integrated onto a mixed signal CMOS analog optical receiver with a digital CMOS microporcessor," *Proc. SPIE* **4089**, 708–714 (2000).
7. C. B. Morrison, R. L. Strijek, and J. H. Bechtel, "New processing approach for grafting optoelectronic devices and applications to multichip modules," *Proc. SPIE* **3005**, 120–127 (1997).

8. T. Yoshimura, J. Roman, Y. Takahashi, M. Lee, B. Chou, S. I. Beilin, W. V. Wang, and M. Inao, "Optoelectronic scalable substrates based on Film/Z-connection and its application to film optical link module (FOLM)," *Proc. SPIE* **3952**, from Photonics West 2000, San Jose, California, 202–213 (2000).
9. Yoshimura, K. Kumai, T. Mikawa, and O. Ibaragi, "Photolithographic packaging with selectively occupied repeated transfer (PL-Pack with SORT) for scalable optical link multi-chip-module (S-FOLM)," *IEEE Trans. Electron. Packag. Manuf.* **25**, 19–25 (2002).
10. T. Yoshimura, M. Ojima, Y. Arai, and K. Asama, "Three-dimensional self-organized micro optoelectronic systems for board-level reconfigurable optical interconnects–performance modeling and simulation–," *IEEE J. Sel. Top. Quantum Electron.* **9**, 492–511 (2003).
11. T. Yoshimura, D. Tamaki, S. Kawamura, Y. Arai, and K. Asama, "A device transfer process "selectively occupied repeated transfer (SORT)" for resource-saving integration of polymer micro optics fabricated by "built-in mask" method," *IEEE Trans. Compon. Packag. Technol.* **27**, 468–471 (2004).
12. T. Yoshimura, T. Kofudo, T. Kashiwazako, K. Naito, K. Ogushi, and Y. Kitabayashi, "A material-saving optical waveguide fabrication process with selective transfers of cores," *Opt. Eng.* **47**, 014601-1-7 (2008).
13. T. Yoshimura, Y. Suzuki, N. Shimoda, T. Kofudo, K. Okada, Y. Arai, and K. Asama, "Three-dimensional chip-scale optical interconnects and switches with self-organized wiring based on device-embedded waveguide films and molecular nanotechnologies," *Proc. SPIE* **6126**, from Photonics West 2006, San Jose, California, 612609-1-15 (2006).
14. T. Yoshimura, J. Roman, W.-C. Wang, M. Inao, and M. T. McCormack, "Device transfer method," U.S. Patent 6,669,801 B2 (2003).
15. C. Stirk, Q. Liu, and M. Ball, "Manufacturing cost analysis of integrated photonic packages," *Proc. SPIE* **3631**, 224–233 (1999).
16. T. Yoshimura, *Trends in Thin Solid Films Research* (ed. A. R. Jost) Ch. 5 "Self-organized growth of polymer wire networks with designed molecular sequences for wavefunction-controlled nano systems." Nova Science Publishers, New York (2007).

Chapter 6

High-Speed/Small-Size Light Modulators and Optical Switches

High-speed/small-size light modulators and optical switches are a challenging core technology for self-organized 3D integrated optical interconnects.

In the present chapter, light modulators and optical switches are classified, and the superiority of flux-controlled devices over phase-controlled devices in large-scale integration is emphasized. As an example of flux-controlled devices, the waveguide prism deflector (WPD) is described, and the predicted performance is compared with that of other types of devices. Materials for light modulators and optical switches are also discussed.

Since in many cases a light modulator and an optical switch have a similar structure, these two are treated as equivalent devices in this chapter.

6.1 Classification of Light Modulators and Optical Switches

Figure 6.1 shows typical device structures of light modulators and optical switches. The Mach–Zehnder light modulator (MZM) shown

Self-Organized 3D Integrated Optical Interconnects with All-Photolithographic Heterogeneous Integration
Tetsuzo Yoshimura

ISBN 978-981-4877-04-6 (Hardcover), 978-1-003-04305-8 (eBook)
www.jennystanford.com

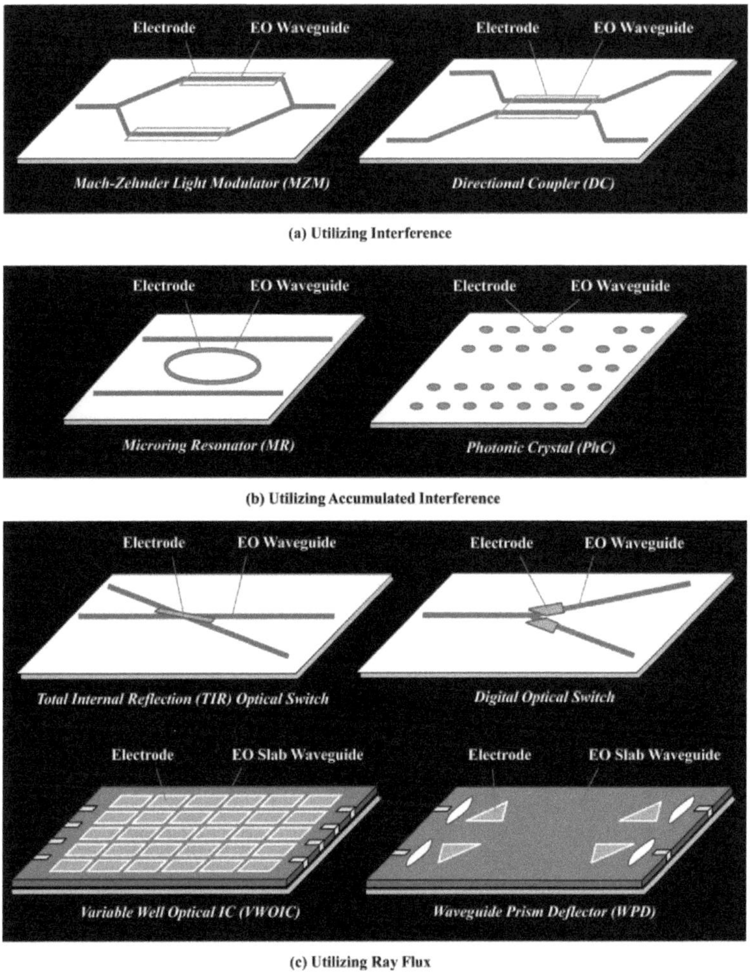

Figure 6.1 Typical device structures of light modulators and optical switches.

in Fig. 6.1a is a most popular device, on which comprehensive work has been done [1–3]. By combining MZMs with directional couplers, optical switch functions are obtained. The MZM is a phase-controlled device because it utilizes the interference of two light beams propagated in branching waveguide arms. The directional coupler (DC) also works as a phase-controlled light modulator and optical switch.

The microring resonator (MR) [4] and the photonic crystal (PhC) [5, 6], shown in Fig. 6.1b, are also applicable to light modulators and optical switches. They utilize accumulated interference. That is, their device functions appear after superposition of all the light beams is completed in the whole device region to reach the saturated state, in which interference is built-up completely. MRs and PhCs exhibit superior modulation/switching capabilities using accumulated interference.

In phase-controlled devices like the MZM, DC, MR, and PhC, however, due to their phase-sensitive characteristics, extremely strict control of waveguide shapes/sizes, refractive indexes, and operation temperatures are required for their stable operation. Furthermore, in the MR and PhC, as described in section 6.5, building-up duration is necessary to settle modulation/switching functions, causing delay time. These drawbacks of phase-controlled devices imply that they are not the final destination for high-speed light modulators/optical switches in self-organized 3D integrated optical interconnects.

On the contrary, the total internal reflection (TIR) optical switch, the digital optical switch, the variable well optical integrated circuit (VWOIC) [7–10], and the WPD [8–16], shown in Fig. 6.1c, are flux-controlled devices that are phase-insensitive. Although these devices require relatively large refractive index changes for their operation, for large-scale integration they are preferable to the MZM, DC, MR, and PhC because flux-controlled devices do not require strict control of waveguide shapes/sizes, refractive indexes, and operation temperatures. Explanations of the WPD as well as a brief explanation of the VWOIC are provided in sections 6.2, 6.3, and 6.4.

To realize high-speed/small-size light modulators and optical switches, materials having enhanced electric-field-induced refractive-index-change characteristics are essential. The candidates are electro-optic (EO) materials, whose refractive index changes by applying electric fields through the Pockels effect or the Kerr effect, and semiconductors, whose refractive index changes by carrier injection or depletion. For these materials, the details are discussed in section 6.3.

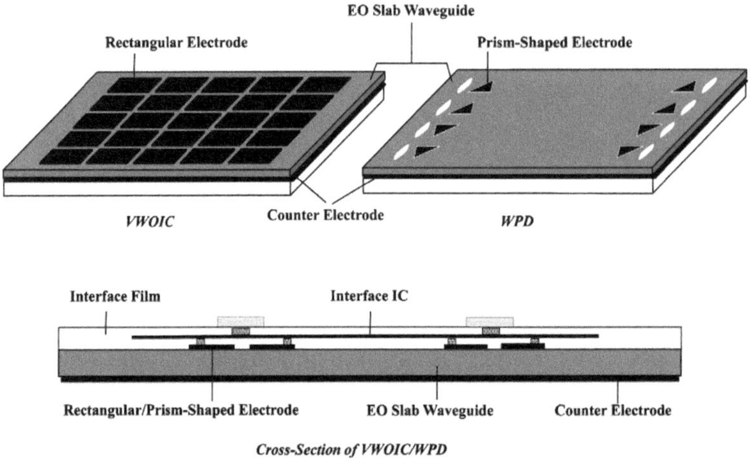

Figure 6.2 Structures of the VWOIC and WPD. Reprinted with permission from Yoshimura et al. [8].

6.2 Variable Well Optical ICs and Waveguide Prism Deflectors

Structures of the VWOIC and WPD are schematically shown in Fig. 6.2. In the former and the latter, rectangular electrodes and prism-shaped electrodes are respectively formed on an EO slab waveguide. The EO slab waveguide consists of three layers of cladding/core/cladding placed between a counter electrode and rectangular or prism-shaped electrodes. The core is made of EO materials, and the refractive index changes by applying voltages between the electrodes. Interface films, in which small/thin-interface IC dies are embedded, are stacked on the surface to apply voltages to the individual electrodes selectively.

In VWOICs, as Fig. 6.3 shows, by selectively applying voltages to the EO slab waveguide through rectangular electrodes, refractive index changes are induced to form potential wells for light beams. Then, electric-field-induced dynamic optical waveguides are constructed all over the slab waveguide to switch guided light beams. In Fig. 6.4, examples of switching characteristics simulated by the beam propagation method (BPM) for light beams of 1.3 μm in

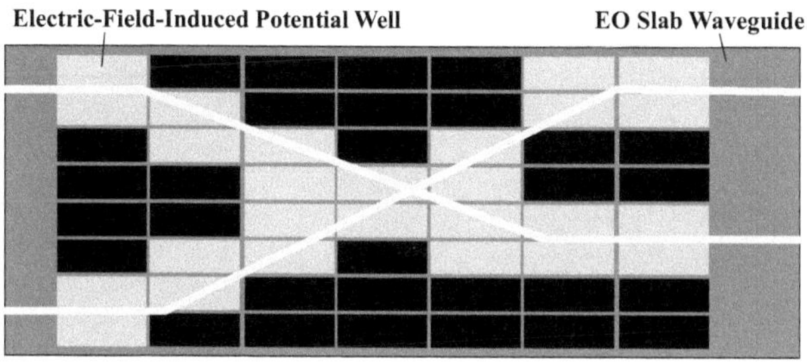

Figure 6.3 Operation principle of the VWOIC. Reprinted with permission from Yoshimura et al. [8].

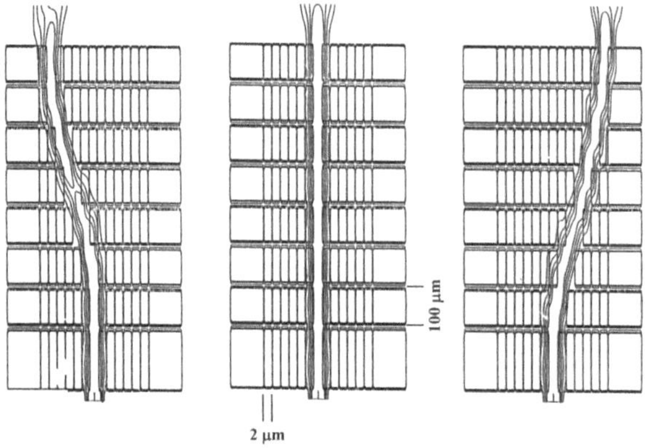

Figure 6.4 BPM simulations of switching characteristics in the VWOIC.

wavelength are shown for a VWOIC with 12 × 7 electrodes of 2 μm × 100 μm. A refractive index change induced by a voltage application is assumed to be 0.005. Two adjacent electrodes are simultaneously selected for voltage application to construct 4 μm wide dynamic waveguides. The selected regions are shifted by 2 μm along the direction perpendicular to the light beam propagation direction in order to achieve quasi-continuous index variations. It can be seen that light beams are switched in a length of 700 μm.

In WPDs, as shown in Fig. 6.5, electric-field-induced dynamic waveguide prisms, which deflect guided light beams, are constructed in the EO slab waveguide, enabling us to switch the light beams through a two-dimensional (2D) free space like

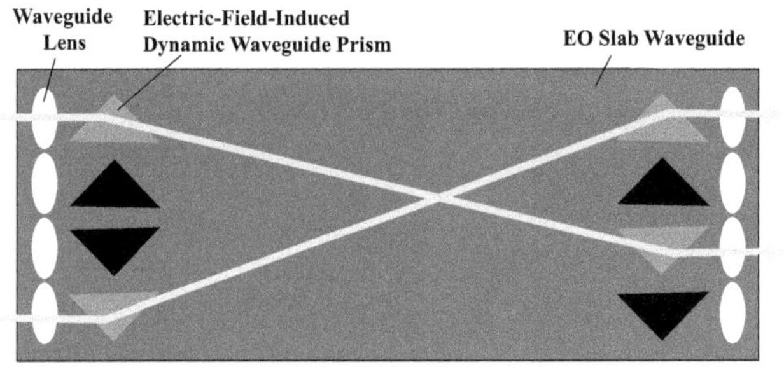

Figure 6.5 Operation principle of the WPD. Reprinted with permission from Yoshimura et al. [8].

"search lights" in the waveguide. Waveguide lenses for light beam collimation are formed to suppress spreading due to diffraction of the propagated light beams in free space. The details of the WPD operation are described in sections 6.3 and 6.4.

It should be noted that VWOICs and WPDs can be used for light modulators as well as optical switches.

Light modulators and optical switches for large-scale integrated optical interconnects and optical switching systems should be high-speed, small, and thin, and at the same time, they should be stable against structural, refractive index, and thermal disturbance. In MZMs, since small phase differences of light beams are utilized for their operation, the influence of temperature and processing accuracy for device fabrication on device operation is substantial. In fact, in a $LiNbO_3$ matrix optical switch, precise individual adjustments are required for each switching element even in an 8 × 8 scale. In MRs and PhCs, the influence is further substantial because they utilize the accumulated interference.

On the contrary, in VWOICs and WPDs, which are phase-insensitive flux-controlled devices, electric-field-induced refractive index differences in a slab waveguide are used to control the optical path directly. The refractive index differences are less sensitive to temperature changes than the refractive index itself. For the processing accuracy, as far as the right edge angles are maintained, the size of each part does not largely affect the device operation. Therefore, VWOICs and WPDs are expected to exhibit more stable operation than MZMs, DCs, MRs, and PhCs, and are suited to large-scale integrated optical interconnects within computers, described in Chapter 8, and three-dimensional micro optical switching systems (3D-MOSSs), described in Chapter 9.

In VWOICs and WPDs, optical waveguides, waveguide lenses, waveguide filters, and waveguide mirrors are integrated with EO thin films for slab waveguides by PL-Pack with SORT, described in Chapter 5, which transplants multiple kinds of small/thin OE dies and materials together onto a substrate. The integration of IC dies into interface films can also be achieved by PL-Pack with SORT.

6.3 Design and Predicted Performance of WPDs

6.3.1 EO Materials for WPDs

Examples of EO material candidates for light modulators and optical switches are listed in Table 6.1. $LiNbO_3$ (LN), III–V semiconductor quantum dot (QD) [17, 18], molecular crystal of styrylpyridinium cyanine dye (SPCD) [19], and polymer multiple quantum dot (polymer MQD) [20–25] are EO materials having the Pockels effect. Lead lanthanum zirconate titanate (PLZT) [26] is an EO material having the Kerr effect. Polymer MQD is an artificial material consisting of polymer wires with MQDs. Because it has not been grown experimentally yet, the EO property was theoretically predicted by simulations based on the molecular orbital method [20] (see Chapter 12).

Table 6.1 Candidates of EO materials for the light modulators and optical switches

	Pockels effect				Kerr effect	
	LN	III-V QD	Molecular crystal (SPCD)	Polymer MQD		PLZT
n_0	2.2	3.5	1.55	2**	n_0	2.48
r(pm/V)	31	26 [17]	430[19]	1000*[20]	$R(m^2/V^2)$	5.7×10^{-16}[26]

*Molecular orbital calculation
**Assumed value

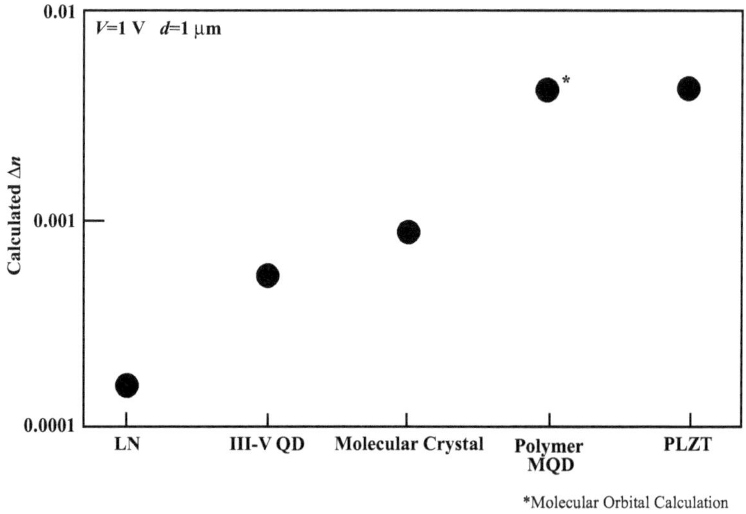

Figure 6.6 Electric-field-induced refractive index change, Δn, calculated for typical EO materials.

To design structures of WPDs and estimate their expected performance, PLZT and polymer MQD were assumed for EO materials.

The electric-field-induced refractive index change, Δn, of EO materials is represented by the following expressions:

$$\text{Pockels effect} \quad \Delta n = \frac{1}{2} n_0^3 r E \tag{6.1}$$

$$\text{Kerr effect} \quad \Delta n = \frac{1}{2} n_0^3 R E^2 \tag{6.2}$$

Here, n_0 and E represent the refractive index of EO materials with no electric fields and the electric fields in the EO materials, respectively. r and R are respectively EO coefficients for the Pockels effect and the Kerr effect.

Figure 6.6 shows Δn calculated using Eqs. 6.1 and 6.2. The thickness of the EO material, namely, the electrode gap distance, d, is 1 μm, and the applied voltage is 1 V. PLZT and polymer MQD are expected to exhibit extremely large Δn. Δn more than 0.01 might be available with applied voltages of a few volts.

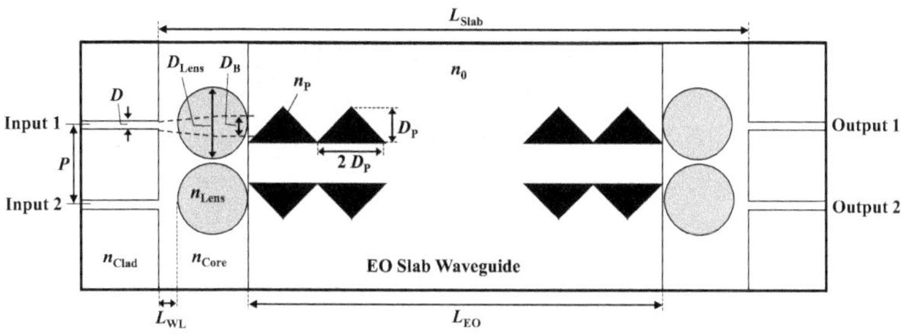

Figure 6.7 Preliminary model of a 2 × 2 WPD optical switch consisting of an EO slab waveguide of PLZT. Reprinted with permission from Yoshimura et al. [15].

6.3.2 Model for 2 × 2 WPD Optical Switch

6.3.2.1 Preliminary model

A preliminary model of a 2 × 2 WPD optical switch is shown in Fig. 6.7. Between input waveguides and output waveguides, an EO slab waveguide, on which prism-shaped electrodes are formed, is placed. Slab waveguides containing waveguide lenses are placed between the input/output waveguides and the EO slab waveguide. When voltages are applied to prism-shaped electrodes, electric-field-induced dynamic waveguide prisms deflect optical signals from input waveguides to switch them through slab waveguides into output waveguides.

The input/output waveguides are channel waveguides with core width, D, of 4 μm, core refractive index, n_{Core}, of 1.52, and cladding refractive index, n_{Clad}, of 1.50, which permits to propagate light beams with the fundamental mode of $m = 0$ and the first excited mode of $m = 1$. The channel waveguide distance P is 20 μm. The total length of the slab waveguide region including the EO slab waveguide and the slab waveguides with lenses is L_{Slab}. The length of the EO slab waveguide, which is placed on a counter electrode, is L_{EO} and the thickness is d. In the present model, the EO slab waveguide is assumed to be a 400 nm thick PLZT film. Right-angle prism-shaped electrodes with an aperture, D_P, of 16 μm and a length of $2D_P$ are formed on the EO slab waveguide. The gap distance, d, between the electrodes and the counter electrode is 400 nm, neglecting the cladding thickness for simplicity. Two-step prism cascading configurations are considered. The refractive index of the EO slab waveguide is n_0 when no electric fields exist. By applying a voltage to the EO slab waveguide through electrodes, electric-field-induced dynamic waveguide prisms with a refractive index of n_P are formed. Waveguide lenses are inserted between the prism-shaped electrodes and the input/output waveguides. The waveguide lens diameter, D_{Lens}, is 40 μm. The lens area refractive index is n_{Lens}. The distance between the lens and the input/output waveguides is L_{WL}. The lens expands the light beam size to D_B. The wavelength, λ, of the probe beams is 1.3 μm.

Reflections from the integrated components are not taken into account. Anti-reflection (AR) coating on the component edge surfaces using atomic layer deposition (ALD) or molecular layer deposition (MLD), which can grow ultrathin conformal films on surfaces with 3D structures, might make the model viable [25].

6.3.2.2 Optimized model for performance evaluation

When designing WPD optical switches, the following two things come about:

(1) Reduction of the prism size and the inter-channel pitch
(2) Suppression of the beam spreading accompanying the free space propagation in the slab waveguide

These two points have a reciprocal relationship. If the prism size is reduced, the prism aperture decreases and the diffraction angle increases. Consequently, the beam width increases. So it is important to find a practical solution for the WPD optical switch structure within this trade-off.

In order to determine an optimized model of the 2 × 2 WPD optical switch for performance evaluation, proper n_{Lens}, L_{Slab}, and L_{EO} were estimated. The detail of the electric-field-induced waveguide prism is shown in Fig. 6.8. A refractive index change, Δn, is induced in the EO slab waveguide by applying a driving voltage, V, according to Eq. 6.1 or 6.2. n_P is given by

$$n_P = n_0 + \Delta n. \tag{6.3}$$

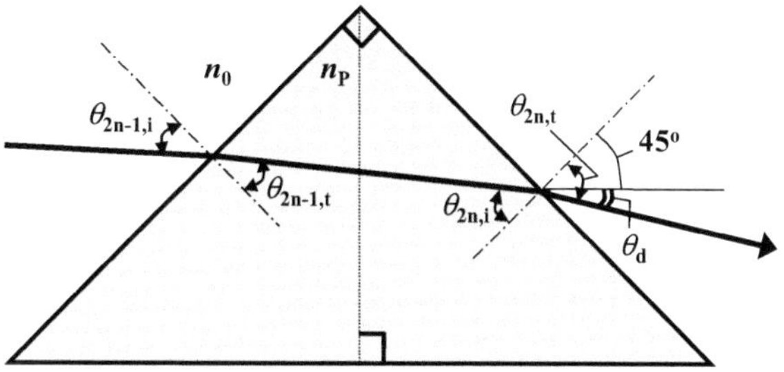

Figure 6.8 Detail of the electric-field-induced waveguide prism in the WPD optical switch. Reprinted with permission from Yoshimura et al. [15].

The deflection angle, θ_d, can be calculated by the following equations:

$$\begin{aligned} &n_0 \sin\theta_{2n-1,i} = n_P \sin\theta_{2n-1,t} \\ &n_P \sin\theta_{2n,i} = n_0 \sin\theta_{2n,t} \\ &\theta_{2n,i} = \pi/2 - \theta_{2n-1,t} \\ &\theta_{2n+1,i} = \pi/2 - \theta_{2n,t} \\ &(n = 1, 2, \cdots, N) \\ &\theta_d = \theta_{2N,t} - \pi/4 \end{aligned} \tag{6.4}$$

Hear, N is the prism count in the multistep prism cascading configurations. The final deflection angle, θ_d, is obtained by performing calculations sequentially for $n = 1, 2, \cdots$, and N. As the 0th approximation, neglecting the sizes of the prism-shaped electrodes and the slab waveguides containing waveguide lenses, L_{Slab}, necessary to achieve the switching operation, is given by

$$L_{Slab} \approx P/\tan\theta_d. \tag{6.5}$$

The spread angles of light beams propagating in the slab waveguide are approximately estimated by $2\theta = \lambda/(\pi n_0 D_B)$. Light beam widths should be less than P at the end of the slab waveguide after propagation. Then, L_{Slab} should satisfy

$$L_{Slab} < \frac{P}{2}/\tan\theta. \tag{6.6}$$

In Fig. 6.9, the driving voltage dependence of L_{Slab} derived from Eq. 6.5 using parameters for PLZT in Table 6.1 is shown for a two-step prism cascading configuration. The diffraction limit boundary derived from Eq. 6.6 is also shown for a D_B of 5 μm. It can be seen

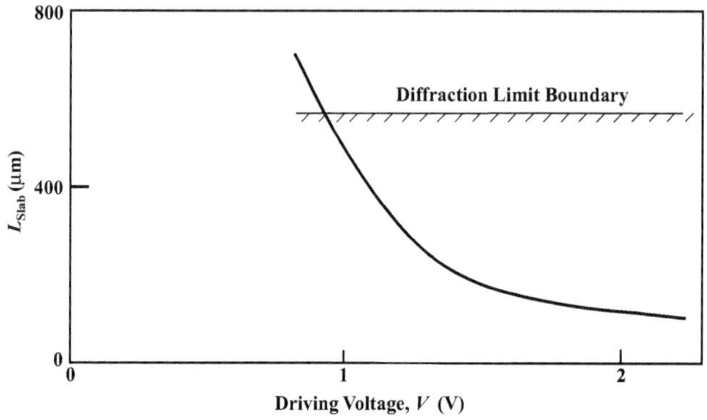

Figure 6.9 Driving voltage dependence of L_{Slab} and diffraction limit boundary for D_B of 5 μm. Reprinted with permission from Yoshimura et al. [15].

Table 6.2 Driving voltages and refractive indexes in 400-nm-thick PLZT and 400-nm-thick polymer MQD

	400-nm-thick PLZT	400-nm-thick polymer MQD
Driving voltage: V	1.2 V	4 V
Prism refractive index (Off): $n_{P\,(Off)}$	2.48	2
Prism refractive index (On): $n_{P(On)}$	2.52	2.04

that the lowest possible driving voltage is around 1 V, predicting a proper L_{Slab} of around 500 μm.

Thus, in the present calculations, as indicated in Table 6.2, the driving voltage is assumed to be $V = 1.2$ V, which generates a waveguide prism refractive index, $n_{P(On)}$, of 2.52 derived from Eqs. 6.2 and 6.3 for 400 nm thick PLZT films. The waveguide prism refractive index for 0 V application, $n_{P(Off)}$, is equal to n_0 of 2.48. In this case, it can be found from Fig. 6.9 that the proper L_{Slab} is around 310 μm.

Based on the parameters mentioned above, n_{Lens}, L_{WL}, and L_{EO} are optimized. Figure 6.10 shows the BPM simulation results for n_{Lens} of 1.8, 2.0, and 2.5. For n_{Lens} of 1.8, collimation of the light beam in the EO slab waveguide is insufficient, while for n_{Lens} of 2.5 the focusing

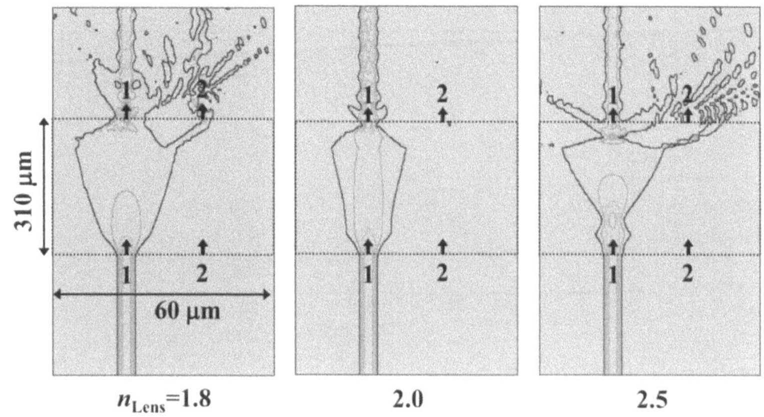

Figure 6.10 BPM simulations of light beam collimation for various n_{Lens}. Reprinted with permission from Yoshimura et al. [15].

effect is too strong, so that large beam size expansion occurs. n_{Lens} of 2.0 is a proper condition to produce collimated light beams, resulting in smooth coupling to the output waveguide in a distance of 310 μm from the input waveguide.

Figure 6.11a shows L_{WL} dependence of transmission efficiency from the input waveguide into the output waveguide. The optimum L_{WL} is found to be 20 μm. For L_{EO}, as shown in Fig. 6.11b, the optimum length is 190–200 μm. From a viewpoint of miniaturization of optical switches, 190 μm should be the optimum L_{EO}. Thus, the total length of the WPD optical switch is calculated to be 310 μm by adding $2(D_{Lens} + L_{WL})$ to L_{EO}.

The design parameters determined for the 2 × 2 WPD optical switch with a 400 nm thick PLZT film are summarized in Table 6.3 and the optimized structure is shown in Fig. 6.12. The waveguide lens refractive index of 2.0 is available by using Ti_xSi_yO, SiO_xN_y, and so on. On the EO slab waveguide of a 400 nm thick PLZT film, right-angle prism-shaped electrodes with an area of 16 × 16 μm^2 are deposited. The size of the main part of the optical switch is 310 μm × 20 μm. When channel separation is widened to 60 μm by bending waveguides of 2.5 mm radius at the input and output ports, the total switch length, L_{SW}, and width, W_{SW}, become 1190 μm and 100 μm, respectively.

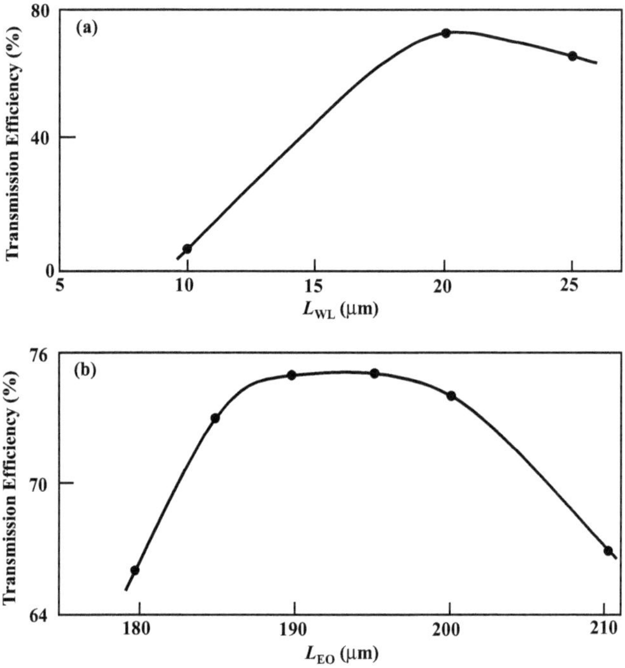

Figure 6.11 (a) L_{WL} dependence and (b) L_{EO} dependence of transmission efficiency from the input waveguide into the output waveguide. Reprinted with permission from Yoshimura et al. [15].

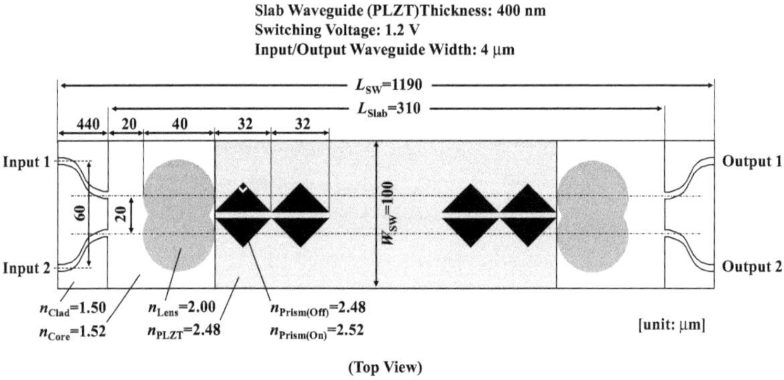

Figure 6.12 Optimized structure of 2 × 2 WPD optical switch with 400-nm-thick PLZT film. Reprinted with permission from Yoshimura et al. [11].

Table 6.3 Design parameters for the 2 × 2 WPD optical switch with 400-nm-thick PLZT film

EO slab waveguide	
Length: L_{EO}	190 μm
Thickness (Electrode gap distance) : d	400 nm
Input/output channel waveguide	
Width: D	4 μm
Channel separation: P	20 μm
Core refractive index: n_{Core}	1.52
Cladding refractive index: n_{Clad}	1.5
Waveguide lens	
Lens-input/output waveguide distance: L_{WL}	20 μm
Diameter: D_{Lens}	40 μm
Refractive index: n_{Lens}	2
Right-angle prism-shaped electrode	
Height: D_P	16 μm
Length: $2D_P$	32 μm
Cascaded prism count: N	2
Total length:	310 μm
Signal light wavelength: λ	1.3 μm

6.3.3 Predicted Performance

Figure 6.13 shows BPM simulation results of the switching characteristics for the 2 × 2 WPD optical switch model utilizing a 400 nm thick PLZT film shown in Fig. 6.12. When no voltages are applied to the prism-shaped electrodes, a 1.3 μm probe beam propagates from Input 1 to Output 1, achieving "Bar" condition. By 1.2 V application to the electrodes, the probe beam propagates from Input 1 to Output 2, achieving "Cross" condition. For both "Bar" and "Cross," cross-talk is less than −20 dB. Transmission efficiency is 80% and 69% for "Bar" and "Cross," respectively. Maximum insertion loss within the WPD optical switch is 1.7 dB, including losses in bending waveguides.

A WPD is regarded as a capacitor with an electrode area of S, thickness of d, and dielectric constant of $\varepsilon_r \varepsilon_0$. Here, ε_0 is the dielectric constant in vacuum. The capacitance of the WPD, C, is given by

$$C = \varepsilon_r \varepsilon_0 S/d. \tag{6.7}$$

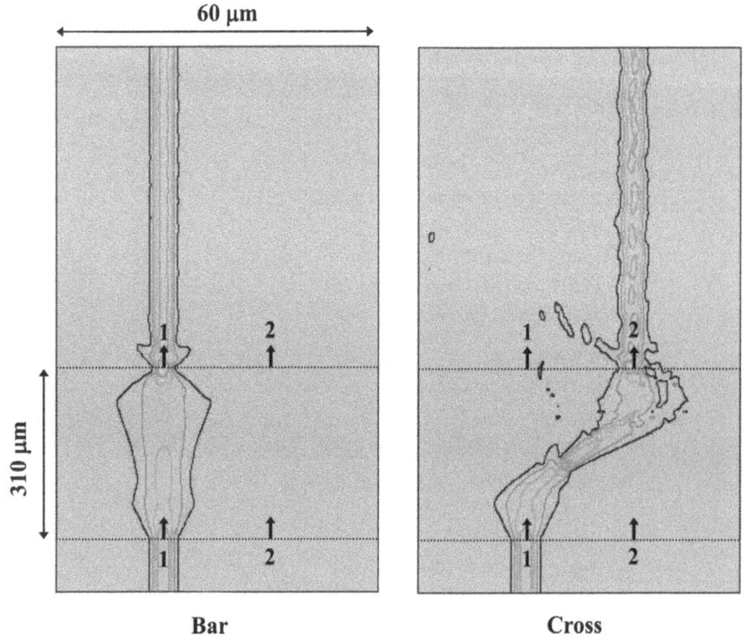

Figure 6.13 BPM simulations of switching characteristics for the 2 × 2 WPD optical switch with 400-nm-thick PLZT film. Reprinted with permission from Yoshimura et al. [11].

The switching speed and power consumption can be estimated based on the charge–discharge model for the capacitor of the WPD as the 0th approximation. The switching speed, τ, is approximately estimated by

$$\tau = R_{\mathrm{ON}}C \tag{6.8}$$

where R_{ON} is the ON resistance of the driver circuit. The energy consumption per bit, P, and power consumption at a switching rate of F, Power, are respectively given by following expressions.

$$P = CV^2 \tag{6.9}$$

$$\text{Power} = FCV^2 \tag{6.10}$$

It is found from Eqs. 6.7–6.10 that low dielectric constant characteristics are preferable for high-speed and low-power operations.

Table 6.4 Predicted electrical performance of 2 × 2 WPD optical switches with 400-nm-thick PLZT and 400-nm-thick polymer MQD

	400-nm-thick PLZT	400-nm-thick polymer MQD
Dielectric constant: ε_r	1000	4*
Capacitance: C [pF]	5.67	0.0227
Switching time (R_{on} =100 Ω): τ [ps]	567	2.27
Energy consumption per bit: P [pJ/bit]	8.17	0.363
Power consumption at $F = 4 \times 10^{10}$: *Power* [mW]	327	14.5

*Assumed value

The predicted electrical performance is summarized in Table 6.4. In the model shown in Fig. 6.12, where $S = 256 \times 10^{-12}$ m^2, $d = 4 \times 10^{-7}$ m, and $\varepsilon_r = 1000$ for PLZT, C is calculated to be 5.67 pF. τ is 567 ps for R_{ON} of 100 Ω, P is 8.17 pJ/bit for $V = 1.2$ V, and Power is 327 mW for $F = 4 \times 10^{10}$ 1/s. When the PLZT is replaced with a 400 nm thick polymer MQD with an assumed ε_r of 4, C is calculated to be 0.0227 pF. τ is 2.27 ps for R_{ON} of 100 Ω, P is 0.363 pJ/bit for $V = 4$ V (see Table 6.2), and Power is 14.5 mW for $F = 4 \times 10^{10}$ 1/s.

In Fig. 6.14a, (Operation Voltage) × (Device Length) characteristics are compared for various devices. For the WPD utilizing PLZT with a structure shown in Fig. 6.12, the length of the main part of the optical switch is 310 μm, and the driving voltage is 1.2 V. Therefore, (Operation Voltage) × (Device Length) is (1.2 V) × (310 × 10^{-4} cm) = 0.0372 V cm. When the PLZT is replaced with polymer MQD, an equivalent refractive index change is available by increasing the driving voltage to 4 V, as can be seen in Table 6.2. Therefore, if the effect of the difference in n_0 on the switching operation of the WPD is neglected, (Operation Voltage) × (Device Length) is calculated to be (4 V) × (310 × 10^{-4} cm) = 0.124 V cm in the WPD utilizing polymer MQD. The values of (Operation Voltage) × (Device Length) for the above-mentioned WPDs are lower than those for MZMs utilizing Si photonics [1] and lithiumniobate (LN) [3], and comparable to the value for the MZM utilizing InGaAsP/Si photonics [2].

In Fig. 6.14b, the energy/bit is compared for various devices. The energy consumption of the MZM utilizing Si photonics and the WPD utilizing PLZT is high. The WPD utilizing polymer MQD and the MZM

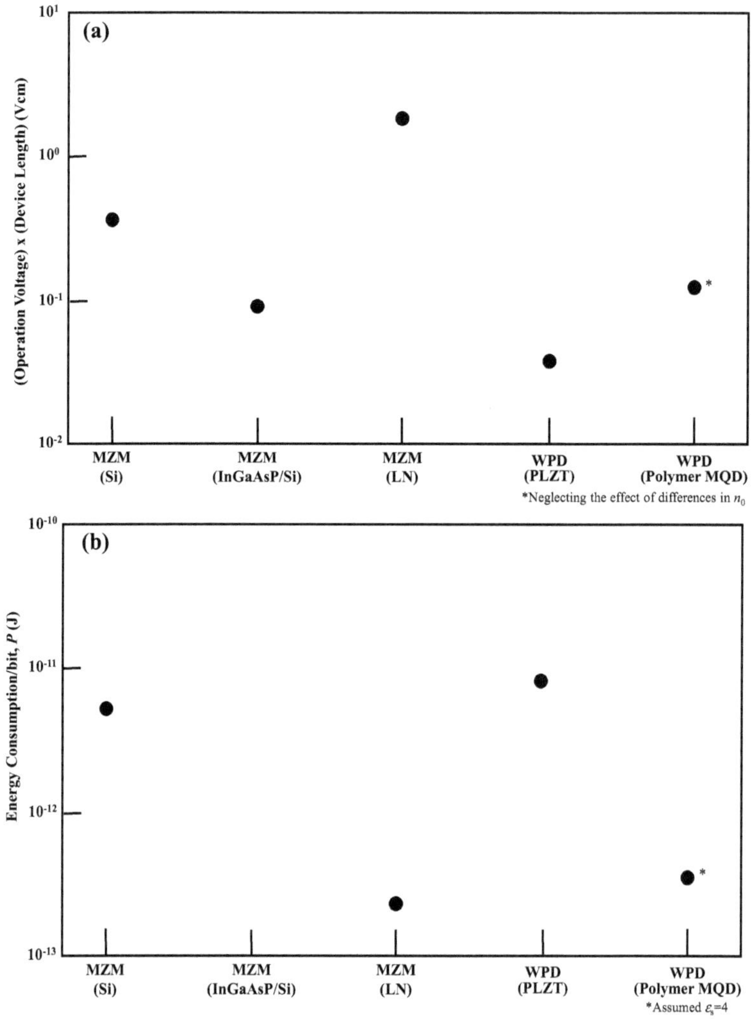

Figure 6.14 (a) (Operation Voltage) × (Device Length) characteristics and (b) energy consumption/bit for various devices.

utilizing LN exhibit extremely low energy consumption, indicating that these devices can reduce the heat generation problem in 3D integrated optical interconnects.

As mentioned in section 6.1, WPDs are phase-insensitive flux-controlled devices, which do not require strict control of waveguide

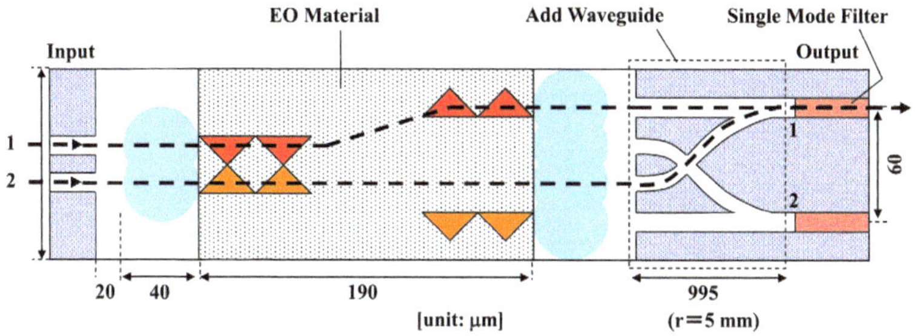

Figure 6.15 Advanced WPD optical switch with ADD functions.

shapes/sizes, refractive indexes, and operation temperature. These arguments suggest that the polymer-MQD WPD is preferable for light modulators/optical switches for 3D integrated optical interconnects from the viewpoint of modulation/switching performance, power consumption, and robustness. It should be noted that the flux-controlled TIR and digital optical switches utilizing the polymer-MQD might also be promising candidates for the 3D integrated optical interconnects.

6.4 Advanced WPDs

6.4.1 WPD Optical Switches with ADD Functions

For optical switching systems, optical switches with ADD functions are sometimes required. Advanced WPDs, which was invented and simulated by Hirotaka Tomita, realize such optical switches as depicted in Fig. 6.15 [27]. The structure is based on the 2 × 2 WPD optical switch. To give ADD functions to optical switch, Add waveguides with a curvature of 5 mm radius are inserted. Single-mode optical filters [15], which extract the fundamental mode from optical signals, are placed after the Output.

By applying driving voltages to four prism-shaped electrodes in the upper-half region, optical signals from Input 1 are connected to Output 1, and at the same time, optical signals from Input 2 are also connected to Output 1, accomplishing the ADD function. Similarly, by applying driving voltages to four electrodes in the lower-half region, optical signals from Input 1 and Input 2 are connected to Output 2

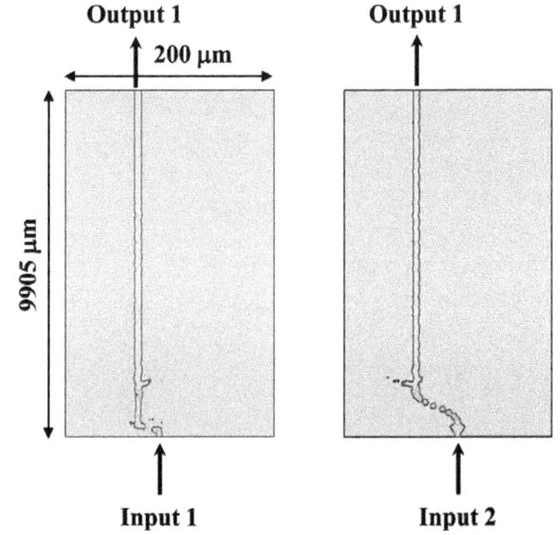

Figure 6.16 Switching characteristics of the advanced WPD optical switch with ADD functions simulated by the BPM.

simultaneously. When driving voltages are applied to all electrodes, optical signals from Input 1 are connected to Output 1, and optical signals from Input 2 to Output 2. When no driving voltages are applied to the electrodes, optical signals from Input 1 are connected to Output 2, and optical signals from Input 2 to Output 1.

Figure 6.16 shows the switching characteristics of the advanced WPD optical switch with ADD functions simulated by BPM. Smooth adding can be seen. The insertion loss and cross-talk are respectively 1.4 dB and −22 dB. After passing through the single-mode optical filter, they respectively become 5.4 dB and −21 dB.

These results suggest a possibility that WPD optical switches will realize various kinds of functional optical circuits.

6.4.2 WPD Optical Switches with MUX/DEMUX Functions

The WPDs can generate MUX/DEMUX functions by using EO materials with large wavelength dispersion characteristics for slab waveguides. Figure 6.17a shows a schematic illustration of a WPD optical switch with DEMUX functions. By applying driving voltages to prism-shaped electrodes, signal beams with wavelengths of λ_1, λ_2, and λ_3 emitted from the input waveguide are respectively

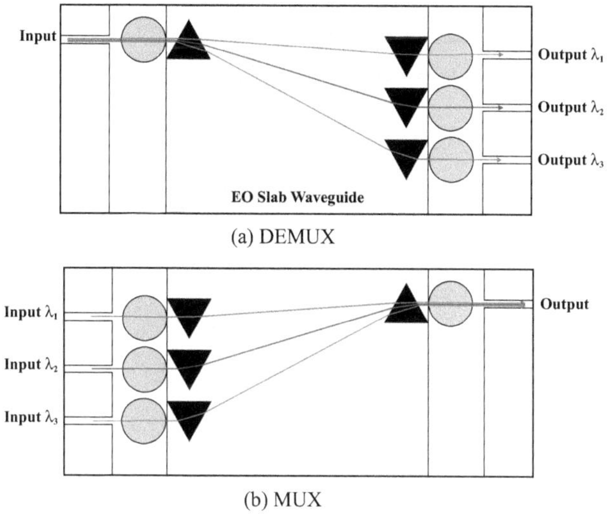

Figure 6.17 WPD optical switches with (a) DEMUX and (b) MUX functions.

guided into output λ_1, output λ_2, and output λ_3. Figure 6.17b shows a WPD optical switch with MUX functions. Signal beams with wavelengths of λ_1, λ_2, and λ_3 respectively emitted from input λ_1, input λ_2, and input λ_3 are guided into the output waveguide. Such MUX/DEMUX functions of WPDs might enable WDM integrated optical interconnects.

6.5 Transient Responses in Microring Resonators and Photonic Crystals

Although MRs and the PhCs provide functions of ultrasmall light modulators/optical switches, they require the building-up duration to complete the interference because they utilize accumulated interference. In the present section, light beam propagation and switching operation in these devices are simulated by the 2D FDTD method and the delay time caused in them is estimated to show the bandwidth limits.

A typical structural model of the MR optical switch is shown in Fig. 6.18. Between two straight waveguides, an EO ring waveguide

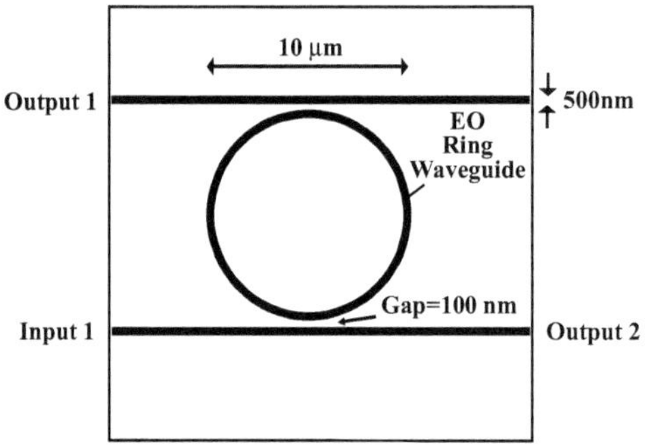

Figure 6.18 Model of an MR optical switch.

with a diameter of 10 μm is inserted. The core width of these optical waveguides is 500 nm. A 100 nm gap exists between straight and ring waveguides. In the case that the core is made of PLZT and cladding is air, the core refractive index, n_{Core}, is 2.48 and the cladding region refractive index, n_{Clad} is 1.0.

As shown in Fig. 6.19, when a light pulse of 1.558 μm in wavelength is introduced from Input 1, the light pulse is partially transferred between the straight waveguides and the ring waveguide. Here, the brightness of the light pulse images is intensified at 0.19 ps to present the light pulse transfer phenomenon clearly.

By tuning the configurations and refractive indexes of optical waveguides to a resonance condition, cw light beams gradually become localized into the ring waveguide, as can be seen in Fig. 6.20. At the same time, the light beams that were guided to Output 2 are switched to Output 1. When a voltage is applied to the ring waveguide, the resonance condition is broken to switch the light beams from Output 1 to Output 2. The resonance condition depends on the wavelength of the guided beams. This means that the resonance wavelength of the MR can be controlled by the driving voltage to achieve tunable wavelength filter operation.

It is found from Fig. 6.20 that to complete the accumulated interference in the MR optical switch, it takes about 4.5 ps. The data rate limit of the MR optical switch is approximately estimated to be $1/(2 \times 4.5\,\mathrm{ps}) \approx 110$ Gbps for the present model. In the case of the WPD optical switch utilizing polymer MQDs, τ is estimated to

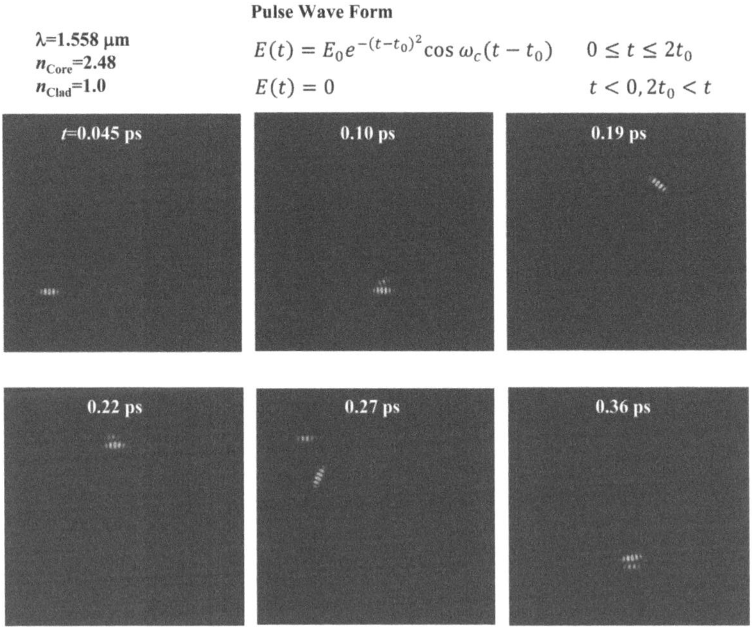

Figure 6.19 FDTD simulations of light pulse propagation in the MR optical switch.

be 2.27 ps, as described in section 6.3.3, and the data rate limit is approximately estimated to be $1/(2 \times 2.27\ \text{ps}) \approx 220$ Gbps. These results indicate that the bandwidth limit of the MR optical switch will become a problem in ultrahigh-speed optical interconnects.

The PhC, whose concept was proposed by Yablonovitch [5, 6], has periodic variations of refractive indexes like dielectric multilayer wavelength filters. As the energy bands for electrons in crystals that have periodic potential variations, energy bands for photons appear in PhCs. When the energy of photons is in the photonic bandgap (PBG) region, light beams cannot exist in PhCs in steady states.

In the case that some imperfections are generated in a PhC, light beams with photon energy within PBG can exist in parts with imperfections. By creating line-shaped imperfections, light beams are confined in the parts with imperfections because they cannot exist in the surrounding perfect PhC parts. When the line width of the imperfection is submicrometers, nanoscale waveguides can be

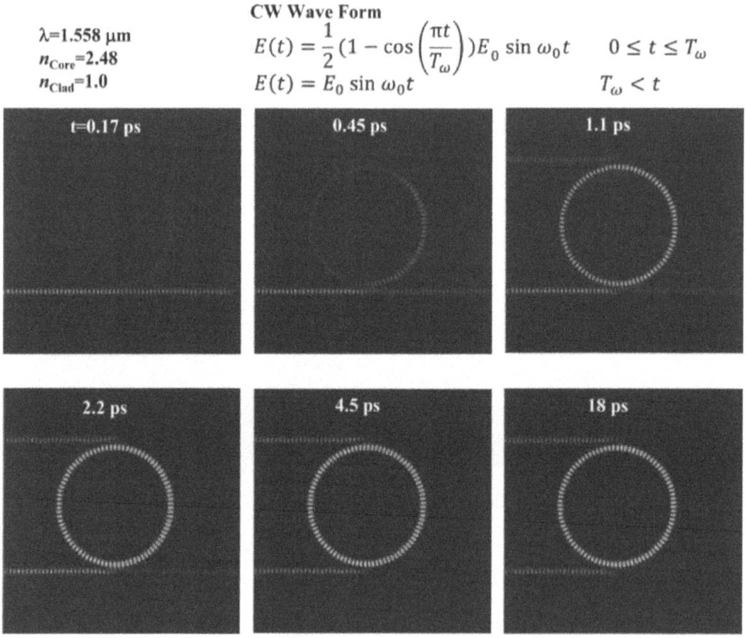

Figure 6.20 FDTD simulations of light beam resonance in the MR optical switch.

constructed in the PhC. By incorporating EO materials in the PhC, ultrasmall optical switches might be available.

Simulations for transient light beam propagation in a two-dimensional PhC waveguide was carried out using the 2D FDTD method. Holes with a radius of 104 nm are arranged with a 521 nm pitch in a medium with a refractive index of 3.38. Light polarization is parallel to the plane for calculations. For a wavelength of 1.30 μm, which corresponds to photon energy near the center of PBG, light beams are strongly confined in the optical waveguide.

Figure 6.21 shows a result for transient light beam propagation in the PhC waveguide [28]. Here, the drawing contrast is adjusted to enhance leaked light beams in the surrounding perfect crystal region. At the initial stage, considerable leakage of light beams into the surrounding perfect crystal region is observed. After 0.6 ps, the leakage is suppressed and a strong confinement of light beams in the optical waveguide is achieved.

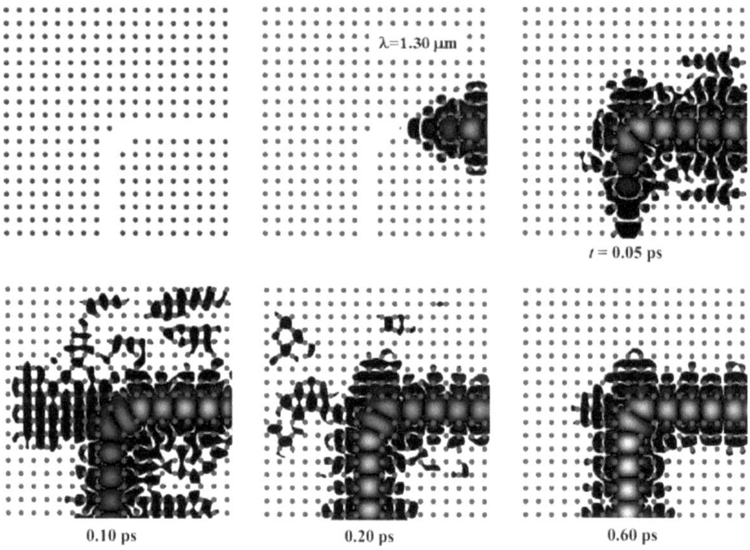

Figure 6.21 FDTD simulations of transient light beam propagation in a PhC waveguide. Reprinted with permission from Yoshimura et al. [28].

Such delay before settling waveguide functions arises from the fact that the PhC device uses interference of light beams at a steady state. In the model shown in Fig. 6.21, at $t = 0.05$ and 0.1 ps, only a part of reflected light beams exist in the PhC—that is to say, the light beam interference in the PhC is not completed yet. At $t = 0.6$ ps, since 0.6 ps corresponds to a light beam propagation distance of ~150 μm, which is 17 times of the PhC size in the present model, the interference is built up completely; that is, the superposition of all the reflected light beams is completed in whole of the PhC, to reach the steady state. Thus, the leakage is suppressed, resulting in perfect light beam confinement in the optical waveguide.

Transient responses of light beam confinement in a PhC waveguide are shown in Fig. 6.22. For a wavelength $\lambda = 1.20$ μm, which corresponds to photon energy at the edge of PBG, a long delay of ~0.5 ps is observed before the confined light intensity reaches its steady state. For a wavelength $\lambda = 1.30$ μm located near the center of PBG, on the other hand, the confined light intensity rapidly increases to its saturated value with a short delay less than 0.2 ps.

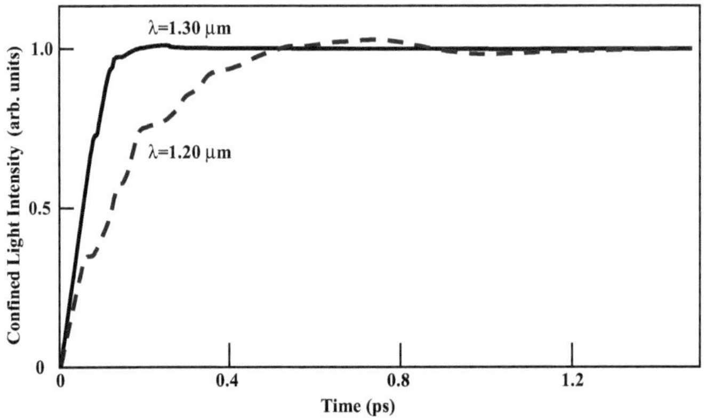

Figure 6.22 FDTD simulations of transient responses of light beam confinement in a PhC waveguide.

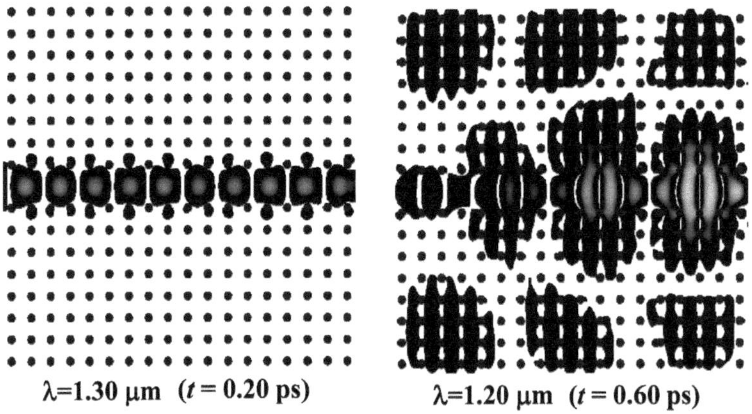

Figure 6.23 FDTD simulations of light beam distribution in PhC waveguides after reaching steady state.

As can be seen in Fig. 6.23, the light beam distribution after reaching a steady state is wider for $\lambda = 1.20\ \mu m$ than for $\lambda = 1.30\ \mu m$. This indicates that in the case of $\lambda = 1.20\ \mu m$ the light beams penetrate into the deep part of the surrounding perfect crystal region, and, consequently, interference occurs over a wide region compared with the case of $\lambda = 1.30\ \mu m$.

References

1. W. M. J. Green, M. J. Rooks, L. Sekaric, and Y. A. Vlasov, "Ultra-compact, low RF power, 10 Gb/s silicon Mach–Zehnder modulator," *Opt. Express* **15**, 17106–17113 (2007).
2. T. Hiraki, T. Aihara, K. Hasebe, K. Takeda, T. Fujii, T. Kakitsuka, T. Tsuchizawa, H. Fukuda, and S. Matsuo, "Heterogeneously integrated III–V/Si MOS capacitor Mach–Zehnder modulator," *Nat. Photonics* **11**, 482–485 (2017).
3. C. Wang, M. Hang, B. Stern, M. Lipson, and M. Loncar, "Nanophotonic lithium niobate electro-optic modulators," *Opt. Express* **26**, 1547–1555 (2018).
4. R. M. Kubacki, "Micro-optic enhancement and fabrication through variable in-plane index of refraction (VIPIR) engineered silicon nanocomposite technology," *Proc. SPIE* **5347**, 233–246 (2004).
5. E. Yablonovitch, "Inhibited spontaneous emission in solid state physics and electronics," *Phys. Rev. Lett.* **58**, 2059–2062 (1987).
6. E. Yablonovitch, T. J. Gmitter, and K. M. Leung, "Photonic band structure: the face-centered-cubic case employing nonspherical atoms," *Phys. Rev. Lett.* **67**, 2295–2298 (1991).
7. T. Yoshimura, S. Tsukada, S. Kawakami, Y. Arai, H. Kurokawa, and K. Asama, "3D micro optical switching system (3D-MOSS)–packaging design-", *2001 IEEE LEOS Annual Meeting Conference Proceedings*, San Diego, California, 499 (2001).
8. T. Yoshimura, S. Tsukada, S. Kawakami, Y. Arai, H. Kurokawa, and K. Asama, "3D micro optical switching system (3D-MOSS) architecture," *Proc. SPIE* **4653**, from Photonics West 2002, San Jose, California, 62–70 (2002).
9. T. Yoshimura, "Waveguide-type electro-optic devices," Japanese Patent Tokukai Hei 4-181231 (1992).
10. T. Yoshimura, "Optical integrated circuit devices," Japanese Patent Tokukai Hei 4-204633 (1992).
11. T. Yoshimura, M. Ojima, Y. Arai, and K. Asama, "Three-dimensional self-organized micro optoelectronic systems for board-level reconfigurable optical interconnects–performance modeling and simulation–," *IEEE J. Sel. Top. Quantum Electron.* **9**, 492–511 (2003).
12. T. Yoshimura, S. Tsukada, S. Kawakami, M. Ninomiya, Y. Arai, H. Kurokawa, and K. Asama, "Three-dimensional micro-optical

switching system architecture using slab-waveguide-based micro-optical switches," *Opt. Eng.* **42**, 439–446 (2003).

13. T. Yoshimura, "Design of organic nonlinear optical materials for electro-optic and all-optical devices by computer simulation," *FUJITSU Sc. Tech. J.* **27**, 115–131 (1991).
14. M. Ninomiya, Y. Arai, T. Yoshimura, H. Kurokawa, and K. Asama, "Characteristics evaluation for waveguide-prism-deflector type micro optical switches (WPD-MOS)," *Electronics and Communications in Japan, Part 2* **86**, 38–48 (2003) (Translated from Denshi Joho Tsushin Gakkai Ronbunshi **J85-C**, 1192–1201 (2002)).
15. T. Yoshimura, M. Ojima, Y. Arai, N. Fujimoto, and K. Asama, "Simulation on cross-talk reduction in multi-mode-waveguide-based micro optical switching systems using single mode filters," *Appl. Opt.* **43**, 1390–1395 (2004).
16. T. Yoshimur and, Y. Arai, "3D optoelectronic micro system," U.S. Patent 7,387,913 B2 (2008).
17. O. Qasaimeh, K. Kamath, P. Bhattacharya, and J. Phikkips, "Linear and quadratic electro-optic coefficients of self-organized $In_{0.4}Ga_{0.6}As/GaAs$ quantum dots," *Appl. Phys. Lett.* **72**, 1275–1277 (1998).
18. T. Yoshimura and T. Futatsugi., "Non-linear optical device using quantum dots," U.S. Patent 6,294,794 (2001)
19. T. Yoshimura, "Characterization of the electro-optic effect in styrylpyridinium cyanine dye thin-film crystals by an ac modulation method," *J. Appl. Phys.* **62**, 2028–2032 (1987).
20. T. Yoshimura, "Enhancing second-order nonlinear optical properties by controlling the wave function in one-dimensional conjugated molecules," *Phys. Rev.* **B 40**, 6292–6298 (1989).
21. T. Yoshimura, S. Tatsuura, W. Sotoyama, A. Matsuura, and T. Hayano, "Quantum wire and dot formation by chemical vapor deposition and molecular layer deposition of one-dimensional conjugated polymer," *Appl. Phys. Lett.* **60**, 268–270 (1992).
22. T. Yoshimura, A. Oshima, D. Kim, and Y. Morita, "Quantum dot formation in polymer wires by three-molecule molecular layer deposition (MLD) and applications to electro-optic/photovoltaic devices," *ECS Trans.* **25**, Issue 4 "Atomic Layer Deposition Applications 5," from the Vienna, Austria Meeting, 15–25 (2009).
23. T. Yoshimura, R. Ebihara, and A. Oshima, "Polymer wires with quantum dots grown by molecular layer deposition of three source molecules for sensitized photovoltaics," *J. Vac. Sci. Technol. A* **9**, 051510 (2011).

24. T. Yoshimura and S. Ishii, "Effect of quantum dot length on the degree of electron localization in polymer wires grown by molecular layer deposition," *J. Vac. Sci. Technol. A* **31**, 031501 (2013).
25. T. Yoshimura, *Thin-Film Organic Photonics: Molecular Layer Deposition and Applications*, CRC/Taylor & Francis, Boca Raton, Florida (2011).
26. T. Aizawa, Furuuchi chemical corporation technical articles for PLZT shutter arrays.
27. N. Fujimoto, H. Tomita, and T. Yoshimura, "A path-independent-loss type optical cross-connect system using high-speed waveguide prism micro-optical switches in WDM networks," *Research Reports of the School of Engineering, Kinki University*, No. 40, 81–85 (2006).
28. T. Yoshimura, Y. Suzuki, N. Shimoda, T. Kofudo, K. Okada, Y. Arai, and K. Asama, "Three-dimensional chip-scale optical interconnects and switches with self-organized wiring based on device-embedded waveguide films and molecular nanotechnologies," *Proc. SPIE* **6126**, from Photonics West 2006, San Jose, California, 612609-1-15 (2006).

Chapter 7

Self-Organized Lightwave Networks

In 3D integrated optical interconnects, in which 3D optical wiring backbones are constructed by connecting constituent OE devices, the following two issues should be solved:

- Enormous effort to assemble a large number of OE devices with strict positional and angular control
- Lack of viable processes to fabricate vertical waveguides for inter-layer optical links

Self-organized lightwave networks (SOLNETs) are expected to solve these problems because of their capabilities to construct self-aligned optical couplings and optical waveguides in free spaces. Details of SOLNETs are presented in a book published in 2018 [1]. In the present chapter, the concept and expected performance of SOLNETs are reviewed.

7.1 Concept of SOLNETs

The inherent lightwave nature of "Go Straight" sometimes causes difficulties. To bring a lightwave from point A to B, strict positional/angular control is necessary. By implementing SOLNETs,

Self-Organized 3D Integrated Optical Interconnects with All-Photolithographic Heterogeneous Integration
Tetsuzo Yoshimura

ISBN 978-981-4877-04-6 (Hardcover), 978-1-003-04305-8 (eBook)
www.jennystanford.com

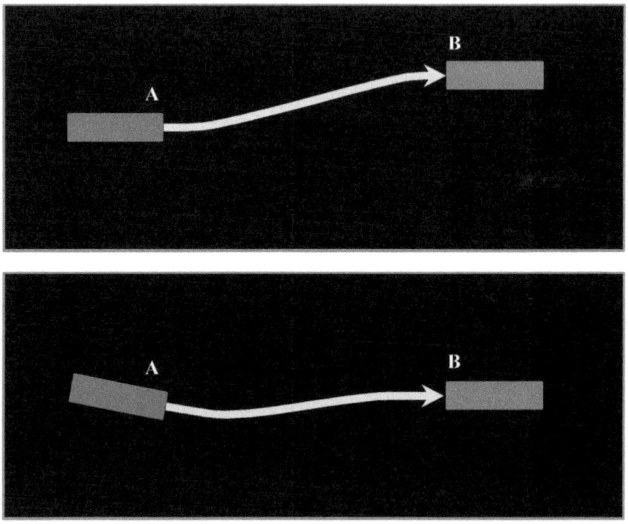

Figure 7.1 Lightwave propagation realized by SOLNETs.

as schematically shown in Fig. 7.1, the lightwave can propagate from point A to B automatically even if there are lateral/angular misalignments between the two points, making strict alignment control unnecessary. This preferable effect is attributed to the following capabilities of SOLNETs [1, 2]:

(1) Fabricating optical solder, which enables self-aligned optical couplings between misaligned OE devices with different core sizes
(2) Fabricating three-dimensional (3D) optical wiring in free spaces
(3) Targeting lightwaves onto specific objects

7.1.1 Types of SOLNETs

The SOLNET is formed by utilizing the interaction between write beams. The interaction generates the capability to fabricate self-aligned coupling waveguides automatically. To date, four types of SOLNETs have been developed, namely, the two-beam-writing SOLNET (TB-SOLNET) [1–8], phosphor SOLNET (P-SOLNET) [1, 7, 8], reflective SOLNET (R-SOLNET) [1, 7–16], and luminescence-

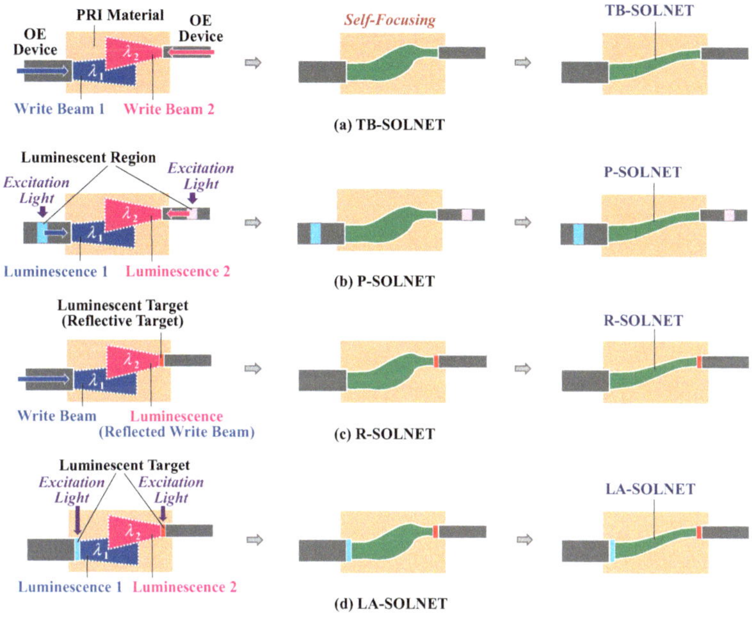

Figure 7.2 Concepts of SOLNETs.

assisted SOLNET (LA-SOLNET) [1, 7, 8]. The concepts are shown in Fig. 7.2. OE devices such as optical waveguides, light modulators, optical switches, wavelength filters, LDs, and PDs are placed in counter-direction arrangements with a photo-induced refractive-index increase (PRI) material filling in between. In the PRI material the refractive index increases by write beam exposure.

The TB-SOLNET is the fundamental type of SOLNET. As Fig. 7.2a shows, a first write beam (write beam 1) of wavelength λ_1 is introduced into a PRI material from an OE device, while a second write beam (write beam 2) of wavelength λ_2 is introduced from another OE device. The refractive index in the region, where the two write beams overlap, increases rapidly compared with that in the surrounding region. Then, because the two write beams tend to be confined in the higher-refractive-index region, they merge through self-focusing to form a self-aligned coupling waveguide of TB-SOLNET automatically between the OE devices, even if misalignments and core size mismatching exist between them.

In the P-SOLNET, as Fig. 7.2b shows, luminescent regions are formed in some parts of OE device cores to generate luminescence within the cores by excitation lights from outside. Luminescence 1 of λ_1 and luminescence 2 of λ_2 respectively correspond to write beam 1 and write beam 2 in the TB-SOLNET. They form a self-aligned coupling waveguide of the P-SOLNET. The same situation is available without luminescent regions in the case that the waveguide cores are emissive intrinsically. For example, in silicon nitride (SiN) and silicon oxynitride (SiON) waveguides having emissive characteristics, luminescence for write beams can be generated within the cores just by exposing the cores to external excitation light. The P-SOLNET is effective when the write beams are difficult to be inserted into cores.

As discussed in previous articles [1, 7], the formation dynamics model for the P-SOLNET is the same as that for the TB-SOLNET. This implies that the results for the TB-SOLNET formation are regarded as those for the P-SOLNET formation.

In the R-SOLNET, as Fig. 7.2c shows, a luminescent target is placed on a front core edge of one of the OE devices. When a write beam of wavelength λ_1 is introduced from the other OE device into the PRI material, the target absorbs the write beam and emits luminescence of wavelength λ_2, which acts as the second write beam, to form a self-aligned coupling waveguide of the R-SOLNET. The same situation is realized without the luminescent target in the case that the waveguide core has emissive characteristics intrinsically, where the core edge plays the role of a luminescent target. The R-SOLNET is effective when the write beams cannot pass through one of the OE devices. It is possible to replace luminescent targets with reflective targets like mirrors and wavelength filters. In this case, a reflected write beam plays the role of the second write beam with $\lambda_2 = \lambda_1$.

During the R-SOLNET formation, the luminescence from the luminescent target (or the reflected write beam from the reflective target) acts just like a trigger to let the incident write beam flow toward the emitting point (or the reflecting point). The luminescence (or the reflected write beam) paves a way to the target location to construct a self-organized waveguide toward the target. We call this phenomenon as the "pulling water" effect.

In the LA-SOLNET, as Fig. 7.2d shows, luminescent targets are placed on front core edges of both OE devices. By introducing excitation lights from outside, the targets respectively emit luminescence 1 of λ_1 and luminescence 2 of λ_2 to form a self-aligned coupling waveguide of the LA-SOLNET. The LA-SOLNET is effective when write beams cannot pass through OE devices.

Usually, SOLNETs mean permanent SOLNETs, which are fixed to remain for a long time once they are formed. By forming SOLNETs in rewritable materials, dynamic SOLNETs, which can be reorganized from time to time, are obtained. SOLNETs are also categorized by the photochemical processes involved in SOLNET formation. One-photon SOLNETs are formed by one-photon photochemistry and two-photon SOLNETs by two-photon photochemistry.

7.1.2 PRI Materials

The requirement for PRI materials is that the refractive index increase upon write beam exposure. There are many kinds of PRI materials for SOLNET formation such as photopolymers [3, 4, 17–23], photo-definable materials, photosensitive glass [24], photosensitive organic/inorganic hybrid materials [25, 26], photorefractive materials [27, 28], and third-order nonlinear optical materials [29].

Figure 7.3a shows a typical mechanism of the refractive index increase in a photopolymer. High-refractive-index photoreactive molecules (high-*n* molecules) and low-refractive-index photoreactive molecules (low-*n* molecules) are mixed. The high-*n* molecules have higher photoreactivity to the write beam than the low-*n* molecules. When the photopolymer is exposed to a write beam, the high-*n* molecules are combined to make dimers, oligomers, or polymers by photochemical reactions. Then, high-*n* molecules diffuse into the exposed region from the surrounding region to compensate for the reduction in the high-*n* molecule concentration in the exposed region, and are combined to make dimers, oligomers, or polymers. This repeated process increases the refractive index of the exposed region to induce inhomogeneous refractive index distributions according to the pattern of the write beam exposure. Curing by blanket light exposure or heating fixes the refractive index distributions to produce permanent SOLNETs.

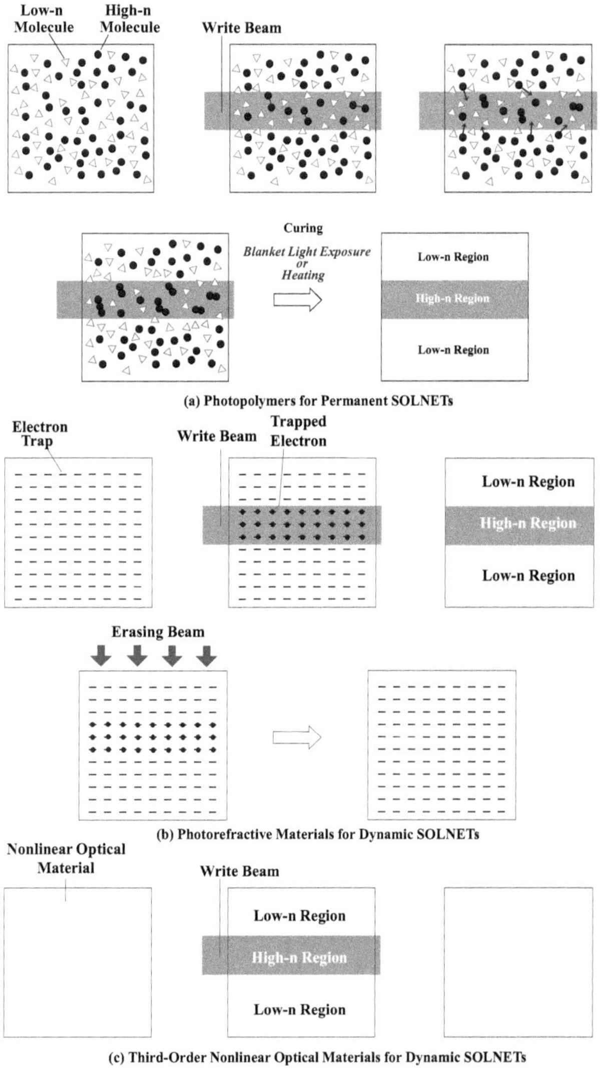

Figure 7.3 Typical mechanisms of refractive index increase in PRI materials.

The spectral response of PRI materials can be adjusted by sensitizers, for which molecules with one-photon or two-photon absorption characteristics can be used. The gamma characteristics, that is, the relationship between refractive index change and write beam exposure, can be controlled by adjusting the refractive index, photoreactivity, and concentration of high/low-n molecules. Precise control of the gamma characteristics might be possible by mixing three or more kinds of molecules with different refractive indexes and photoreactivity.

Figure 7.3b shows a refractive index increase mechanism in a photorefractive material such as lithium niobate (LN) crystals and photorefractive polymers. When a photorefractive material is exposed to a write beam, electrons excited by the beam are captured in electron traps, generating local electric fields around the electrons. This gives rise to refractive index changes due to the EO effect to induce inhomogeneous refractive index distributions according to the pattern of the write beam exposure. Blanket exposure of erasing beams releases the trapped electrons to clear the refractive index distributions. This makes the material ready for rewriting, enabling dynamic SOLNETs. Fazio et al. proposed transient interconnection network in a photorefractive material as a novel paradigm for integrated photonics circuits [27].

Figure 7.3c shows a case in which a third-order nonlinear optical material such as poly-diacethylene and compound semiconductors are used for the PRI material. When the material is exposed to a write beam, inhomogeneous refractive index distributions are induced according to the pattern of the write beam exposure due to the optical Kerr effect. The refractive index distributions disappear just after the write beam exposure stops to reset the material. This enables dynamic SOLNETs.

7.1.3 One-Photon and Two-Photon SOLNETs

In the one-photon SOLNET, PRI materials exhibiting one-photon photochemistry are used. In this case, as shown in Fig. 7.4a, the λ_1 write beam excites electrons from S_0 to S_n state in sensitizing molecules to induce chemical reactions, specifically an increase in the refractive index of the PRI material. In parallel, the λ_2 write beam excites electrons from S_0 to $S_{n'}$ state to induce an increase in the refractive index. Therefore, the refractive index increase rate R is

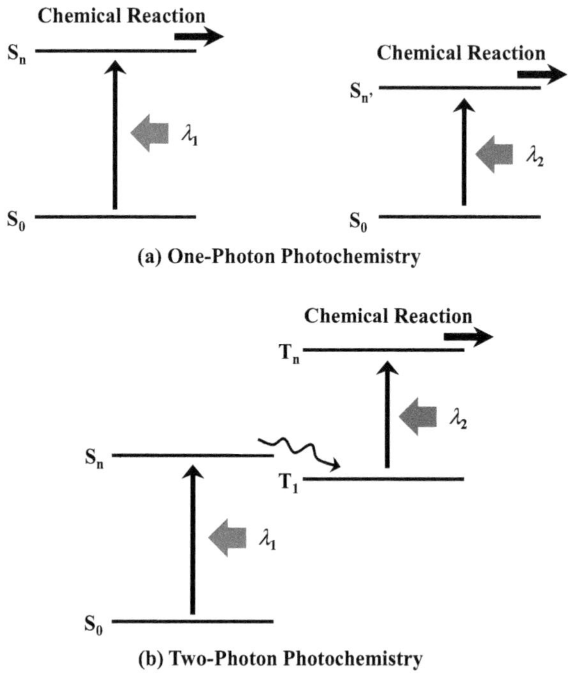

Figure 7.4 Energy-level schemes for sensitization with one-photon and two-photon photochemistry.

expressed as

$$R \propto \gamma_1 I_1 + \gamma_2 I_2, \tag{7.1}$$

where I_1 and I_2 are respectively the intensity of the λ_1 write beam and the λ_2 write beam, and γ_1 and γ_2 are respectively the PRI material sensitivity to the λ_1 write beam and the λ_2 write beam. In the case that $\gamma_1 \approx \gamma_2 = \gamma$, Eq. 7.1 is simplified as

$$R \propto \gamma (I_1 + I_2). \tag{7.2}$$

The one-photon SOLNET formation mechanism works in both cases, where, $\lambda_1 \neq \lambda_2$ and $\lambda_1 = \lambda_2$.

In the two-photon SOLNET [1, 6–8], PRI materials exhibiting two-photon photochemistry are used. Two-photon photosensitive materials were reported by Brauchle et al. for holography [30]. Two-photon absorption molecules such as biacetyl (BA) and

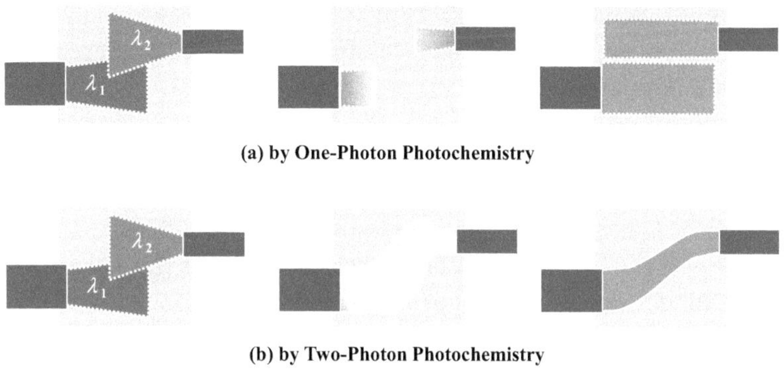

Figure 7.5 Schematic illustrations of one-photon and two-photon SOLNET formation for cases with extremely large misalignments.

camphorquinone (CQ) were incorporated in the materials for hologram formation to induce two-photon photochemical reactions.

The two-photon sensitization mechanism is shown in Fig. 7.4b. Electrons excited from S_0 to S_n state by the λ_1 write beam transfer to T_1 state, and are further excited to T_n state by the λ_2 write beam, resulting in chemical reactions to increase the refractive index. In this case, because two-step electron excitation occurs in series, the refractive index increase rate is expressed as

$$R \propto I_1 \; I_2. \tag{7.3}$$

The above-described difference in the write beam intensity dependence of the refractive index increase rate causes differences in formation dynamics between one-photon SOLNET and two-photon SOLNET. In Fig. 7.5, SOLNET formation is schematically illustrated for cases of large misalignments. When SOLNET formation by one-photon photochemistry is attempted, the refractive index increases with a rate proportional to $\gamma_1 I_1 + \gamma_2 I_2$. Because I_1 and I_2 are respectively large near front edges of the left-hand-side and right-hand-side OE devices, the refractive index tends to increase rapidly in the PRI material near the front edges of the OE devices, as shown in Fig. 7.5a. This prevents the write beams from expanding into the write-beam-overlapping region in the middle, resulting in two separated optical waveguides stretching from individual OE devices. This means that one-photon photochemistry cannot form SOLNETs when the misalignment is extremely large.

When SOLNET formation by two-photon photochemistry is attempted, on the other hand, the refractive index increases with a rate proportional to $I_1 I_2$, which implies that photochemical reactions occur only if the λ_1 write beam and the λ_2 write beam coexist. Then, as shown in Fig. 7.5b, the increase in the refractive index near the front edges of the OE devices becomes less notable because I_2 is very small near the front edge of the left-hand-side OE device and I_1 is very small near the front edge of the right-hand-side OE device. In the write-beam-overlapping region in the middle, where both of I_1 and I_2 are not so small, the refractive index increase develops to make the two write beams merge to form a coupling optical waveguide of SOLNET. Thus, in the two-photon SOLNET the write-beam-overlapping effect is enhanced to extend tolerances to misalignments.

The two-photon SOLNET also has an advantage in that it can suppress the waveguide broadening caused by over-exposure, as illustrated in Fig. 7.5, because the refractive index increase rate proportional to $I_1 I_2$ results in high-contrasting refractive-index images.

7.1.4 Fabrication Processes of Luminescent Targets and Luminescent Regions

To form R-SOLNETs and LA-SOLNETs, luminescent targets should be deposited just on the core regions of OE device edges. The luminescent targets can be fabricated by the light curing process or the selective deposition process [16].

For OE devices consisting of optical waveguides, in which curing light can be transmitted, the light curing process shown in Fig. 7.6a is applicable. First, an edge of an OE device is coated with a luminescent material like a phosphor-doped photopolymer. Then, in order to cure the region just on the core, the luminescent material is exposed to curing light that propagates in the core of the OE device toward the device edge. The luminescent target is finally obtained by etching the non-cured region. In principle, the target size is determined by the beam size of the propagated curing light.

In the selective deposition process shown in Fig. 7.6b, different surface properties are given to the core surface and the cladding

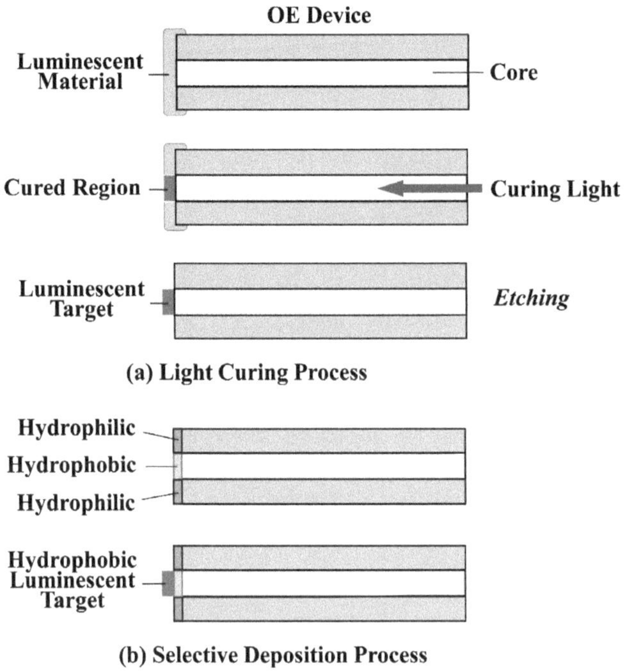

Figure 7.6 Processes for luminescent target fabrication. Reprinted with permission from Yoshimura [16].

surface of an OE device edge. For example, the core surface and the cladding surface are respectively hydrophobic and hydrophilic, which is the case of a Si waveguide with SiO_2 cladding. When a hydrophobic luminescent material is put on the surface, it is deposited on the hydrophobic core surface selectively to form a luminescent target. The luminescent material can be polymer, glass, or thin films. The selective deposition process can also be achieved by surface treatments such as self-assembled monolayer (SAM) formation with designated patterns (see Chapter 12).

Luminescent thin films can be grown by vacuum deposition [14], organic chemical vapor deposition [31, 32], atomic layer deposition (ALD) [33], or molecular layer deposition (MLD) [31, 34, 35]. MLD enables selective conformal growth of organic thin films with designated molecular arrangements. Poly-azomethine (poly-AM) thin films grown by MLD emit blue-green luminescence

[35], and they can selectively be grown on hydrophilic surfaces [31, 32, 35]. When hydrophilic and hydrophobic treatments are respectively applied on the core and cladding surfaces, a poly-AM thin film is deposited on the core surface selectively to form a luminescent target. This process is suitable for submicrometer-scale target fabrication. Similarly, wavelength filters for reflective targets can be deposited using ALD or MLD.

Formation of luminescent regions for P-SOLNETs can be performed by thermal diffusion of phosphor molecules in polymer waveguides. In the case that the waveguide cores exhibit emissive characteristics intrinsically, write beams can be generated within the cores just by exposing the cores to excitation light without any additional processes. For example, in SiON waveguides and sol–gel material waveguides, which are fabricated by using starting materials with dopants for luminescent centers, write beams can be generated within the cores just by excitation light exposure.

7.2 Performance of SOLNETs Predicted by Computer Simulations

Comprehensive theoretical study on SOLNETs has been carried out to predict their performance [1, 3–7, 9–14, 16]. In the present section, computer simulations of SOLNET formation and optical coupling efficiency are reviewed for four types of SOLNETs. The features of one-photon SOLNETs and two-photon SOLNETs are compared for couplings between nanoscale waveguides and couplings between microscale and nanoscale waveguides.

7.2.1 Simulation Models

Models for computer simulations of SOLNETs between nanoscale waveguides are shown in Fig. 7.7 [1, 7]. An input waveguide and an output waveguide are placed with a gap distance of 32 μm. A PRI material is placed in the gap region. The waveguide core width is 600 nm. The refractive index of the waveguide cores is 2.0 and that of the cladding region is 1.5. Such refractive index distributions are available by using, for example, SiN, SiON [36],

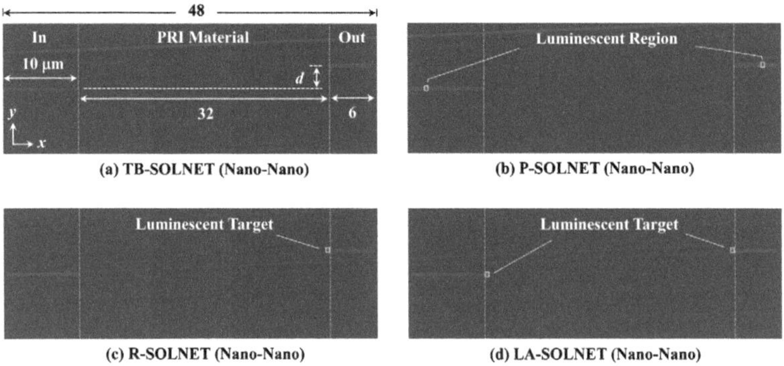

Figure 7.7 Models for 2D FDTD simulations of SOLNETs between nanoscale waveguides.

or sol–gel material based on silicon/titanium oxide (SiTiO) for the cores and a polymer or glass for the cladding region. The refractive index of the PRI material increases from 1.5 to 1.7 upon write beam exposure. Such a photosensitive property is available in the PRI sol–gel material (Nissan Chemical Industries Ltd.) [25, 26, 37, 38], which was found to be suitable for high-refractive-index-contrast SOLNET fabrication [38] as well as nanoscale waveguide fabrication [37]. The lateral misalignment between the input and output waveguides is expressed by d.

In the TB-SOLNET (Fig. 7.7a), a λ_1 write beam is emitted from the input waveguide and a λ_2 write beam from the output waveguide. Thirty point light sources for the λ_1 write beam are distributed across the core along the y direction with a pitch of 20 nm at $x = 1.4$ μm in the input waveguide, and 30 point light sources for the λ_2 write beam are distributed with a pitch of 20 nm at $x = 47.4$ μm in the output waveguide. Here, the origin of the x–y coordinate is set at the left-hand-side bottom corner of the model. The electric field of the point light source is expressed as follows.

$$\left.\begin{array}{ll} E(t) = \frac{1}{2}\left(1 - \cos\left(\dfrac{\pi t}{T_\omega}\right)\right) E_0 \sin\omega_0 t & 0 \le t \le T_\omega \\ E(t) = E_0 \sin\omega_0 t & T_\omega \le t \end{array}\right\} \quad (7.4)$$

Here, ω_0 is the angular frequency of the lightwave and $\omega_0 T_\omega = 4\pi$. The electric-field amplitude of the write beam at the point light source location is $E_0 = 6$ mV/m.

In the P-SOLNET (Fig. 7.7b), by excitation lights from outside, luminescence 1 for the λ_1 write beam and luminescence 2 for the λ_2 write beam are respectively generated from luminescent regions at $x = 1.4\,\mu m$ in the input waveguide and at $x = 47.4\,\mu m$ in the output waveguide. In this model, the point light source locations and arrangements in the luminescent regions are the same as those for the write beams in the TB-SOLNET. This implies that the model for the P-SOLNET is exactly the same as the model for the TB-SOLNET, and consequently, that results for P-SOLNETs and TB-SOLNETs are the same. In the present chapter, therefore, the results for TB-SOLNETs mean the results for both TB-SOLNETs and P-SOLNETs.

As mentioned in section 7.1.1, the same situation for P-SOLNET formation is available without luminescent regions when the waveguide cores are emissive intrinsically. In this case, the luminescent region in the model corresponds to the waveguide core portion, which is exposed to the excitation light.

In the R-SOLNET (Fig. 7.7c), a luminescent target is placed on the front edge of the output waveguide. When a λ_1 write beam is introduced into the PRI material from the input waveguide, the luminescent target is excited by it and emits luminescence of λ_2, which acts as the λ_2 write beam from the output waveguide. The point light source arrangements in the luminescent target are the same as those for the write beam in the TB-SOLNET. The electric-field amplitude of the emitted luminescence is determined by assuming that it is proportional to the temporary time average of absolute values of the λ_1 write-beam electric field at the location of the point light source. Efficiency for the luminescence generation is assumed 0.7.

The same situation for R-SOLNET formation is available without the luminescent target when the waveguide core has emissive characteristics. In this case, the luminescent target in the model is regarded as the luminescent surface of the output waveguide core edge.

In the LA-SOLNET (Fig. 7.7d), by excitation lights from outside, the λ_1 write beam and the λ_2 write beam are respectively generated from luminescent targets on the front edges of the input and output waveguides. The point light source arrangements in the luminescent targets are the same as those for the write beam in the TB-SOLNET.

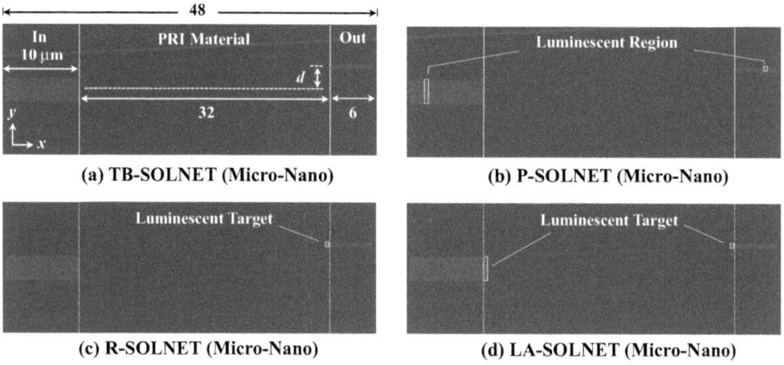

Figure 7.8 Models for 2D FDTD simulations of SOLNETs between a microscale waveguide and a nanoscale waveguide.

To assess the coupling efficiency between the input and output waveguides, a probe beam of 650 nm in wavelength is propagated from the input waveguide to the output waveguide. The point light source locations and arrangements for the probe beam are the same as those for the λ_1 write beam.

Brauchle et al. reported that the absorption spectra of CQ, which is a two-photon sensitizing molecule, for λ_1 and λ_2 are respectively located in wavelength regions of 400–520 nm and 580–1100 nm (see Fig. 7.40) [30]. According to these spectral data, we performed simulations for a case of $\lambda_1 = 500$ nm, $\lambda_2 = 600$ nm.

In Fig. 7.8, models for computer simulations of SOLNETs between a microscale waveguide and a nanoscale waveguide are shown. Widths of the input and output waveguides are respectively 3 μm and 600 nm. The basic structures of the models are the same as those shown in Fig. 7.7 except for the point light source distributions in the 3 μm wide input waveguide, 3 μm wide luminescent region in the input waveguide, and 3 μm wide luminescent target on the front edge of the input waveguide. In the 3 μm wide input waveguide, 75 point light sources for the λ_1 write beam generation are distributed across the core along the y direction with a pitch of 40 nm at $x = 1.4$ μm. In the 3 μm wide luminescent region, similarly, 75 point light sources for λ_1 luminescence 1 generation are distributed at $x = 1.4$ μm. In the 3 μm wide luminescent target the 75 point light sources for the λ_1 write beam generation are distributed at the front edge of the input waveguide.

7.2.2 Simulation Procedures

The computer simulations were performed by the 2D FDTD method [1, 7]. In the simulations the light absorption term in PRI materials is neglected. This implies that the simulations are for models under the small-absorption-limit condition, which is preferable for SOLNET formation.

For calculations of SOLNET formation, after inputting the initial refractive index distributions, the electric and magnetic fields are calculated using absorption boundary conditions. Because the refractive index of the PRI material changes with time, its distribution is rewritten at each time step n.

The energy density of the λ_1 write beam exposure and that of the λ_2 write beam exposure on the PRI material over a time interval for a step, Δt, are respectively $(1/2)\varepsilon E_1^2 v \Delta t$ and $(1/2)\varepsilon E_2^2 v \Delta t$, where $\varepsilon = \varepsilon_r \varepsilon_0$ is the dielectric constant and v is the velocity of the lightwave. E_1 and E_2 denote the respective electric field amplitudes for the λ_1 write beam and the λ_2 write beam.

Then, for the one-photon SOLNET, since the rate of the refractive index increase is proportional to $\gamma_1 I_1 + \gamma_2 I_2$ according to Eq. 7.1, the refractive index change Δn during Δt is expressed as

$$\Delta n = \frac{1}{2}\varepsilon v \left(\gamma_1 E_1^2 + \gamma_2 E_2^2\right) \Delta t. \tag{7.5}$$

In the present calculations, for simplicity, it is assumed that the refractive index changes in proportional to write beam exposure until it reaches a saturation limit Δn_{sat}, and γ_1 and γ_2 are set to the same value, γ. In this case, Eq. 7.5 is simplified with a constant of proportionality:

$$\Delta n = C_{1\mathrm{Photon}} \left(E_1^2 + E_2^2\right) \Delta t, \tag{7.6}$$

$$C_{1\mathrm{Photon}} = \frac{1}{2}\gamma \varepsilon v \tag{7.7}$$

For the two-photon SOLNET, since the rate of the refractive index increase is proportional to $I_1\, I_2$ according to Eq. 7.3, Δn during Δt can be expressed as

$$\Delta n = C_{2\mathrm{Photon}} \left(E_1^2 E_2^2\right) \Delta t. \tag{7.8}$$

Table 7.1 Constants of proportionality in 2D FDTD simulations for SOLNETs

Simulation model	$C_{1\text{Photon}}$ [$\times 10^{14}$ (m/V)^2/s]	$C_{2\text{Photon}}$ [$\times 10^{18}$ (m/V)^4/s]
TB-SOLNET/P-SOLNET	1.5	1.5
R-SOLNET	0.83	1.5
LA-SOLNET	2.6	9.1

The values of $C_{1\text{Photon}}$ and $C_{2\text{Photon}}$ are adjusted so that duration of SOLNET formation for two-photon SOLNETs is comparable to that for one-photon SOLNETs. This means adjusting the PRI material sensitivity. In the present study, we performed the simulations with constants of proportionality shown in Table 7.1.

The mesh sizes are $\Delta x = \Delta y = 20$ nm, and $\Delta t = 0.0134$ fs, which were chosen to satisfy the convergence condition. The polarization direction is along the z axis that is perpendicular to the calculated plane. Because PRI materials are usually isotropic, polarization effects are likely to be negligible.

In real systems, the typical PRI material response time for SOLNET formation is over 1 second as can be seen, for example, in Figs. 7.25, 7.34, 7.36, 7.41, 7.45, 7.47 in sections 7.3 and 7.4. In the 2D FDTD method, however, it is difficult to perform calculations with such long time spans because Δt should be on a sub-fs scale. We therefore rescaled the time parameter in the results as follows: $\Delta t' = \Delta t \times 10^{15}$ and $\gamma' = \gamma \times 10^{-15}$. This corresponds to the FDTD calculations with $\Delta t = 0.0134$ fs referring to a PRI material with very high sensitivity.

7.2.3 SOLNETs between Nanoscale Waveguides

7.2.3.1 TB-SOLNET/P-SOLNET

As mentioned in section 7.2.1, because the TB-SOLNET and the P-SOLNET are calculated by the same model, the results for TB-SOLNETs are, at the same time, those for P-SOLNETs.

Figure 7.9a shows one-photon TB-SOLNET formation between 600 nm wide waveguides by write beams of $\lambda_1 = 500$ nm and $\lambda_2 = 600$ nm for a lateral misalignment of 3000 nm. The left, middle,

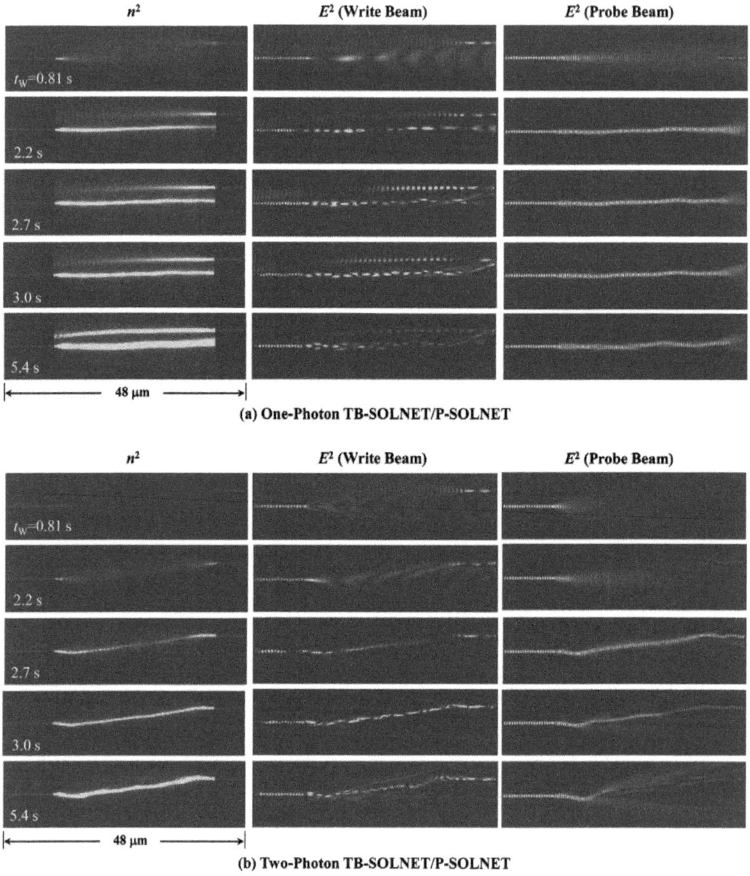

Figure 7.9 One-photon and two-photon TB-SOLNET/P-SOLNET formation between 600 nm wide waveguides with a lateral misalignment of 3000 nm.

and right columns respectively represent the dielectric constant, ε_r, i.e., the square of the refractive index, n^2, the intensity of the write beams, E^2 (Write Beam), and the intensity of the probe beams, E^2 (Probe Beam). A 650 nm probe beam is introduced from the input waveguide.

In the one-photon TB-SOLNET, a coupling waveguide is unable to form. Instead, optical waveguides extend from the input and output waveguides with writing time, t_W, to construct completely separated

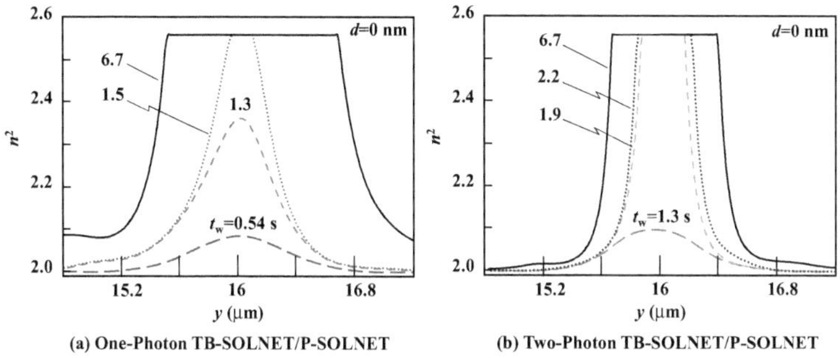

Figure 7.10 Cross-sectional n^2 profiles of one-photon and two-photon TB-SOLNETs/P-SOLNETs at $x = 36$ μm.

two optical waveguides. At the same time, considerable waveguide broadening is observed.

In the two-photon TB-SOLNET, on the other hand, as shown in Fig. 7.9b, the two write beams begin to merge at 2.2 s to rapidly form a self-aligned coupling waveguide of two-photon TB-SOLNET. The probe beam is smoothly guided toward the output waveguide at 2.7 s. At 5.4 s, due to over-exposure, zigzag shapes appear in the SOLNET, resulting in considerable probe beam leakage.

Figure 7.10 shows n^2 profiles across TB-SOLNETs (along the y direction in Fig. 7.7) at $x = 36$ μm for no lateral misalignments. Here, $x = 36$ μm corresponds to a position of 6 μm from the front edge of the output waveguide. The position of $y = 16$ μm corresponds to the core axis location of the input waveguide. Graded-index-like profiles are observed at the initial stage of SOLNET formation, and later change to step-index-like profiles. The waveguide broadening is suppressed in the two-photon SOLNET, while considerable broadening occurs in the one-photon SOLNET. The refractive index increase rate proportional to $I_1 I_2$ produces the steep SOLNET formation characteristics in two-photon SOLNET.

In Fig. 7.11a one-photon TB-SOLNETs for various lateral misalignments are summarized. For a lateral misalignment of 600 nm, a self-aligned coupling waveguide is formed, and the probe beam is guided to the output waveguide. For lateral misalignments of 1800 and 3000 nm, normal SOLNETs cannot be formed. Instead, branching or parallel waveguides are formed, indicating that the

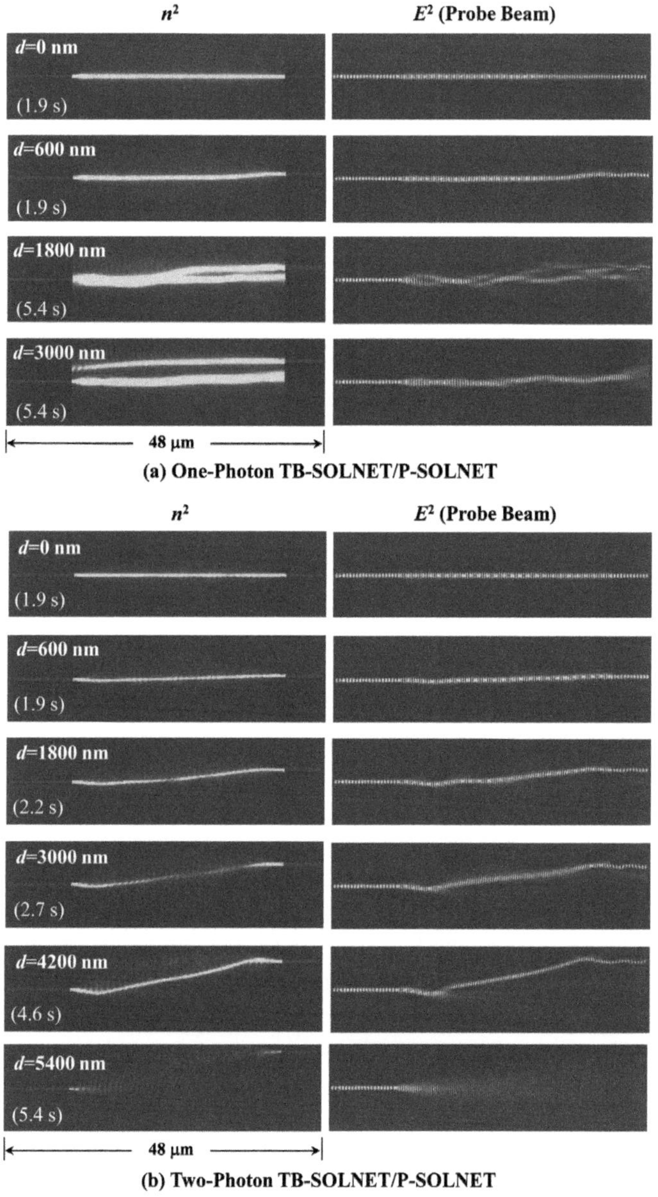

Figure 7.11 One-photon and two-photon TB-SOLNETs/P-SOLNETs between 600 nm wide waveguides at optimum writing time indicated in brackets for various lateral misalignments.

tolerance to the lateral misalignment is ~600 nm. In two-photon TB-SOLNETs, as shown in Fig. 7.11b, the lateral misalignment tolerance is drastically widened up to 4200 nm. The waveguide broadening is suppressed, and consequently, the meandering of probe beams in the output waveguides is suppressed, indicating that the two-photon SOLNETs induce small mode disturbances.

Figure 7.12 shows dependence of coupling efficiency at 650 nm in wavelength on the writing time in the TB-SOLNETs. For one-photon SOLNETs, when no lateral misalignments exist, the coupling efficiency increases to 94% with writing time, and then becomes unstable, varying between 47% and 88%. This instability of the coupling efficiency is due to the waveguide broadening. For a lateral misalignment of 600 nm, the same trend is found in the coupling efficiency. For lateral misalignments of 1800 nm and 3000 nm, the coupling efficiency becomes very low, corresponding to the appearance of branching or separated waveguides with little sign of self-aligning (see Fig. 7.11a).

For two-photon SOLNETs, as shown in Fig. 7.12b, when no lateral misalignments exist, the coupling efficiency rises with writing time to 94%, and is maintained above 73% until 8.9 s. The instability of the coupling efficiency, which is observed in one-photon SOLNETs, is suppressed. This is attributed to the suppression of the waveguide broadening in two-photon SOLNETs. When the lateral misalignment increases to 3000 nm, still relatively high coupling efficiency of 53% is maintained. For a misalignment of 4200 nm, the maximum coupling efficiency is 40%, which is much higher than the coupling efficiency of less than 5% in one-photon SOLNETs. These results confirm that the two-photon TB-SOLNET achieves greater lateral misalignment tolerances than the one-photon TB-SOLNET.

It is found from Fig. 7.12b that with increasing lateral misalignments the optimum writing time required to reach the maximum coupling efficiency tends to increase, and writing time windows keeping high coupling efficiency becomes narrow. For practical use of the SOLNET optical solder, it is desirable that the optimum writing time is constant or the writing time windows are wide regardless of the lateral misalignments so that SOLNETs can be formed at fixed writing time without monitoring the coupling efficiency. Solutions for the problem should be found by future work.

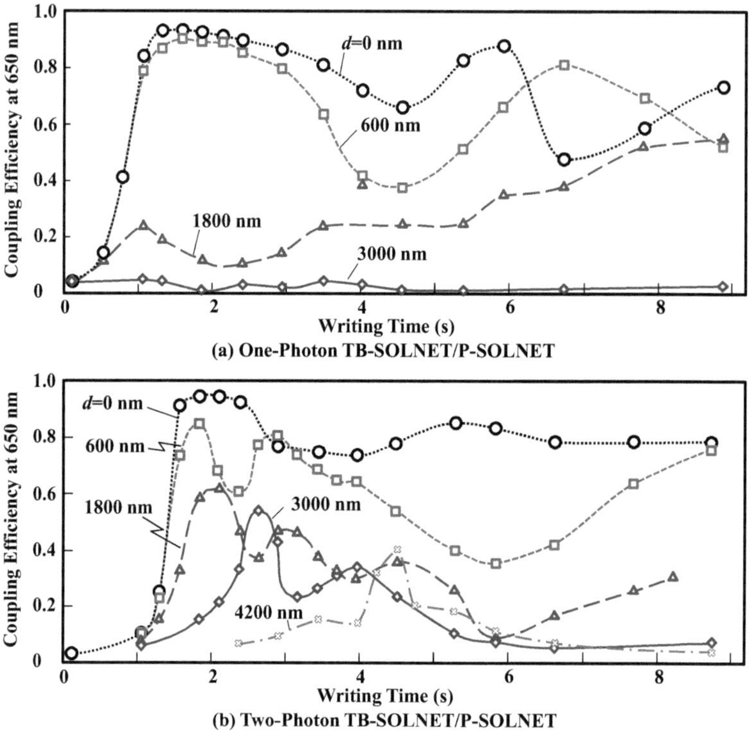

Figure 7.12 Dependence of coupling efficiency on writing time in TB-SOLNETs/P-SOLNETs formed between 600 nm wide waveguides.

Dependence of maximum coupling efficiency on lateral misalignments is shown in Fig. 7.13 for TB-SOLNETs. Peak values of coupling efficiency in Fig. 7.12 are plotted as a function of lateral misalignments. By introducing SOLNETs, the efficiency greatly increases from that in butt joints. In accordance with the results shown in Fig. 7.11, the lateral misalignment tolerance is wider in two-photon TB-SOLNETs than in one-photon TB-SOLNETs. The wider misalignment tolerances of two-photon SOLNETs are attributed to their capability to form high-contrasting refractive-index images as mentioned in section 7.1.3 (see Fig. 7.5). Because the rate of the refractive index increase is proportional to $I_1 I_2$ in two-photon SOLNETs, the refractive index increase occurs only if both of a λ_1 write beam and a λ_2 write beam coexist, resulting in

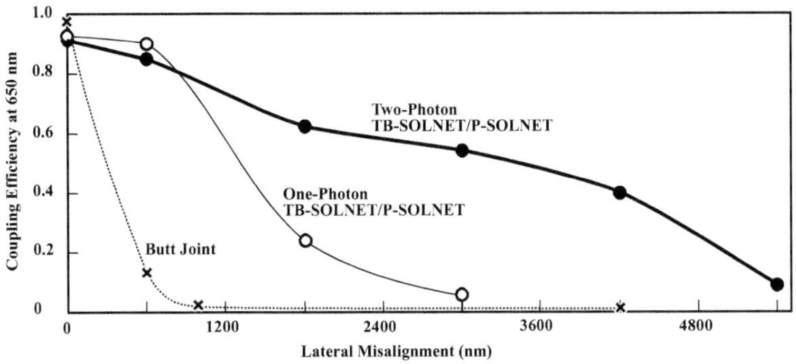

Figure 7.13 Dependence of maximum coupling efficiency on lateral misalignments in TB-SOLNETs/P-SOLNETs formed between 600 nm wide waveguides.

the enhancement of the write-beam-overlapping effect to extend the misalignment tolerances.

7.2.3.2 R-SOLNET

Two-photon R-SOLNET formation between 600 nm wide waveguides with a lateral misalignment of 1800 nm is shown in Fig. 7.14. Here, the middle columns represent the intensity of the write beam E^2 (Write Beam) and the intensity of the luminescence E^2

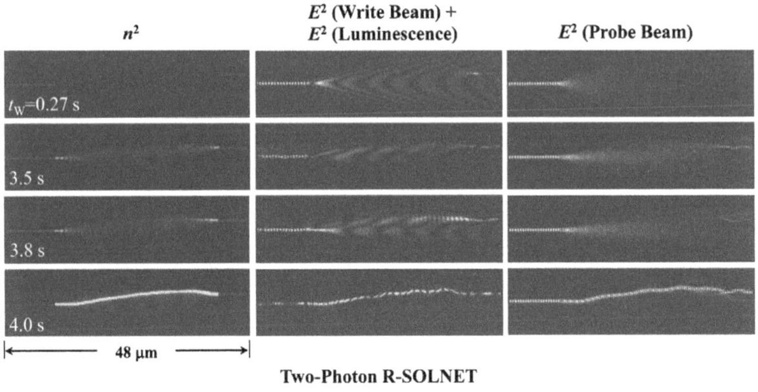

Figure 7.14 Two-photon R-SOLNET formation between 600 nm wide waveguides with a lateral misalignment of 1800 nm.

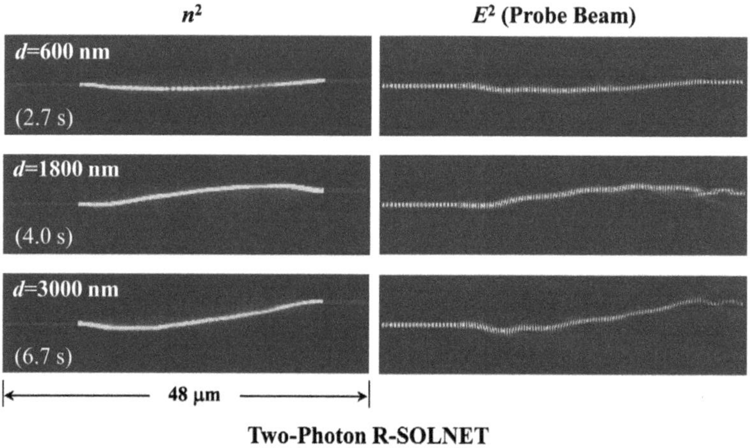

Figure 7.15 Two-photon R-SOLNETs between 600 nm wide waveguides at optimum writing time indicated in brackets for various lateral misalignments.

(Luminescence). When a 500 nm write beam is introduced from the input waveguide on the left, a luminescent target on the front edge of the output waveguide is excited by the write beam and emits 600 nm luminescence. The write beam and the luminescence merge to form a self-aligned coupling waveguide at 4.0 s.

As Fig. 7.15 shows, two-photon R-SOLNETs form a self-aligned coupling waveguide even for a lateral misalignment of 3000 nm, exhibiting a wide misalignment tolerance.

7.2.3.3 LA-SOLNET

Two-photon LA-SOLNET formation between 600 nm wide waveguides with a lateral misalignment of 3000 nm is shown in Fig. 7.16. 500 nm luminescence 1 is emitted from the luminescent target on the input waveguide front edge and 600 nm luminescence 2 from the target on the output waveguide front edge to form a self-aligned coupling waveguide.

As Fig. 7.17 shows, two-photon LA-SOLNETs form a self-aligned coupling waveguide even for a lateral misalignment of 5400 nm, exhibiting an extremely wide misalignment tolerance.

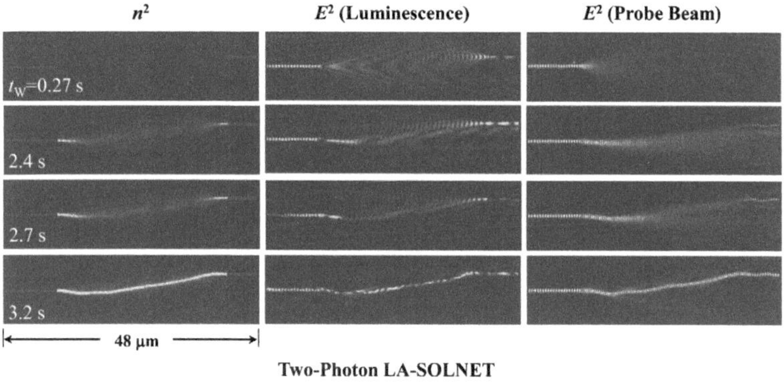

Figure 7.16 Two-photon LA-SOLNET formation between 600 nm wide waveguides with a lateral misalignment of 3000 nm.

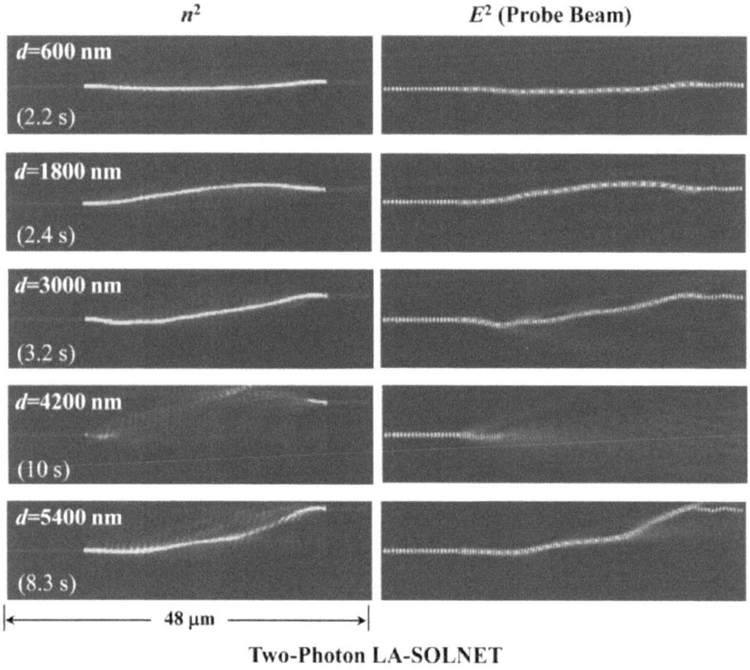

Figure 7.17 Two-photon LA-SOLNETs between 600 nm wide waveguides at optimum writing time indicated in brackets for various lateral misalignments.

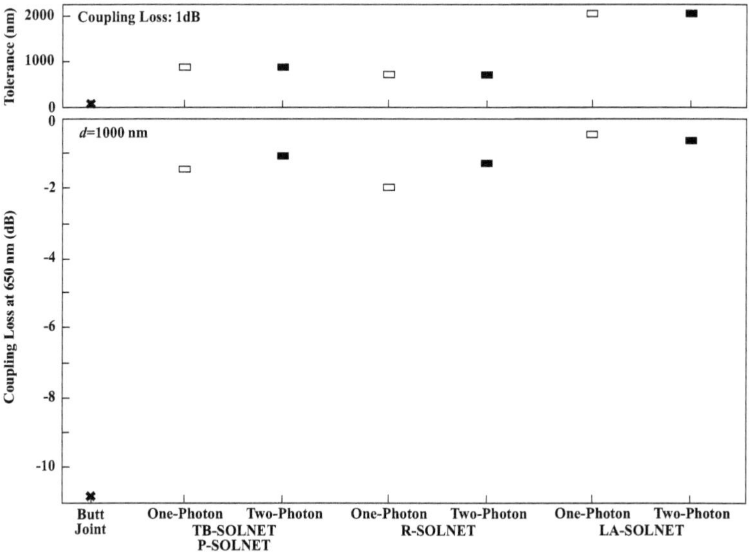

Figure 7.18 Performance of optical couplings between 600 nm wide waveguides for the butt joint and SOLNETs. The upper represents lateral misalignment tolerances for 1 dB coupling loss, and the lower represents coupling losses at 1000 nm lateral misalignment.

7.2.3.4 Performance of couplings

In Fig. 7.18, performance of optical couplings between 600 nm wide waveguides is compared. In the butt joint, the lateral misalignment tolerance for 1 dB coupling loss and the coupling loss at 1000 nm lateral misalignment are 80 nm and 11 dB, respectively.

In the TB-SOLNET and P-SOLNET, the tolerance is widened by one order of magnitude to be 870 nm, and the coupling losses are reduced to 1.4 and 1.1 dB for one-photon and two-photon SOLNETs, respectively. In the R-SOLNET the tolerance is 710 nm, and the coupling loss is 2.0 and 1.2 dB for one-photon and two-photon SOLNETs, respectively. In the LA-SOLNET the tolerance is widened up to 2050 nm, and the coupling loss is reduced to 0.4 and 0.6 dB for one-photon and two-photon SOLNETs, respectively.

These results indicate that LA-SOLNETs for nano–nano couplings exhibit lateral misalignment tolerances wide enough to exceed the die bonding accuracy of ~1 μm in cases that coupling losses of 0.4–0.6 dB are permitted. The TB-SOLNET, P-SOLNET, and R-SOLNET are also applicable in cases that coupling losses of 1.1–2.0 dB are permitted, allowing us to widen the choice variation for SOLNET formation processes.

7.2.4 SOLNETs between Microscale and Nanoscale Waveguides

7.2.4.1 TB-SOLNET/P-SOLNET

As mentioned in section 7.2.3, the results for TB-SOLNETs described in the present section mean those for TB-SOLNETs and for P-SOLNETs.

Figure 7.19a shows one-photon TB-SOLNET formation between 3 μm wide and 600 nm wide waveguides with no lateral misalignments by write beams of $\lambda_1 = 500$ nm and $\lambda_2 = 600$ nm. With writing time, a coupling waveguide of TB-SOLNET is formed to guide a 650 nm probe beam. However, broadening of the coupling waveguide due to over-exposure causes mode disturbances in probe beam propagation. In one-photon SOLNETs, because the rate of the refractive index increase is proportional to $\gamma_1 I_1 + \gamma_2 I_2$, the refractive index increase can develop by one write beam, namely, λ_1 write beam or λ_2 write beam. This causes limitless broadening of the SOLNET throughout the gap region, preventing the expected tapered feature from being formed between the microscale and nanoscale waveguides.

Two-photon TB-SOLNET formation is shown in Fig. 7.19b for no lateral misalignments. With writing time, a coupling waveguide of SOLNET is formed, and at 1.1 s it combines the input and output waveguides. Although the SOLNET width is close to the width of the 600 nm wide waveguide and is much narrower than the width of the 3 μm wide waveguide to give a straight-line-shaped coupling waveguide, the probe beam is efficiently guided from the 3 μm wide waveguide to the 600 nm wide waveguide. At 3.0 s, a zigzag shape appears in the SOLNET. This causes a probe beam to meander, resulting in probe beam leakage.

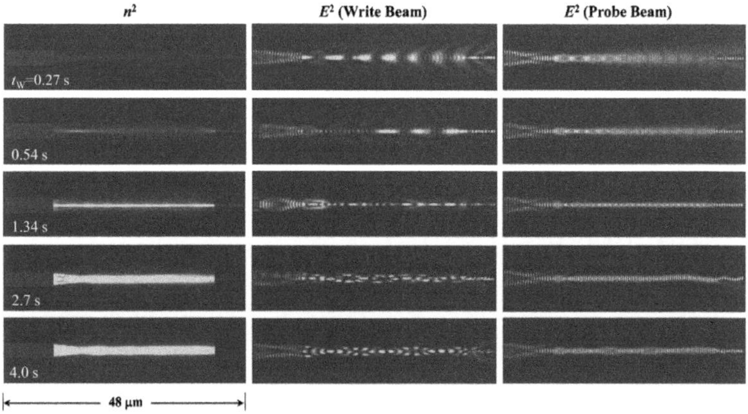

(a) One-Photon TB-SOLNET/P-SOLNET

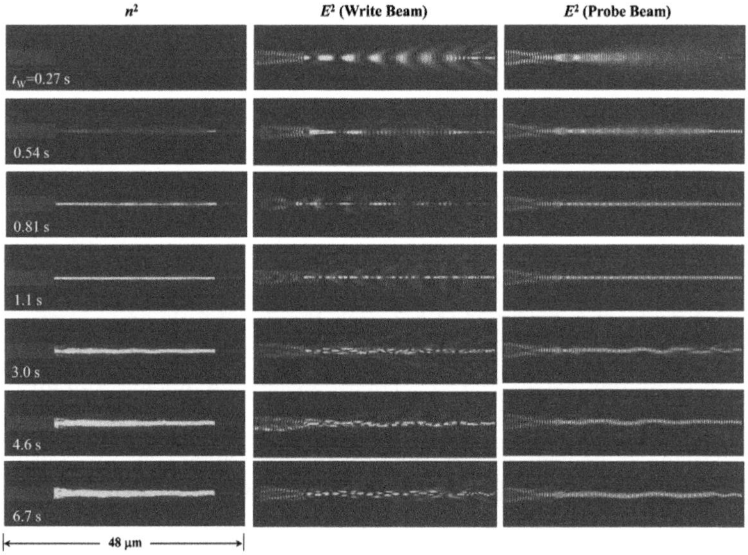

(b) Two-Photon TB-SOLNET/P-SOLNET

Figure 7.19 One-photon and two-photon TB-SOLNET/P-SOLNET formation between 3 μm wide and 600 nm wide waveguides with no lateral misalignments.

After 3.0 s, broadening of the SOLNET occurs mainly on the 3 μm wide waveguide side. At 4.6–6.7 s a coupling waveguide with a tapered shape is formed to confine the probe beam strongly and guide it into the 600 nm wide waveguide efficiently. In two-photon SOLNETs, because the rate of the refractive index increase is proportional to $I_1 I_2$, the refractive index increase develops by the two-write-beam overlapping. This suppresses the broadening of the SOLNET near the nanoscale waveguide, producing the taper feature.

As shown in Fig. 7.20, in the one-photon TB-SOLNET the misalignment limit is ~600 nm. In the two-photon TB-SOLNET a self-aligned coupling waveguide is still formed when the misalignment is 3000 nm. The probe beam is guided into the output waveguide smoothly for lateral misalignments of 0 nm, 600 nm, and 1800 nm. Thus, in the micro–nano couplings similarly as in the nano-nano couplings, the two-photon TB-SOLNET achieves a wider lateral misalignment tolerance than the one-photon TB-SOLNET.

The probe beam leakage occurring near the output waveguide in the two-photon TB-SOLNET with a lateral misalignment of 3000 nm is due to the small bending radius of the SOLNET.

7.2.4.2 R-SOLNET

As Fig. 7.21 shows, two-photon R-SOLNETs form a self-aligned coupling waveguide between 3 μm wide and 600 nm wide waveguides with 1800 nm misalignment. When the lateral misalignment increases to 3000 nm, however, no traces of SOLNET formation are observed.

It should be noted that for the nano–nano couplings two-photon R-SOLNETs form a coupling waveguide even when 3000 nm misalignment exists as shown in Fig. 7.15. The limited lateral misalignment tolerances in the micro–nano couplings can be explained from a view point of the write beam divergence in the PRI material as follows. In nano–nano couplings the spread angle of the λ_1 write beam emitted from the input waveguide is large. Consequently, even when the lateral misalignment increases to 3000 nm, the luminescent target emits strong luminescence, enabling two-photon R-SOLNET formation. On the contrary, in

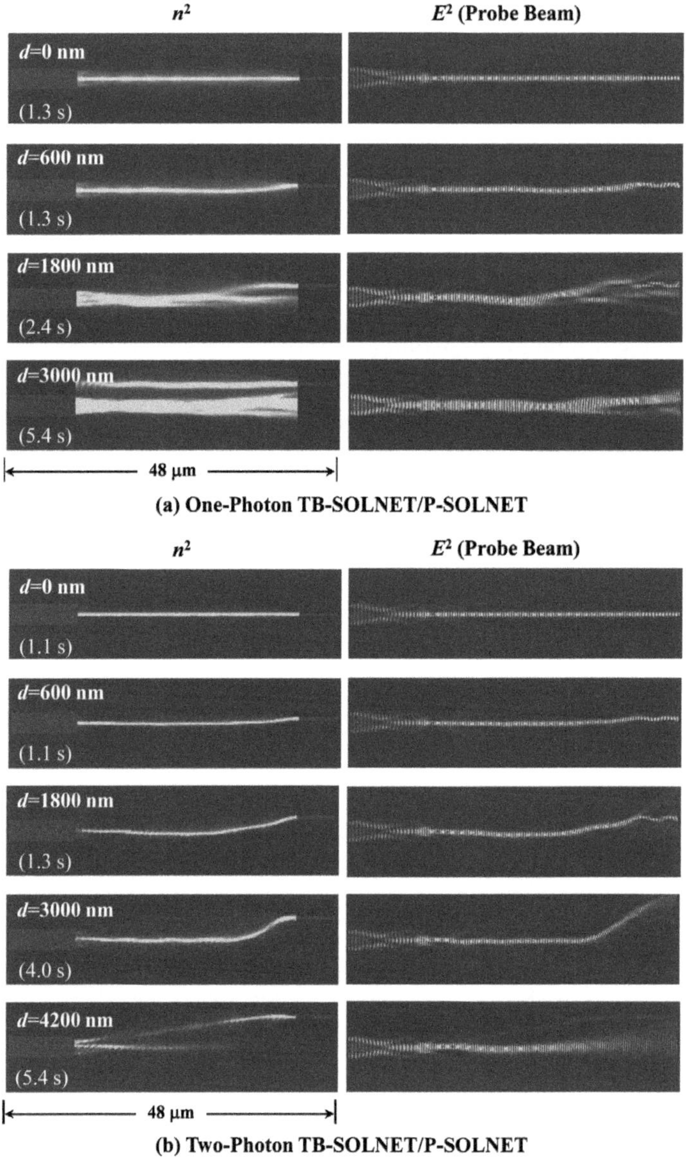

Figure 7.20 One-photon and two-photon TB-SOLNETs/P-SOLNETs between 3 μm wide and 600 nm wide waveguides at optimum writing time indicated in brackets for various lateral misalignments.

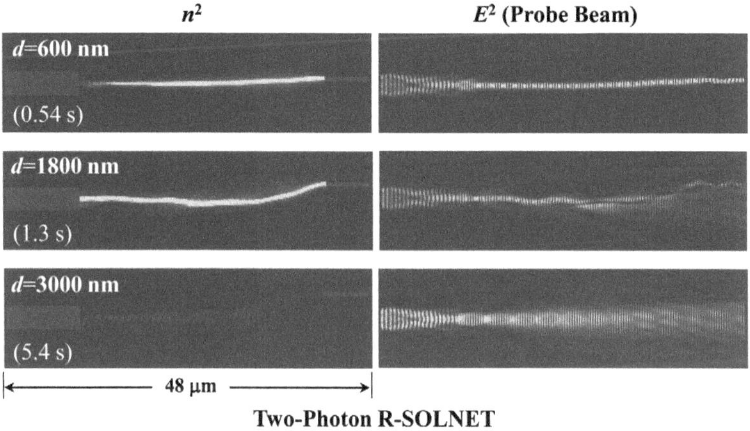

Figure 7.21 Two-photon R-SOLNETs between 3 μm wide and 600 nm wide waveguides at optimum writing time indicated in brackets for various lateral misalignments.

micro–nano couplings, because the spread angle of the λ_1 write beam is small, the luminescent target cannot receive enough write beam to emit strong luminescence when the lateral misalignment is 3000 nm, preventing the R-SOLNET formation.

These results suggest that the lateral misalignment tolerance of the two-photon R-SOLNET will be widened by increasing the spread angle of the λ_1 write beam emitted from the input waveguide by, for example, adjusting the waveguide edge structures or placing waveguide lenses.

7.2.4.3 LA-SOLNET

As Fig. 7.22 shows, two-photon LA-SOLNETs form a self-aligned coupling waveguide between 3 μm wide and 600 nm wide waveguides by 500 nm luminescence 1 and 600 nm luminescence 2 when the lateral misalignment increases to 3000 nm.

One concern in the two-photon LA-SOLNETs for micro–nano couplings is the branching structures that appear in SOLNETs for lateral misalignments of 1800 nm and 3000 nm. The branching might be attributed to the wavefront shape of the luminescence. In nano-nano couplings, luminescence with smooth wavefronts

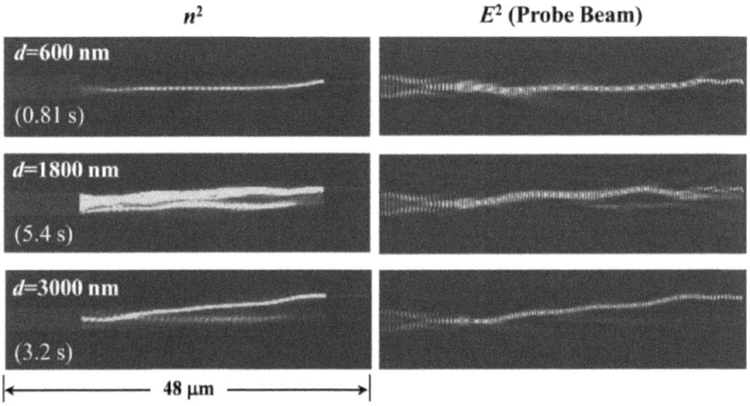

Figure 7.22 Two-photon LA-SOLNETs between 3 μm wide and 600 nm wide waveguides at optimum writing time indicated in brackets for various lateral misalignments.

is emitted from nanoscale luminescent targets, propagating in a manner of a light emitted from a single-mode optical fiber. Consequently, smooth two-photon LA-SOLNETs without branching structures are available as shown in Fig. 7.17. On the contrary, in micro–nano couplings, luminescence with a complicated wavefront is emitted from the microscale luminescent target, propagating in a plurality of directions in a manner of a light emitted from a multimode optical fiber. This might cause the branching structures. It is suggested from these results that two-photon LA-SOLNETs without branching structures might be obtained by adjusting luminescent target shapes to generate luminescence with smooth wavefronts.

7.2.4.4 Performance of couplings

In Fig. 7.23, performance of optical couplings between 3 μm wide and 600 nm wide waveguides is compared. In the butt joint, the coupling loss is over 1 dB in all cases. The coupling loss at 1000 nm lateral misalignment is 7.7 dB.

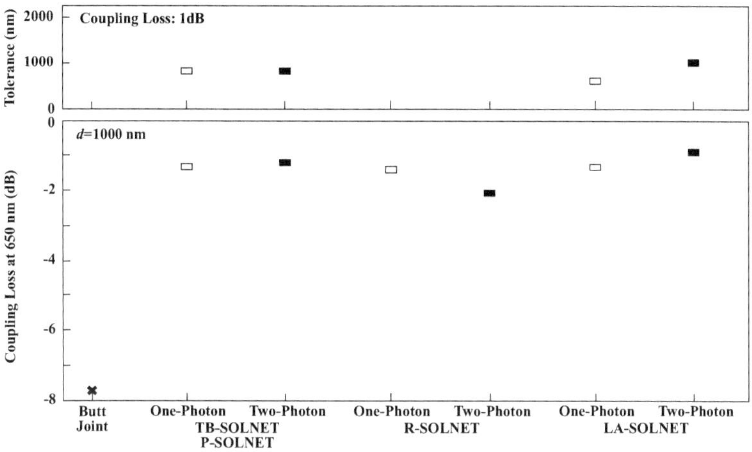

Figure 7.23 Performance of optical couplings between 3 μm wide and 600 nm wide waveguides for the butt joint and SOLNETs. The upper represents lateral misalignment tolerances for 1 dB coupling loss, and the lower represents coupling losses at 1000 nm lateral misalignment.

In the TB-SOLNET and P-SOLNET, the lateral misalignment tolerance for 1 dB coupling loss is 810 nm, and the coupling losses at 1000 nm lateral misalignment are 1.4 and 1.2 dB for one-photon and two-photon SOLNETs, respectively. In the R-SOLNET the coupling loss is over 1 dB in all cases. The coupling loss at 1000 nm lateral misalignment is 1.4 and 2.1 dB for one-photon and two-photon SOLNETs, respectively. In the LA-SOLNET, the lateral misalignment tolerance for 1 dB coupling loss is ~600 nm and ~1000 nm for one-photon and two-photon SOLNETs, respectively, and the coupling loss at 1000 nm lateral misalignment is 1.4 and 0.9 dB for one-photon and two-photon SOLNETs, respectively.

These results suggest that the SOLNETs enable self-aligned optical couplings with a mode size conversion function. LA-SOLNETs are applicable to the optical solder that allows 1 μm lateral misalignment when a coupling loss of 0.9 dB is permitted. TB-SOLNETs and P-SOLNETs are applicable when a coupling loss of 1.2 dB is permitted. R-SOLNETs are also applicable when a coupling loss of 1.4 dB is permitted.

7.3 Experimental Demonstrations of One-Photon SOLNETs

In the present section, experimental demonstrations of one-photon SOLNETs are presented. The experiments were carried out using following three kinds of optical fibers.

- 50 μm optical fibers: optical fibers with a core diameter of ~50 μm (multimode for 1.3 μm in wavelength)
- 9 μm optical fibers: optical fibers with a core diameter of ~9 μm (single-mode for 1.3 μm)
- 3 μm optical fibers: optical fibers with a core diameter of ~3 μm (single-mode for 405 nm)

7.3.1 One-Photon TB-SOLNETs

Figure 7.24 shows an experimental demonstration for one-photon TB-SOLNET [5]. Two 9 μm optical fibers are put on a glass substrate. The gap between the optical fibers is filled with a PRI material

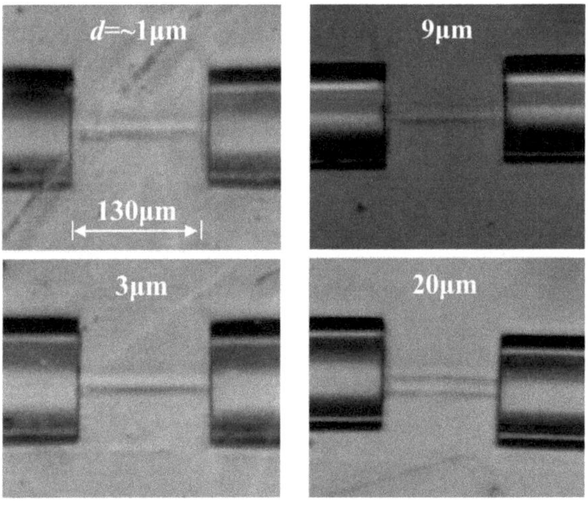

Figure 7.24 One-photon TB-SOLNET formation between 9 μm optical fibers for various lateral misalignments in the monomer/monomer-type photopolymer. Writing beam intensity is 1.3 W/cm^2.

of a monomer/monomer-type photopolymer developed by JSR Corporation [21]. It consists of a high-n acrylic component and a low-n epoxy component with a mixing ratio of 50:50 in wt%. The former exhibits larger reactivity than the latter upon irradiation of light in the blue-violet regions, giving rise to a maximum refractive index change around 0.02. For lateral misalignments of $1 \sim 9\ \mu m$, self-aligned coupling waveguides of one-photon TB-SOLNETs are formed. When the lateral misalignment increases to 20 μm, optical waveguides grown from the two optical fibers do not merge, but remain as separated optical waveguides.

In Fig. 7.25a, dependence of coupling efficiency between two optical fibers on writing time in a one-photon TB-SOLNET is shown for a lateral misalignment of ~1 μm. The probe beam wavelength is 1.3 μm. With writing time, the coupling efficiency increases up to 85% (coupling loss of 0.7 dB), then decreases slightly to some extent. This tendency of the coupling efficiency seems similar to that predicted by the 2D FDTD simulation for one-photon TB-SOLNETs with lateral misalignments of 0 and 600 nm (see Fig. 7.12a). In Fig. 7.25b, the dependence of maximum coupling efficiency on lateral misalignments is shown for one-photon TB-SOLNETs and butt joints. TB-SOLNETs exhibit a wider tolerance to misalignment than butt joints.

These results confirm that the one-photon TB-SOLNET acts as optical solder for a self-aligned coupling waveguide.

7.3.2 One-Photon R-SOLNETs with Micromirrors Formed by Free-Space Write Beams

Figure 7.26a depicts an experimental setup for formation of one-photon R-SOLNET with micromirrors by free-space write beams [9]. A PRI material of the monomer/monomer-type photopolymer is inserted between two glass substrates placed in parallel with a gap distance of ~200 μm. On the upper glass substrate, an Al mask with a circle window is deposited. The diameter of the window is 100 μm. On the lower glass substrate, an Al micromirror with a square or circle pattern is formed. A free-space UV write beam from a high-pressure mercury lamp comes through the circle window into the

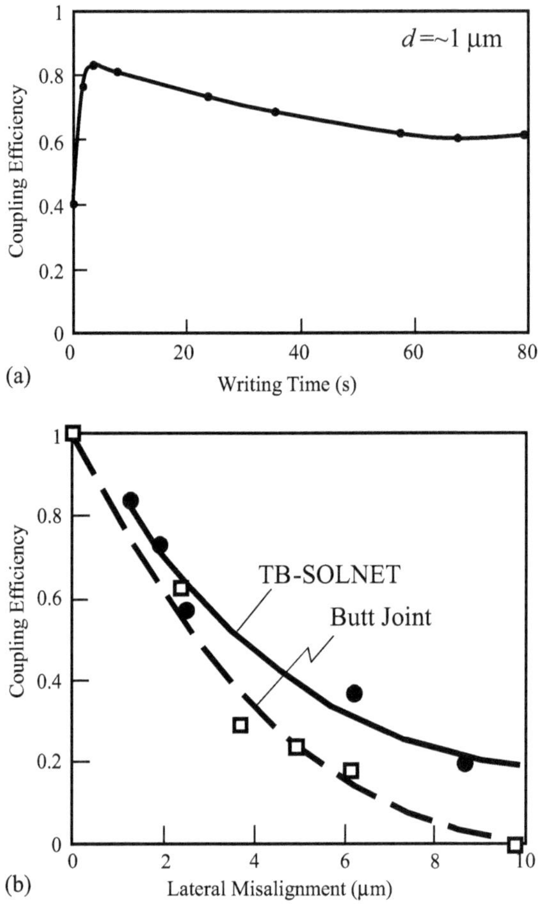

Figure 7.25 (a) Dependence of coupling efficiency between 9 µm optical fibers on writing time in a one-photon TB-SOLNET for a lateral misalignment of ~1 µm. (b) Dependence of maximum coupling efficiency on lateral misalignments in one-photon TB-SOLNETs and butt joints between 9 µm optical fibers. Reprinted with permission from Yoshimura et al. [5].

photopolymer and is reflected from the micromirror to form a one-photon R-SOLNET as shown in Fig. 7.26b.

To observe the SOLNET formation, photographs were taken from the lower substrate side by detaching the lower substrate after the write beam exposure.

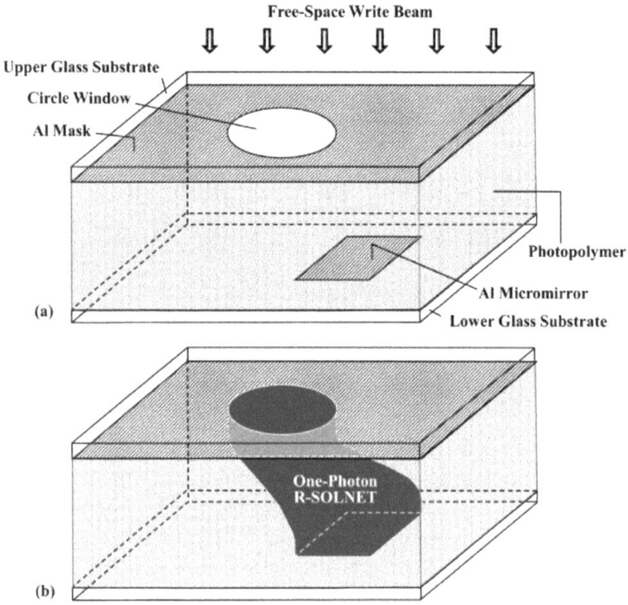

Figure 7.26 (a) Experimental setup for formation of one-photon R-SOLNET with micromirrors by free-space write beams. (b) Schematic illustration of an expected structure of the one-photon R-SOLNET. Reprinted with permission from Yoshimura et al. [9].

In the case that the lower glass substrate was not placed during write beam exposure, as Fig. 7.27a shows, a straight optical waveguide of SWW grown from the circle window on the upper glass substrate is observed.

Figure 7.27b shows a one-photon R-SOLNET formed between the circle window on the upper glass substrate and a 100 μm diameter circle micromirror deposited on the lower glass substrate with no misalignments. A coupling optical waveguide with a circle bottom and a circle top are observed. Figure 7.27c shows a one-photon R-SOLNET formed between the circle window and a square micromirror with no misalignments. A coupling optical waveguide with a circle bottom and a small square top is observed. Figure 7.27d shows a one-photon R-SOLNET formed between the circle window and the square micromirror, which is deposited with 30 μm misalignment. A coupling optical waveguide with a circle

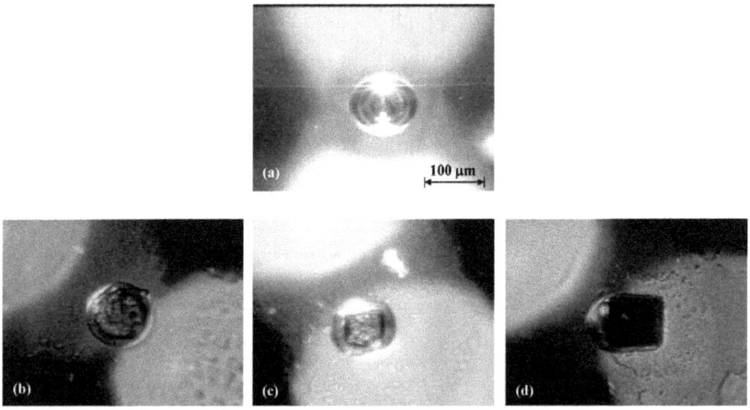

Figure 7.27 (a) Straight optical waveguide of SWW grown from the circle window on the upper glass substrate. (b) One-photon R-SOLNET between the circle window and a 100 μm diameter circle micromirror on the lower glass substrate with no misalignments, (c) one-photon R-SOLNET between the circle window and a square mirror, which is smaller than the circle window, with no misalignments, and (d) one-photon R-SOLNET between the circle window and a square mirror with 30 μm misalignment. Reprinted with permission from Yoshimura et al. [9].

bottom and a square top can be seen. A displacement exists between the top and the bottom according to the misalignment between the circle window and the square micromirror.

These results confirm that R-SOLNETs with micromirrors can be formed using free-space write beams.

7.3.3 One-Photon R-SOLNETs with Micromirrors

Figure 7.28 shows an experimental demonstration for a one-photon R-SOLNET with micromirrors [11]. The setup to form a one-photon R-SOLNET between a 50 μm optical fiber and an Al micromirror deposited on an optical fiber edge is depicted in Fig. 7.28a. They are placed with a gap distance of ~800 μm in the monomer/monomer-type photopolymer. An angular misalignment of 3° exists between the fiber direction and the surface-normal direction of the micromirror.

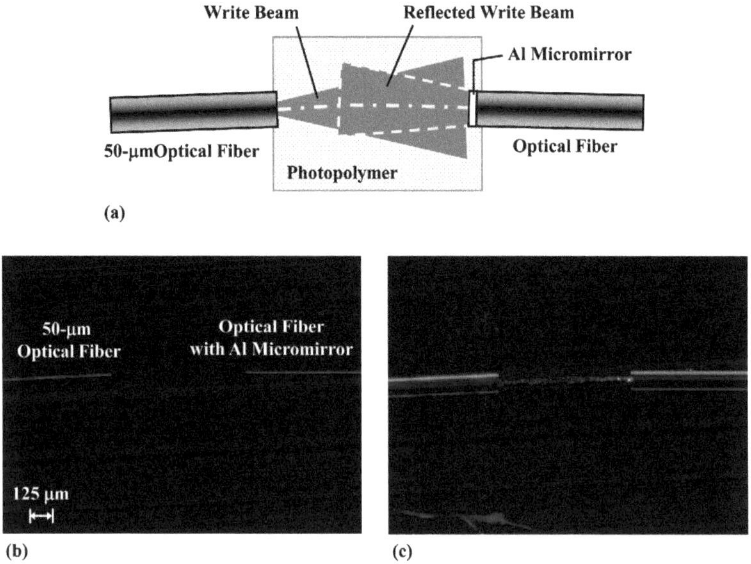

Figure 7.28 One-photon R-SOLNET formation between a 50 μm optical fiber and an Al micromirror on an optical fiber edge with an angular misalignment of 3°. (a) Experimental setup, (b) the 50 μm optical fiber and the optical fiber with an Al micromirror on the edge surface in a photopolymer before R-SOLNET formation, and (c) 650 nm probe beam propagation in the bow-shaped self-aligned coupling waveguide of one-photon R-SOLNET.

Figure 7.28b shows a photograph before R-SOLNET formation. By introducing a UV write beam from the 50 μm optical fiber into the photopolymer, a one-photon R-SOLNET was attempted to form. If the "pulling water" effect described in section 7.1.1 did not exist, two separated optical waveguides with a directional difference of 6° would be observed stretching from the micromirror, namely, one grown by the incident write beam and the other by the reflected write beam. However, actually, as Fig. 7.28c shows, when the write beam is introduced, a bow-shaped one-photon R-SOLNET is formed between the optical fiber and the micromirror by the "pulling water" effect, and a 650 nm probe beam from the 50 μm optical fiber propagates in the R-SOLNET. The core width of the R-SOLNET is about 50 μm, which is close to that of the 50 μm optical fiber. Further

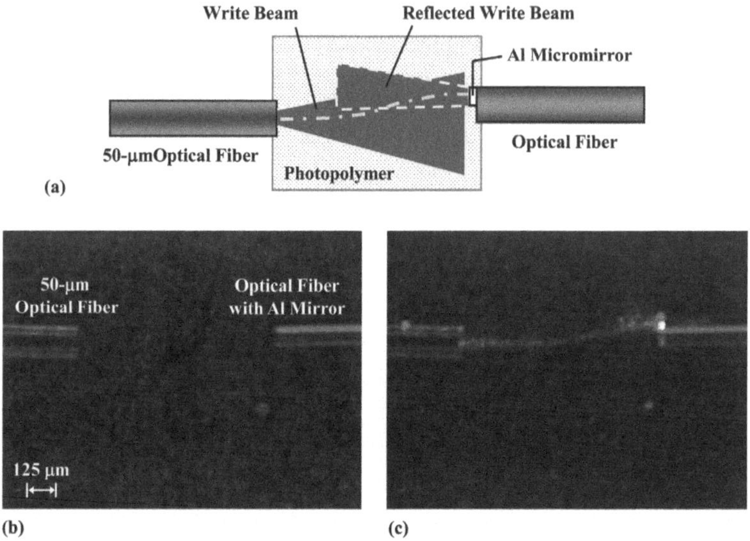

Figure 7.29 One-photon R-SOLNET formation between a 50 μm optical fiber and an Al micromirror on an optical fiber edge with a lateral misalignment of ~60 μm. (a) Experimental setup, (b) the 50 μm optical fiber and the optical fiber with an Al micromirror on the edge surface in a photopolymer before R-SOLNET formation, and (c) 650 nm probe beam propagation in the S-shaped self-aligned coupling waveguide of one-photon R-SOLNET.

investigation revealed that bow-shaped R-SOLNETs can be formed when the angular misalignment is less than 6°.

The setup to form a one-photon R-SOLNET between a 50 μm optical fiber and a Al micromirror with a lateral misalignment of ~60 μm is depicted in Fig. 7.29a. A photograph before R-SOLNET formation is shown in Fig. 7.29b. When a UV write beam is introduced from the 50 μm optical fiber into the photopolymer, as Fig. 7.29c shows, an S-shaped one-photon R-SOLNET is formed, connecting the 50 μm optical fiber to the misaligned micromirror, and a 650 nm probe beam propagates in the S-shaped R-SOLNET with a core width of ~50 μm.

These results confirm that the one-photon R-SOLNET can be formed using reflected write beams from micromirrors, enabling self-aligned optical couplings.

7.3.4 One-Photon R-SOLNETs with Reflective Objects

One-photon R-SOLNET formation with reflective objects was demonstrated using the PRI sol–gel material described in section 4.6. The experimental setup is illustrated in Fig. 7.30 [38]. On a Si substrate with a 2000 nm thick SiO_2 film for an undercladding layer, a PRI sol–gel thin film of approximately 230 nm in thickness is deposited by spin-coating and prebaking. A silver paste droplet is put as a reflective object in the PRI sol–gel thin film. A 405 nm write beam with power of 0.5 mW is introduced from a 3 μm optical fiber into the PRI sol–gel thin film under heating at 200°C to form a one-photon R-SOLNET. As Fig. 7.31 shows, at writing time of 25 s, the

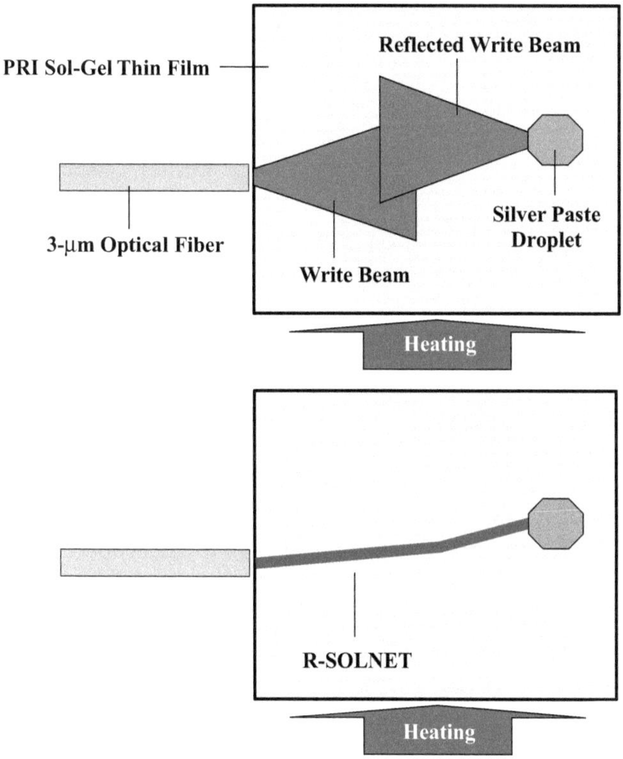

Figure 7.30 Experimental setup for formation of one-photon R-SOLNET with reflective objects in a PRI sol-gel thin film. Reprinted with permission from Ono et al. [38].

Figure 7.31 Formation of one-photon R-SOLNET between a 3 μm optical fiber and a silver paste droplet in a PRI sol–gel thin film. Reprinted with permission from Ono et al. [38].

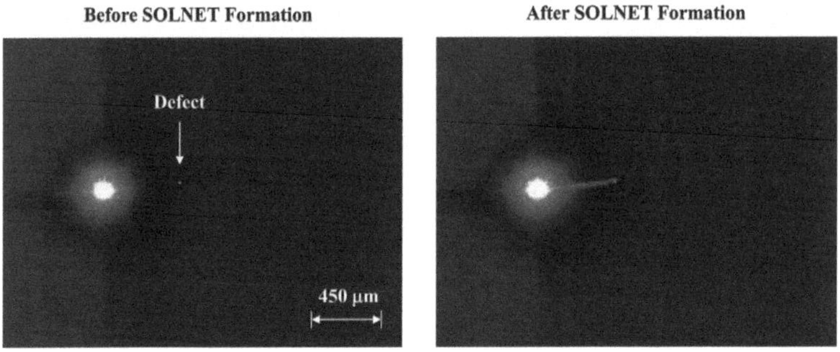

Figure 7.32 Formation of one-photon R-SOLNET between a 3 μm optical fiber and a reflective defect in a PRI sol–gel thin film. Reprinted with permission from Ono et al. [38].

R-SOLNET is slightly drawn to the silver paste droplet. At 50 s, the R-SOLNET completely forms to be drawn to the droplet.

Figure 7.32 shows another example of one-photon R-SOLNET formation in the PRI sol–gel thin film. A 405 nm write beam with power of 0.5 mW is introduced into the PRI sol–gel thin film for 30 s under heating at 200°C. Then, a one-photon R-SOLNET is formed toward a defect, which acts as a reflective object in the PRI sol–gel thin film.

These results suggest that the one-photon R-SOLNET has the capability to find locations of reflective objects and to guide light beams there.

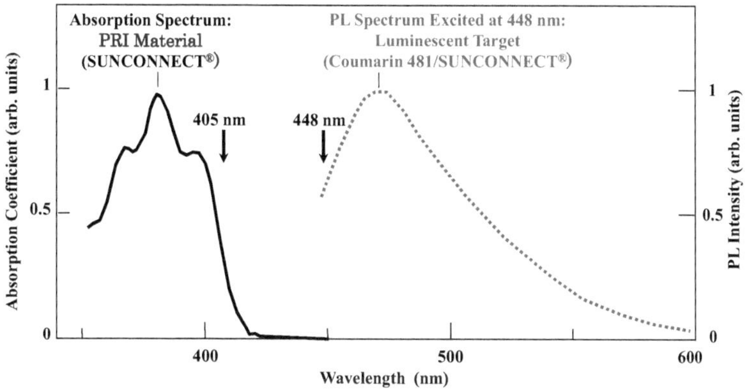

Figure 7.33 Absorption spectrum of PRI material of SUNCONNECT® and PL spectrum of a coumarin 481/SUNCONNECT® luminescent target excited by a write beam at 448 nm.

7.3.5 One-Photon R-SOLNETs with Luminescent Targets

7.3.5.1 Coumarin 481 luminescent targets

Formation of one-photon R-SOLNETs with luminescent targets was demonstrated in a PRI material of photosensitive organic/inorganic hybrid material, SUNCONNECT® [15], using luminescent targets made of a laser dye, coumarin 481, dissolved in SUNCONNECT® (coumarin 481/SUNCONNECT®.

To enhance the targeting effect, the sensitivity of the PRI material to the luminescence should be comparable with that to the write beam. As Fig. 7.33 shows, the absorption spectrum of the PRI material of SUNCONNECT® has a peak in UV region, and the absorption coefficient decreases at longer wavelengths, indicating that the sensitivity decreases with increasing the wavelength. Since the photoluminescence (PL) spectrum of the coumarin 481/SUNCONNECT® luminescent target has a peak around 470 nm, the wavelength separation between the PL peak and the write beam wavelength is 65 nm for a 405 nm write beam, and 22 nm for a 448 nm write beam. The 448 nm write beam is therefore preferable for reducing the sensitivity unbalance between the write beam and the luminescence.

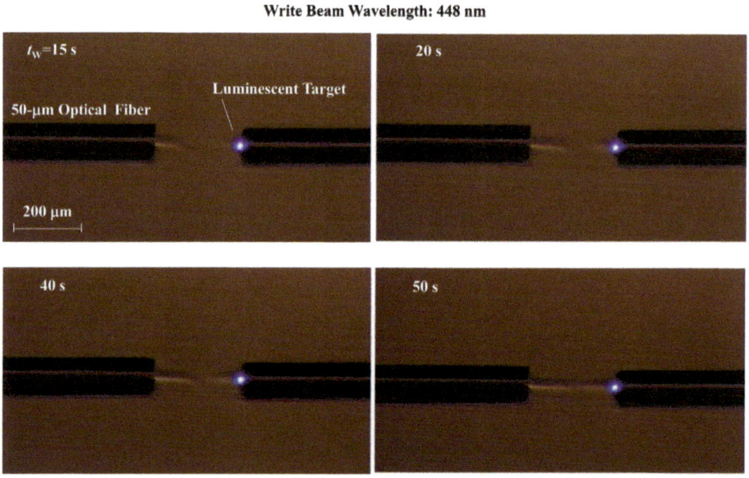

Figure 7.34 Formation of one-photon R-SOLNET between a 50 μm optical fiber and a luminescent target on the core edge of another 50 μm optical fiber with 20 μm lateral misalignment in SUNCONNECT® by a 448 nm write beam with power of 20 μW.

To deposit a luminescent target on the core region of an optical fiber edge, the light curing process shown in Fig. 7.6a was used. The fiber edge was coated with a luminescent material of coumarin 481 dissolved in SUNCONNECT® at concentration of ~0.1 wt%. The coating was exposed to a UV light from the fiber core, curing the region just on the core. The luminescent target was obtained by wet-etching the coating to remove the non-cured part using a mixture of isopropyl alcohol/4-methyl-2-pentanone.

Figure 7.34 shows a one-photon R-SOLNET formed by a 448 nm write beam between a 50 μm fiber and a luminescent target on the core edge of another 50 μm fiber in a PRI material of SUNCONNECT®. The two optical fibers are placed with a gap distance of 250 μm and a lateral misalignment of 20 μm. A 448 nm write beam with power of 20 μW is introduced from the optical fiber on the left. Bright luminescence from the luminescent target is observed. As the write beam irradiation progresses, a trace of optical waveguide growth from the core of the optical fiber on the left is observed. At the same time, a trace of optical waveguide growth

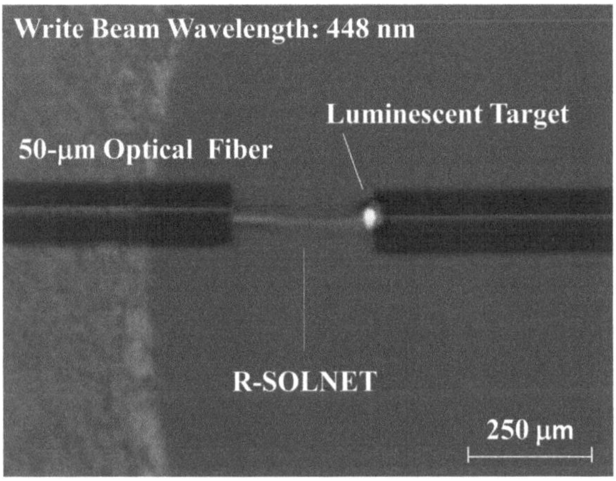

Figure 7.35 Another example of one-photon R-SOLNET formation between a 50 μm optical fiber and a luminescent target on the core edge of another 50 μm optical fiber. Reprinted with permission from Yoshimura [16].

from the luminescent target on the right is observed. The trace from the target is induced by luminescence emitted from the target. The two traces finally merge to form a single self-aligned coupling waveguide to complete a one-photon R-SOLNET with luminescent targets between the two 50 μm optical fibers. In Fig. 7.35, another example of one-photon R-SOLNET formation is presented. A self-aligned coupling waveguide is observed.

Figure 7.36 shows a one-photon R-SOLNET formed between a 9 μm fiber and a luminescent target on the core edge of another 9 μm fiber. A 448 nm write beam with power of 20 μW is introduced from the left-hand-side optical fiber. A trace of optical waveguide growth from the optical fiber on the left and another trace of optical waveguide growth from the luminescent target are observed. The two traces finally merge to form a single self-aligned coupling waveguide.

Figure 7.37 shows the coupling efficiency between 50 μm optical fibers connected by one-photon R-SOLNETs with luminescent targets for a gap distance of 300 μm. The probe beam wavelength is 850 nm. The coupling efficiency decreases with the lateral misalignment between optical fibers. By forming R-SOLNETs the coupling

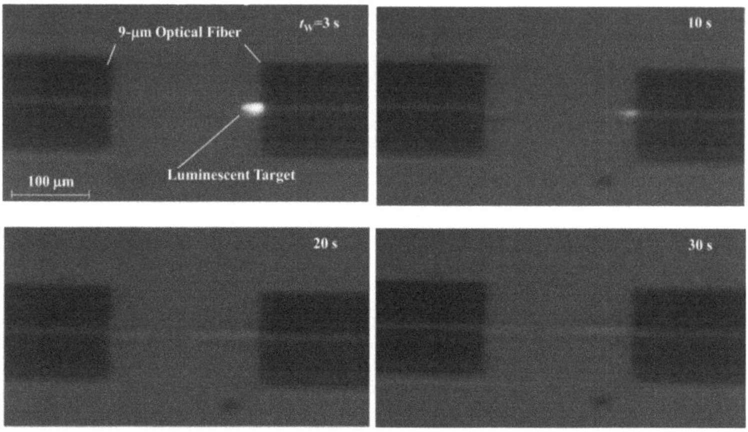

Figure 7.36 Formation of one-photon R-SOLNET between a 9 μm optical fiber and a luminescent target on the core edge of another 9 μm optical fiber with 9 μm lateral misalignment in SUNCONNECT® by a 448 nm write beam with power of 20 μW.

efficiency increases by 10–30%, and the lateral misalignment tolerance increases. Coupling efficiency of ~80% (coupling loss: ~1 dB) is maintained even when the lateral misalignment increases to 20 μm.

These results indicate that the one-photon R-SOLNET with luminescent targets has the potential to realize the optical solder function for in-plane couplings, and, in principle, for surface-normal couplings such as optical Z-connections and couplings between waveguide gratings and optical fibers. The coupling efficiency might increase by optimizing the luminescent target structures.

7.3.5.2 Alq3 luminescent targets

Formation of one-photon R-SOLNETs with luminescent targets was also demonstrated using tris(8-hydroxyquinolinato) aluminum (Alq3) [13], which is known as phosphor for green emission in organic light emitting diodes (LEDs). A luminescent target of Alq3-dispersed polyvinyl alcohol (Alq3/PVA) was placed on an optical

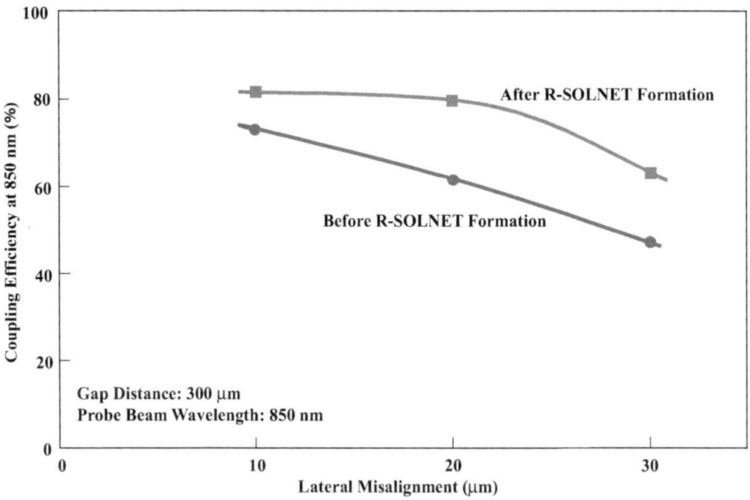

Figure 7.37 Dependence of maximum coupling efficiency on lateral misalignments in one-photon R-SOLNETs with luminescent targets between 50 μm optical fibers.

fiber edge. The PRI material was NOA81/NOA65, which is a mixture of acrylic materials of NOA81 (NORLAND) with a refractive index of 1.56 and NOA65 with a refractive index of 1.52. The mixing weight ratio of NOA81:NOA65 was 2:1. On a slide glass, a 50 μm optical fiber and an optical fiber with the luminescent target were placed in the PRI material with a gap distance of ~4 mm.

As shown in Fig. 7.38a, when a 405 nm write beam is introduced from the left-hand-side 50 μm optical fiber, the luminescent target emits green/blue luminescence to pull the write beam to the target to form a one-photon R-SOLNET that connects the fiber core and the luminescent target. It is found from Fig. 7.38b that the R-SOLNET width gradually expands from the fiber core diameter to the width of the Alq3/PVA luminescent target, constructing a tapered R-SOLNET. This indicates that the one-photon R-SOLNET with luminescent targets acts as optical solder with a mode size conversion function.

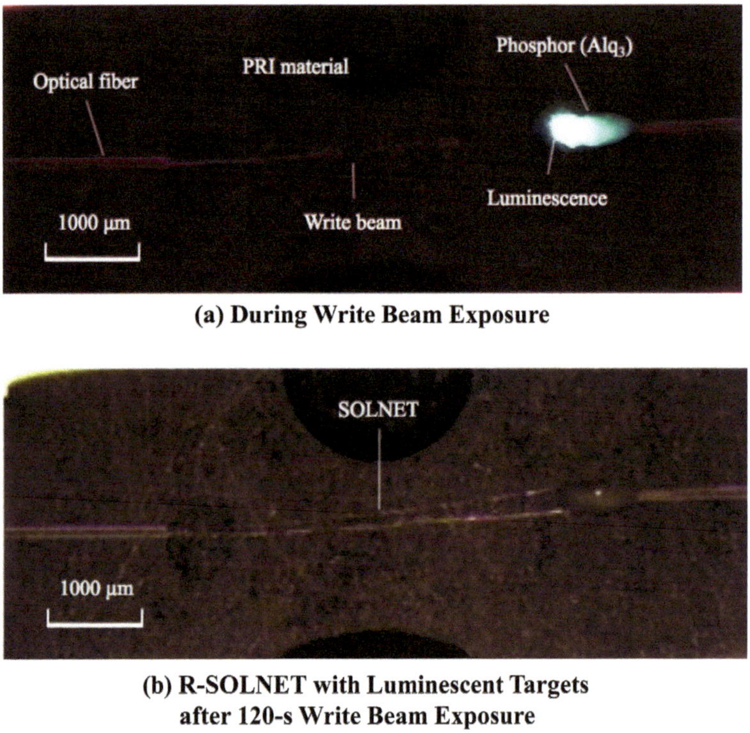

Figure 7.38 Formation of one-photon R-SOLNET with luminescent targets between a 50 μm optical fiber and a large Alq3/PVA target. The R-SOLNET has a function of a mode size converter. Reprinted with permission from Seki and Yoshimura [13].

Figure 7.39 shows one-photon R-SOLNET formation using an Alq3 thin-film luminescent target deposited on an optical fiber edge by the vacuum evaporation [14]. By introducing a 405 nm write beam from the left-hand-side 50 μm optical fiber into the PRI material of NOA81/NOA65, green/blue luminescence is generated from the Alq3 thin film (Fig. 7.39b). The write beam is pulled toward the Alq3 thin-film target to form a one-photon R-SOLNET with luminescent targets (Fig. 7.39c).

These results confirm that the one-photon R-SOLNET can be formed using luminescence emitted from luminescent targets.

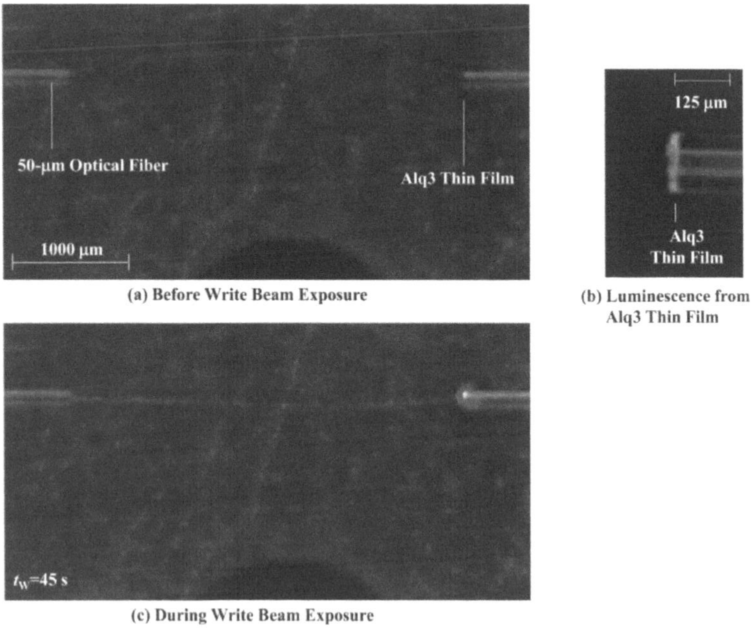

Figure 7.39 Formation of one-photon R-SOLNET with luminescent targets between a 50 μm optical fiber and an Alq3 thin-film target deposited on the edge of an optical fiber. Reprinted with permission from Yoshimura and Seki [14].

7.4 Experimental Demonstrations of Two-Photon SOLNETs

In the present section, experimental demonstrations of two-photon SOLNETs are described. The experiments were carried out using 50 μm optical fibers.

7.4.1 Two-Photon TB-SOLNETs

Two-photon TB-SOLNET was formed in two-photon PRI materials made of a host material of New-SUNCONNECT® and two-photon absorption molecules such as camphorquinone (CQ) and biacetyl (BA) [8]. The two-photon absorption molecules induce two-photon photochemical reactions [30]. New-SUNCONNECT® is a photosen-

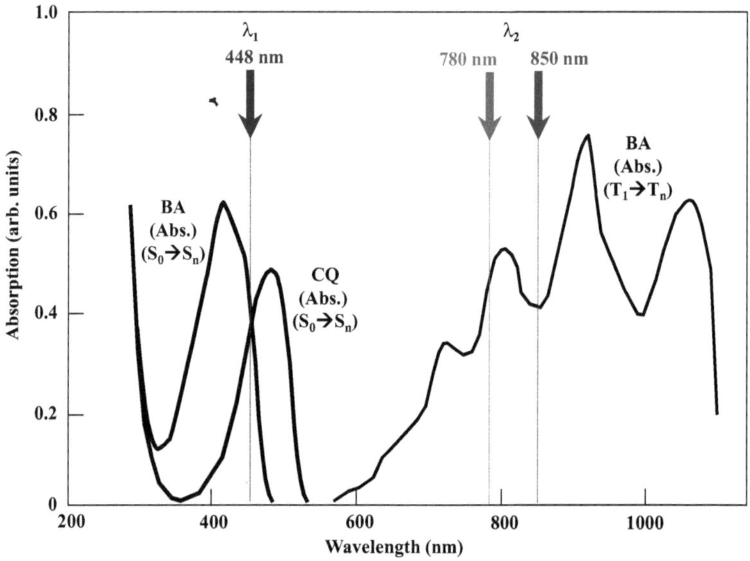

Figure 7.40 Absorption spectra of two-photon absorption molecules, BA and CQ.

sitive organic/inorganic hybrid material, whose light absorption is smaller than that of SUNCONNECT® [25, 26] in blue wavelength regions. The small light absorption of the host material is preferable to feature the effect of the two-photon absorption molecules.

Absorption spectra of CQ and BA are presented in Fig. 7.40 for the $S_0 \rightarrow S_n$ transition and the $T_1 \rightarrow T_n$ transition [30]. Here, although a $T_1 \rightarrow T_n$ absorption spectrum is drawn only for BA, Brauchle et al. reported that the $T_1 \rightarrow T_n$ absorption spectrum of CQ is almost the same as that of BA. For the λ_1 write beam that induces the $S_0 \rightarrow S_n$ transition, a wavelength of 448 nm is chosen because both of CQ and BA absorb the 448 nm write beam. For the λ_2 write beam, a wavelength of 850 nm or 780 nm, which is within the wavelength region for the $T_1 \rightarrow T_n$ transition, is chosen.

Figure 7.41 shows two-photon TB-SOLNET formation between 50 μm optical fibers placed with a gap distance of 500 μm in a two-photon PRI material containing 20wt%-CQ (20wt%-CQ/New-SUNCONNECT®). A 448 nm write beam from the left-hand-side optical fiber and an 850 nm write beam from the right-hand-side

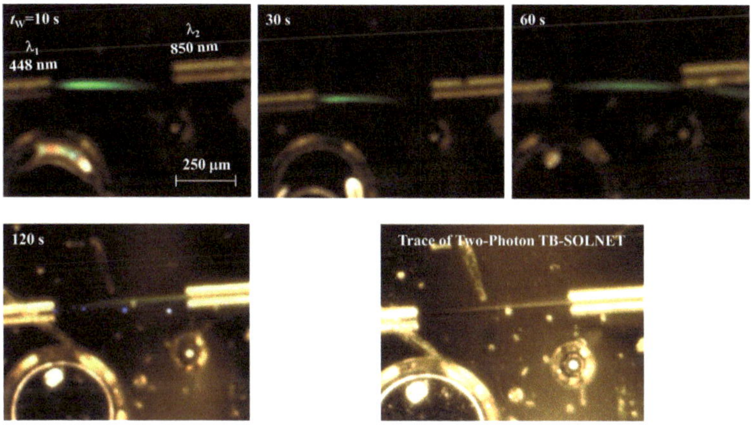

Figure 7.41 Formation of two-photon TB-SOLNET between 50 μm optical fibers and a trace of the two-photon TB-SOLNET in 20wt%-CQ/New-SUNCONNECT®. Reprinted with permission from Yoshimura et al. [8].

optical fiber are introduced into the two-photon PRI material. CQ is excited by the 448 nm write beam to emit green luminescence, which is probably accompanied with the $S_n \rightarrow S_0$ transition or the $T_1 \rightarrow S_0$ transition. Since the intensity of the green luminescence is proportional to that of the 448 nm write beam, the green luminescence is enhanced with an increase in the 448 nm write beam confinement. Therefore, the green luminescence can be used to monitor the SOLNET formation dynamics. A green luminescence line, which corresponds to a trace of two-photon TB-SOLNET, stretches from the left-hand-side fiber core, and is initially deviated from the opposing right-hand-side fiber core. With writing time the luminescence line is gradually pulled toward the right-hand-side fiber core location. Finally, at 120 s the luminescence line connects the left-hand-side and right-hand-side fiber cores, indicating the two-photon TB-SOLNET formation. In the photograph, which is taken after the write beam is turned off, a trace of the two-photon TB-SOLNET connecting the two optical fibers is observed. The two-photon TB-SOLNET keeps connecting the two optical fibers even after the left-hand-side optical fiber moves as Fig. 7.42 shows.

Figure 7.43 shows coupling efficiency of the two-photon TB-SOLNET. The wavelength of probe beams is 448 nm. The probe

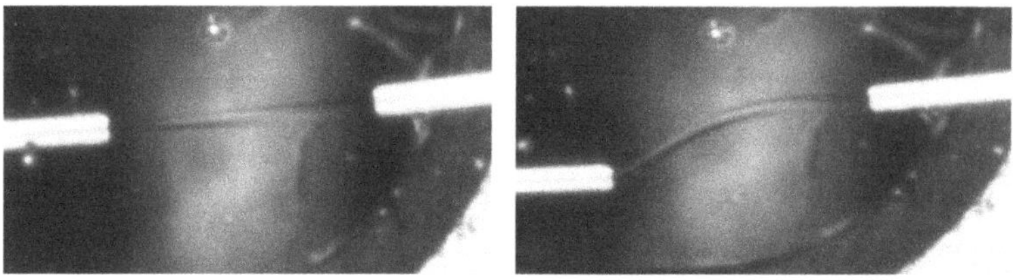

Figure 7.42 Photographs of the two-photon TB-SOLNET before and after the left-hand-side optical fiber is moved. Reprinted with permission from Yoshimura et al. [8].

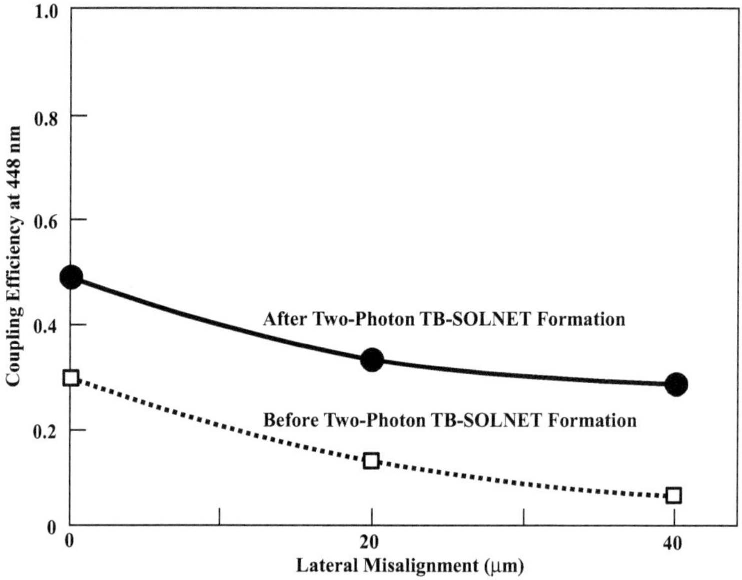

Figure 7.43 Dependence of maximum coupling efficiency on lateral misalignments in two-photon TB-SOLNETs formed between 50 μm optical fibers in CQ/New-SUNCONNECT®. λ_1 and λ_2 are respectively 448 nm and 850 nm, and gap distance is 500 μm. Reprinted with permission from Yoshimura et al. [8].

beams propagate from the left-hand-side optical fiber to the right-hand-side optical fiber. By the two-photon TB-SOLNET formation, the coupling efficiency rises. The coupling losses for 0 μm misalignment and 40 μm misalignment are respectively ~3 dB and

~5.3 dB after the SOLNET formation, indicating that the two-photon TB-SOLNET can overcome misalignments of 40 μm with ~2.3 dB penalty.

These results provide an experimental proof-of-concept for the two-photon TB-SOLNET formation. Since the formation dynamics for P-SOLNETs and TB-SOLNETs are the same as mentioned in section 7.1.1, the experimental demonstration of TB-SOLNETs hints to the potential feasibility of P-SOLNETs.

7.4.2 Two-Photon R-SOLNETs

Two-photon R-SOLNET was formed using luminescent targets of SUNCONNECT® containing a laser dye called styryl 7 or DCM [8]. In Fig. 7.44a an absorption spectrum and a PL spectrum of styryl 7 in methanol [39] are shown together with the $S_0 \rightarrow S_n$ absorption spectrum of CQ and the $T_1 \rightarrow T_n$ absorption spectrum of BA, which is almost the same as that of CQ. Styryl 7 absorbs the 448 nm write beam and emits luminescence in a wavelength region of 600–800 nm, which can be absorbed in the $T_1 \rightarrow T_n$ transition by CQ. Therefore, styryl 7 is a suitable material for luminescent targets in the two-photon R-SOLNET. It is found from Fig. 7.44b that DCM is another suitable material because DCM absorbs the 448 nm write beam and emits luminescence, which can be absorbed in the $T_1 \rightarrow T_n$ transition by CQ.

The light curing process was used to deposit a luminescent target on the core region of a 50 μm optical fiber edge. After the fiber edge was coated with a luminescent material of styryl 7 or DCM dissolved in SUNCONNECT®, the coating was exposed to a 448 nm light from the fiber core, curing the region just on the core. The luminescent target was obtained by wet-etching the coating to remove the non-cured part using a mixture of isopropyl alcohol/4-methyl-2-pentanone.

In Fig. 7.45 formation of two-photon R-SOLNET with a luminescent target of 13wt%-styryl 7/SUNCONNECT® between 50 μm optical fibers is shown. The two-photon PRI material is the CQ/New-SUNCONNECT®. When a 448 nm write beam with power of 130 μW is emitted from the left-hand-side optical fiber, a green luminescence line monitoring the SOLNET formation is gradually pulled toward

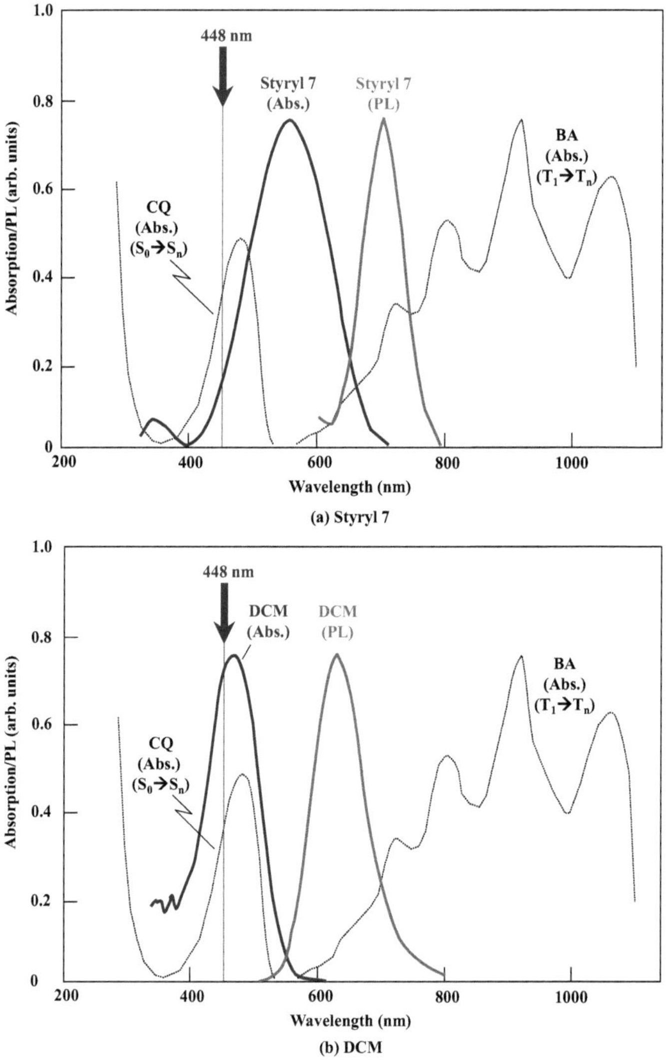

Figure 7.44 Absorption spectra of BA and CQ, and absorption/PL spectra of (a) styryl 7 and (b) DCM in methanol.

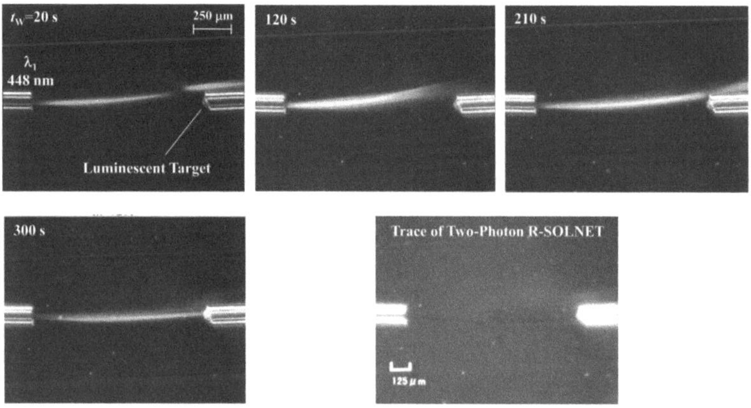

Figure 7.45 Formation of two-photon R-SOLNET with luminescent targets between 50 μm optical fibers and a trace of the two-photon R-SOLNET in CQ/New-SUNCONNECT®. The luminescent target is styryl 7/SUNCONNECT®. 448 nm write beam is introduced from the left-hand-side optical fiber. Reprinted with permission from Yoshimura et al. [8].

the luminescent target with writing time by the write beam and luminescence from the luminescent target. Finally, the green luminescence line reaches the luminescent target. A trace of the two-photon R-SOLNET connecting the two optical fibers is observed.

Figure 7.46 shows coupling efficiency of the two-photon R-SOLNET. 448 nm probe beams propagate from the left-hand-side optical fiber to the right-hand-side optical fiber. Although coupling efficiency is found to be low (about 20–40%), the decrease in the coupling effect is suppressed by the two-photon R-SOLNET. In the case that the 448 nm write beam power is 130 μW, the coupling losses for 0 μm misalignment and 25 μm misalignment are respectively ~6.3 dB and ~7.7 dB, indicating that the two-photon R-SOLNET can overcome misalignments of 25 μm with ~1.4 dB penalty. Optimization of the fabrication conditions such as write beam power and luminescent target structures might improve the efficiency.

When telecom-relevant wavelengths like 850 nm, 1310 nm, and 1550 nm, which are longer than 448 nm, are used for probe beams, it is expected that the coupling loss is reduced because the host material of the PRI material, SUNCONNECT®, is a waveguide material

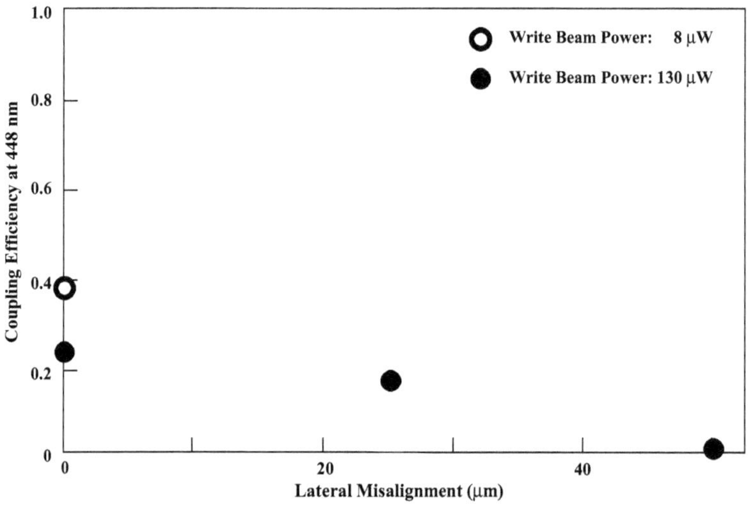

Figure 7.46 Dependence of maximum coupling efficiency on lateral misalignments in two-photon R-SOLNETs with luminescent targets formed between 50 μm optical fibers in CQ/New-SUNCONNECT®. The gap distance is 500 μm. Reprinted with permission from Yoshimura et al. [8].

designed to exhibit low-loss characteristics in the telecom relevant wavelength regions [37, 38]. Actually, as mentioned in section 7.3.5, one-photon R-SOLNETs formed in the SUNCONNECT®-based PRI material enable low-loss couplings between 50 μm optical fibers at 856 nm.

Two-photon R-SOLNETs formed between 50 μm optical fibers are expected to be multimode because the SOLNET width is close to that of the 50 μm optical fiber cores. When two-photon R-SOLNETs are formed between single-mode optical fibers, it is expected that SOLNETs become single mode.

Figure 7.47 shows the formation of a two-photon R-SOLNET with a luminescent target of 1wt%-DCM/SUNCONNECT® between 50 μm optical fibers in the CQ/New-SUNCONNECT®. The diameter of the right-hand-side optical fiber seems much larger than that of the left-hand-side optical fiber. This indicates that the right-hand-side optical fiber is located at a position closer to the objective lens. In other words, the two optical fibers are arranged with misalignments along both of the horizontal and the vertical

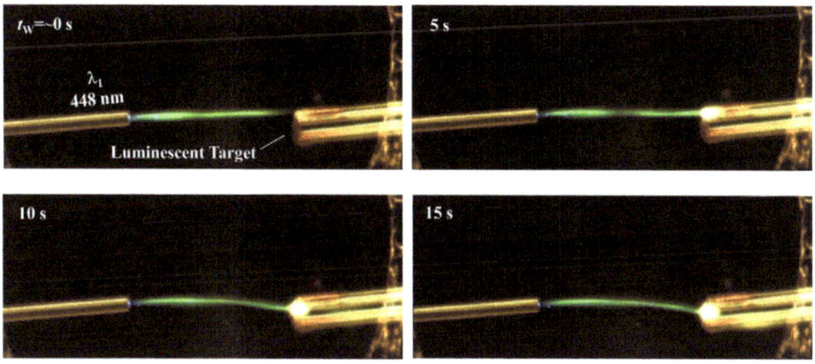

Figure 7.47 Formation of two-photon R-SOLNET with luminescent targets between 50 μm optical fibers with horizontal and vertical misalignments in CQ/New-SUNCONNECT®. The luminescence target is DCM/SUNCONNECT®. Reprinted with permission from Yoshimura et al. [8].

directions. When a 448 nm write beam is introduced from the left-hand-side optical fiber, a green luminescence line is gradually pulled toward the luminescent target on the core edge of the right-hand-side optical fiber with writing time, confirming that a two-photon R-SOLNET with luminescent targets is formed even when such a large misalignment exists.

These results provide an experimental proof-of-concept for the two-photon R-SOLNET formation. The experimental demonstration of R-SOLNET using placed luminescent targets hints to the feasibility of LA-SOLNET.

References

1. T. Yoshimura, *Self-Organized Lightwave Networks: Self-Aligned Coupling Optical Waveguides*, CRC/Taylor & Francis, Boca Raton, Florida (2018).
2. T. Yoshimura, W. Sotoyama, K. Motoyoshi, T. Ishitsuka, K. Tsukamoto, S. Tatsuura, H. Soda, and T. Yamamoto, "Method of producing optical waveguide system, optical device and optical coupler employing the same, optical network and optical circuit board," U.S. Patent 6,081,632 (2000).
3. T. Yoshimura, J. Roman, Y. Takahashi, W. V. Wang, M. Inao, T. Ishitsuka, K. Tsukamoto, K. Motoyoshi, and W. Sotoyama, "Self-organizing waveguide

coupling method "SOLNET" and its application to film optical circuit substrates," *Proc. 50th Electronic Components & Technology Conference (ECTC)*, 962–969 (2000).

4. T. Yoshimura, J. Roman, Y. Takahashi, W. V. Wang, M. Inao, T. Ishituka, K. Tsukamoto, K. Motoyoshi, and W. Sotoyama, "Self-organizing lightwave network (SOLNET) and its application to film optical circuit substrates," *IEEE Trans. Compon. Packag. Technol.* **24**, 500–509 (2001).
5. T. Yoshimura, A. Hori, Y. Yoshida, Y. Arai, H. Kurokawa, T. Namiki, and K. Asama, "Coupling efficiencies in reflective self-organized lightwave network (R-SOLNET) simulated by the beam propagation method," *IEEE Photonics Technol. Lett.* **17**, 1653–1655 (2005).
6. T. Yoshimura, D. Takeda, T. Sato, Y. Kinugasa, and H. Nawata, "Polymer waveguides self-organized by two-photon photochemistry for self-aligned optical couplings with wide misalignment tolerances," *Opt. Commun.* **362**, 81–86 (2016).
7. T. Yoshimura and H. Nawata, "Micro/nanoscale self-aligned optical couplings of the self-organized lightwave network (SOLNET) formed by excitation lights from outside," *Opt. Commun.* **383**, 119–131 (2017).
8. T. Yoshimura, S. Yasuda, H. Yamaura, Y. Yamada, M. Takashima, R. Ito, and T. Hamazaki, "Self-aligned coupling waveguides experimentally formed by two-photon photochemistry for 3D integrated optical interconnects," *Opt. Commun.* **430**, 284–292 (2019).
9. T. Yoshimura, T. Inoguchi, T. Yamamoto, S. Moriya, Y. Teramoto, Y. Arai, T. Namiki, and K. Asama, "Self-organized lightwave network based on waveguide films for three-dimensional optical wiring within boxes," *J. Lightwave Technol.* **22**, 2091–2100 (2004).
10. H. Kaburagi and T. Yoshimura, "Simulation of micro/nanometer-scale self-organized lightwave network (SOLNET) using the finite difference time domain method," *Opt. Commun.* **281**, 4019–4022 (2008).
11. T. Yoshimura and H. Kaburagi, "Self-organization of optical waveguides between misaligned devices induced by write-beam reflection," *Appl. Phys. Express* **1**, 062007 (2008).
12. T. Yoshimura, K. Wakabayashi, and S. Ono, "Analysis of reflective self-organized lightwave network (R-SOLNET) for Z-connections in three-dimensional optical circuits by the finite difference time domain method," *IEEE J. Sel. Top. Quantum Electron.* **17**, 566–570 (2011).
13. M. Seki and T. Yoshimura, "Reflective self-organizing lightwave network using a phosphor," *Opt. Eng.* **51**, 074601 (2012).

14. T. Yoshimura and M. Seki, "Simulation of self-organized parallel waveguides targeting nanoscale luminescent objects," *J. Opt. Soc. Am. B* **30**, 1643–1650 (2013).
15. T. Yoshimura, M. Iida, and H. Nawata, "Self-aligned optical couplings by self-organized waveguides toward luminescent targets in organic/inorganic hybrid materials," *Opt. Lett.* **39**, 3496–3499 (2014).
16. T. Yoshimura, "Simulation of self-aligned optical coupling between micro- and nanoscale devices using self-organized waveguides," *J. Lightwave Technol.* **33**, 849–856 (2015).
17. S. J. Frisken, "Light-induced optical waveguide uptapers," *Opt. Lett.* **18**, 1035–1037 (1993).
18. A. S. Kewitsch and A. Yariv, "Self-focusing and self-trapping of optical beams upon photopolymerization," *Opt. Lett.* **21**, 24–26 (1996).
19. M. Kagami, T. Yamashita, and H. Ito, "Light-induced self-written three-dimensional optical waveguide," *Appl. Phys. Lett.* **79**, 1079–1081 (2001).
20. K. Dorkenoo, O. Crégut, L. Mager, and F. Gillot, "Quasi-solitonic behavior of self-written waveguides created by photopolymerization," *Opt. Lett.* **27**, 1782–1784 (2002).
21. F. Huang, H. Takase, Y. Eriyama, and T. Ukachi, "Single-mode optical fiber interconnection by using self-written waveguides," *Proc. 12th Int. POF. Conf.*, Seattle, USA, 199–202 (2003).
22. N. Hirose and O. Ibaragi, "Optical component coupling using self-written waveguides," *Proc. SPIE* **5355**, 206–214 (2004).
23. S. Jradi, O. Soppera, and D. J. Lougnot, "Fabrication of polymer waveguides between two optical fibers using spatially controlled light-induced polymerization," *Appl. Opt.* **47**, 3987–3993 (2008).
24. T. M. Monro, C. M. de Sterke, and L. Poladian, "Investigation of waveguide growth in photosensitive germanosilicate glass," *J. Opt. Soc. Am. B* **13**, 2824–2832 (1996).
25. T. Sato, "Novel organic-inorganic hybrid materials for optical interconnects," *Proc. SPIE* **7944**, 79440M (2011).
26. H. Nawata, "Organic-inorganic hybrid material for on-board optical interconnection and its applications in optical coupling," in *IEEE CPMT Symposium Japan*, 121–124 (2013).
27. E. Fazio, A. Belardini, L. Bastiani, M. Alonzo, M. Chauvet, N. I. Zheludev, and C. Soci, "Novel paradigm for integrated photonics circuits: transient interconnection network," *Proc. SPIE* **10130**, 1013006 (2017).

28. E. Fazio, A. Zaltron, A. Belardini, N. Argiolas, and C. Sada, "Influence of iron doping on spatial soliton formation and fixing in lithium niobate crystals," *Opt. Mater.* **37**, 175–180 (2014).
29. Y. R. Shen, *The Principles of Nonlinear Optics*, John Wiley & Sons, New York (1984).
30. C. Brauchle, U. P. Wild, D. M. Burland, G. C. Bjorkund, and D. C. Alvares, "Two-photon holographic recording with continuous-wave lasers in the 750–1100-nm range," *Opt. Lett.* **7**, 177–179 (1982).
31. T. Yoshimura, S. Ito, T. Nakayama, and K. Matsumoto, "Orientation-controlled molecule-by-molecule polymer wire growth by the carrier-gas-type organic chemical vapor deposition and the molecular layer deposition," *Appl. Phys. Lett.* **91**, 033103 (2007).
32. T. Yoshimura, N. Terasawa, H. Kazama, Y. Naito, Y. Suzuki, and K. Asama, "Selective growth of conjugated polymer thin films by the vapor deposition polymerization," *Thin Solid Films* **497**, 182–184 (2006).
33. M. Pessa, R. Makela, and T. Suntola, "Characterization of surface exchange reactions used to grow compound films," *Appl. Phys. Lett.* **31**, 131–133 (1981).
34. T. Yoshimura, S. Tatsuura, and W. Sotoyama, "Polymer films formed with monolayer growth steps by molecular layer deposition," *Appl. Phys. Lett.* **59**, 482–484 (1991).
35. T. Yoshimura and S. Ishii, "Effect of quantum dot length on the degree of electron localization in polymer wires grown by molecular layer deposition," *J. Vac. Sci. Technol. A* **31**, 031501 (2013).
36. F. E. Doany, B. G. Lee, S. Assefa, W. M. J. Green, M. Yang, C. L. Schow, C. V. Jahnes, S. Zhang, J. Singer, V. I. Kopp, J. A. Kash, and Y. A. Vlasov, "Multichannel high-bandwidth coupling of ultradense silicon photonic waveguide array to standard-pitch fiber array," *J. Lightwave Technol.* **29**, 475–482 (2011).
37. S. Ono, T. Yoshimura, T. Sato, and J. Oshima, "Fabrication and evaluation of nano-scale optical circuits using sol–gel materials with photo-induced refractive index variation characteristics," *J. Lightwave Technol.* **27**, 1229–1235 (2009).
38. S. Ono, T. Yoshimura, T. Sato, and J. Oshima, "Fabrication of self-organized optical waveguides in photo-induced refractive index variation sol–gel materials with large index contrast," *J. Lightwave Technol.* **27**, 5308–5313 (2009).
39. Catalogue of Laserdyes, Acros Organics.

Chapter 8

Self-Organized 3D Integrated Optical Interconnects: Model Proposals

The overall concept of the self-organized 3D integrated optical interconnects are described in Chapter 2. In the present chapter, their precise models are proposed [1–3].

8.1 3D Integrated Optical Interconnects with P- and R-SOLNETs

Figure 8.1 shows a model for self-organized 3D integrated optical interconnects, where optical waveguides are transparent in visible and near-infrared wavelength regions and write beams can propagate in them as in the case of polymers, SiN, SiON, and SiTiO waveguides.

Firstly, a 3D preform is prepared (Fig. 8.1a). OE-Films, which contain embedded small/thin OE dies for light modulators and PDs as well as nanoscale waveguides with vertical mirrors, are stacked together with OE-Films containing thinned LSIs. Here, the light modulators could be, e.g., WPD, TIR, or digital optical switches, which are made of, e.g., EO polymers like the polymer MQW, PLZT, LN, Si, or III–V materials like GaAs and InP. Vertical mirrors in

Self-Organized 3D Integrated Optical Interconnects with All-Photolithographic Heterogeneous Integration
Tetsuzo Yoshimura

ISBN 978-981-4877-04-6 (Hardcover), 978-1-003-04305-8 (eBook)
www.jennystanford.com

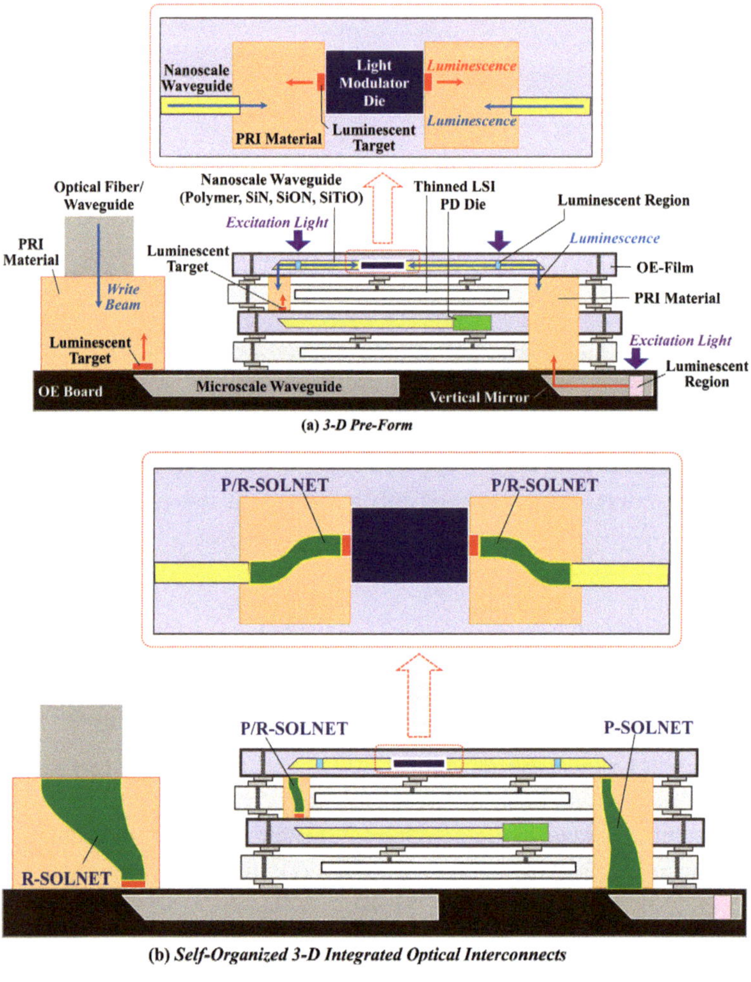

Figure 8.1 Model for self-organized 3D integrated optical interconnects built by two-photon P-SOLNETs and R-SOLNETs. Reprinted with permission from Yoshimura et al. [2].

the waveguides can be formed by the built-in mask method, laser ablation, and oblique-exposure reactive ion etching.

OE-die embedding into OE-Films and OE-Film stacking are performed by PL-Pack with SORT described in Chapter 5 and by the Film/Z-connection technology as follows. The small/thin OE dies are picked up on supporting substrates using the ELO. These OE dies are transferred onto pads on a film substrate at one time by using aligners. It is also possible to transfer nanoscale waveguide cores with vertical mirrors onto the film substrate by PL-Pack with SORT for waveguide transfer. After covering the OE dies with a polymer layer, surface planarization and metallization for vias, pads, and electrical lines are carried out to complete the OE-Films. The OE-Films are stacked by using aligners and are connected by, e.g., the solder bonding or metal bonding for mechanical/electrical joints to construct the 3D preform.

For in-plane couplings between nanoscale waveguides and light modulators with nanoscale waveguides, luminescent targets are deposited on the light modulator edges, and PRI materials are placed between the waveguides and the light modulators. For vertical couplings between nanoscale waveguides in the 3D structure and between nanoscale waveguides in the 3D structure and microscale waveguides in the OE board, luminescent regions and luminescent targets are distributed. For couplings between optical fibers (or waveguides) and microscale waveguides in the OE board, luminescent targets are deposited at vertical mirror sites on the microscale waveguides. Similar optical coupling structures are available if the vertical mirrors are replaced with grating couplers.

Then, external excitation lights are introduced into the 3D preform from outside by blanket exposure, or by focusing optics or optical fibers. As a result, as illustrated in Fig. 8.1b, self-aligned in-plane couplings of P/R-SOLNET between nanoscale waveguides and light modulators, self-aligned vertical couplings of P/R-SOLNET between nanoscale waveguides, and self-aligned vertical couplings of P-SOLNET between nanoscale waveguides and microscale waveguides are formed. Here, P/R-SOLNET means R-SOLNET formed by write beams from luminescent regions. For couplings between optical fibers (or waveguides) and microscale waveguides, R-SOLNETs are formed by introducing write beams from the optical fibers (or waveguides). Thus, self-organized 3D integrated optical interconnects are built. For in-plane couplings between nanoscale waveguides and light modulators, it might also be possible to form P/R-SOLNETs before OE-Film stacking.

Light modulators in the 3D structure transmit optical signals by modulating cw or pulse-train lights from light sources like high-power LDs and mode-locked LDs. Since the light sources generate considerable heat, it is preferable for thermal managements not to place the light sources within the 3D structure. The cw or pulse-train lights can be inserted into the 3D structure by introducing them from the optical fibers (or waveguides) into the waveguides in the OE board via vertical mirrors. They can also be inserted into the 3D structure by introducing them from light sources mounted on the waveguides in the OE board via the vertical mirrors. By using light sources with a plurality of wavelengths and by replacing the vertical mirrors with wavelength filters, WDM interconnects can be performed.

Figure 8.2 shows an example of the PL-Pack with SORT process for integration of small/thin light modulator dies and nanoscale waveguides with vertical mirrors. As shown in Fig. 8.2a, the light modulator dies are picked up on Supporting Substrate-I by the ELO, and luminescent targets are deposited on the core edges of the dies by the selective deposition process. The nanoscale waveguides with vertical mirrors are picked up on Supporting Substrate-II. These dies and nanoscale waveguides are selectively transferred to final substrates, as Fig. 8.2b shows. In the case that PRI materials are placed between the light modulator dies and the nanoscale waveguides, gap spaces are inserted between them.

For fabrication of luminescent regions, partial thermal diffusion of phosphor molecules is a possible process. For example, in polymer waveguides, phosphor molecules are expected to diffuse into the waveguide cores selectively using a mask with apertures distributed at the luminescent region locations. For SiN and SiON waveguides, they have emissive characteristics intrinsically. For example, a-SiN_x:H emits PL, and the peaks of the PL spectra vary from 620 nm to 310 nm with an increase in the N-composition x from 0.85 to 1.24 [4]. This implies that the wavelength of luminescence emitted from the luminescent regions can be adjusted by controlling x in the luminescent regions. In such cases, luminescence for write beams can be generated just by exposing the waveguide cores to external excitation lights without preparing the luminescent regions. To find optimum conditions in excitation light absorption and PL emission

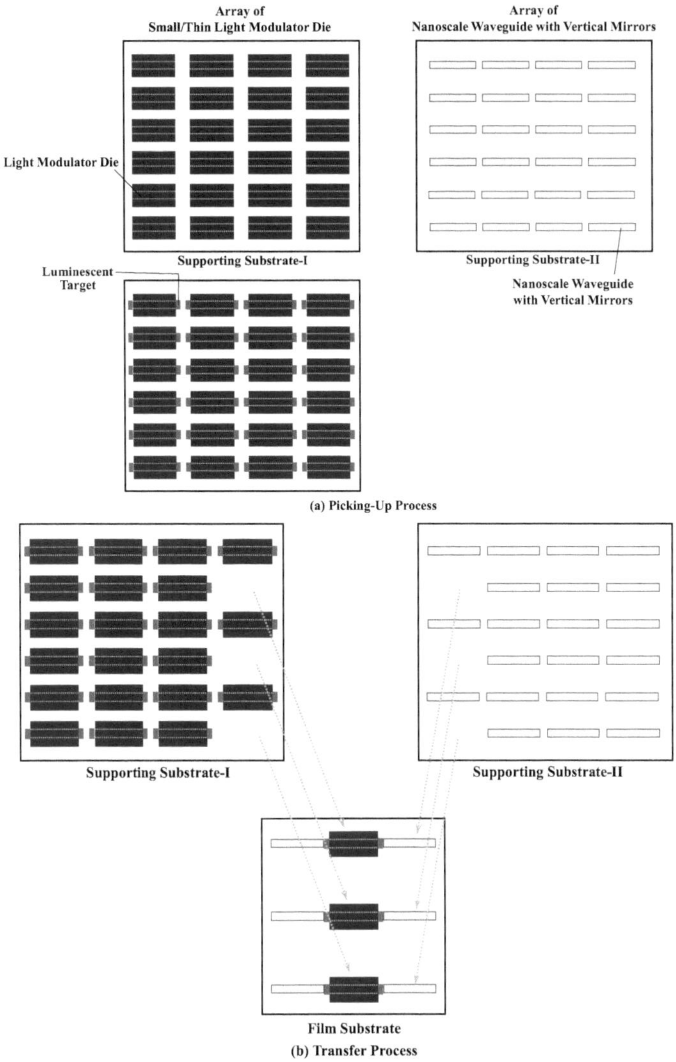

Figure 8.2 Example of the PL-Pack with SORT process for integration of small/thin light modulator dies and nanoscale waveguides with vertical mirrors. In the case that PRI materials are placed between the light modulator dies and the nanoscale waveguides, gap spaces are inserted between them.

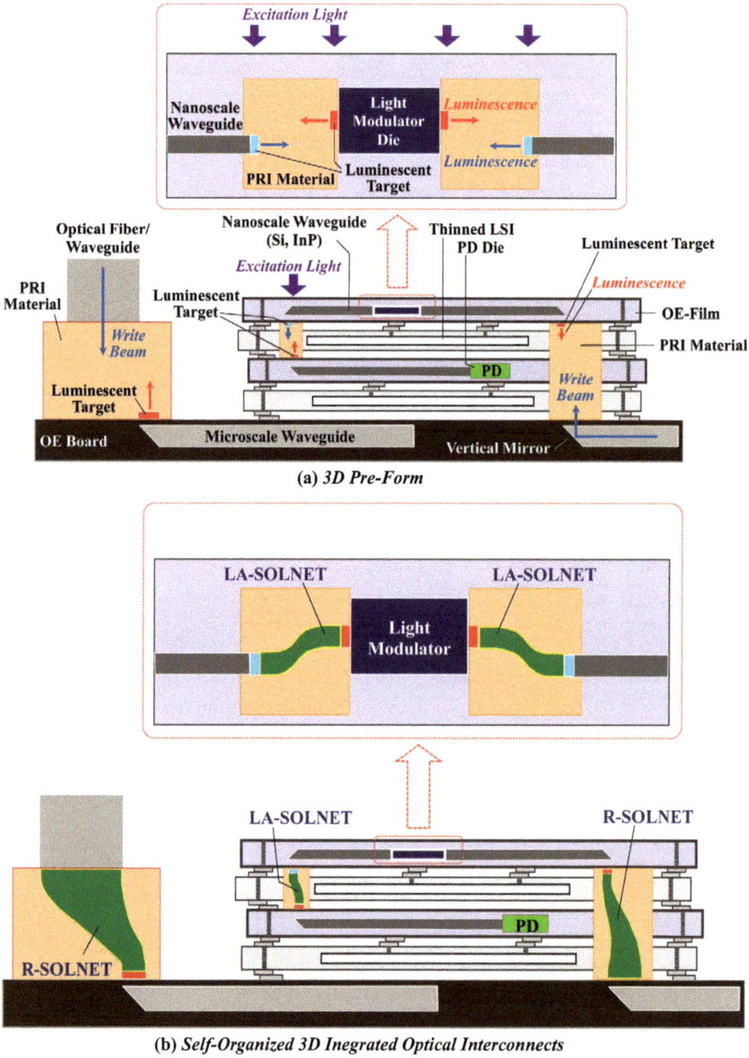

Figure 8.3 Model for self-organized 3D integrated optical interconnects built by two-photon LA-SOLNETs and R-SOLNETs. Reprinted with permission from Yoshimura et al. [2].

for luminescent regions, detailed investigations into the material composition dependence of absorption/PL spectra for SiN and SiON will be required.

8.2 3D Integrated Optical Interconnects with LA- and R-SOLNETs

In the model shown in Fig. 8.3, nanoscale waveguides are not transparent in visible wavelength regions as in the case of Si and InP, in which write beams cannot propagate. For in-plane couplings between nanoscale waveguides and light modulators, luminescent targets are deposited on the edges of the nanoscale waveguides and the light modulators, as shown in Fig. 8.3a. Because curing lights cannot propagate in waveguides, the selective deposition process should be used. For vertical couplings between nanoscale waveguides in the 3D structure and between nanoscale waveguides in the 3D structure and microscale waveguides in the OE board, luminescent targets are deposited at vertical mirror sites on the nanoscale waveguides.

By introducing external excitation light into the 3D preform from outside, luminescent targets emit write beams. Then, as illustrated in Fig. 8.3b, self-aligned in-plane couplings of LA-SOLNET between nanoscale waveguides and light modulators, and self-aligned vertical couplings of LA-SOLNET between nanoscale waveguides are formed to build self-organized 3D integrated optical interconnects. Vertical couplings of R-SOLNET between nanoscale waveguides and microscale waveguides in the OE board are formed by introducing write beams from the microscale waveguides. Couplings of R-SOLNET between optical fibers (or waveguides) and microscale waveguides are formed by introducing write beams from optical fibers (or waveguides).

References

1. T. Yoshimura, *Self-Organized Lightwave Networks: Self-Aligned Coupling Optical Waveguides*, CRC/Taylor & Francis, Boca Raton, Florida (2018).

2. T. Yoshimura, S. Yasuda, H. Yamaura, Y. Yamada, M. Takashima, R. Ito, and T. Hamazaki, "Self-aligned coupling waveguides experimentally formed by two-photon photochemistry for 3D integrated optical interconnects," *Opt. Commun.* **430**, 284–292 (2019).
3. T. Yoshimura, S. Yasuda, H. Yamaura, and Y. Yamada, "Self-organized lightwave network (SOLNET) formed by two-photon photochemistry for 3D integrated optical interconnects," *2018 IEEE Optical Interconnects Conference (OI)*, Santa Fe, New Mexico, 33–34 (2018).
4. K. Hiranaka, T. Yoshimura, and T. Yamaguchi, *J. Appl. Phys.* **62**, 2197–2200 (1987).

Chapter 9

Self-Organized 3D Micro Optical Switching Systems: Model Proposals and Predicted Performance

In the present chapter, as an example of S-FOLM-based advanced OE systems, precise models of the 3D switching system called "Three-Dimensional Micro Optical Switching System (3D-MOSS)" is proposed. The advantage of the SOLNET implementation is discussed. It is predicted that the 3D-MOSS has potential to achieve high-speed massive optical switching in the 3D integrated optical interconnects with cost, size and power dissipation reduction [1–3].

9.1 Advantages of 3D-MOSS

High-speed massive optical switching systems, which are required to enhance network traffic, have been developed using optical switches based on thermo-optics (TO), the microelectromechanical system (MEMS), and bubble reflection. These optical switches, however, have a switching speed limit around milliseconds. Meanwhile, waveguide-type optical switches, which utilize EO materials such as

Self-Organized 3D Integrated Optical Interconnects with All-Photolithographic Heterogeneous Integration
Tetsuzo Yoshimura

ISBN 978-981-4877-04-6 (Hardcover), 978-1-003-04305-8 (eBook)
www.jennystanford.com

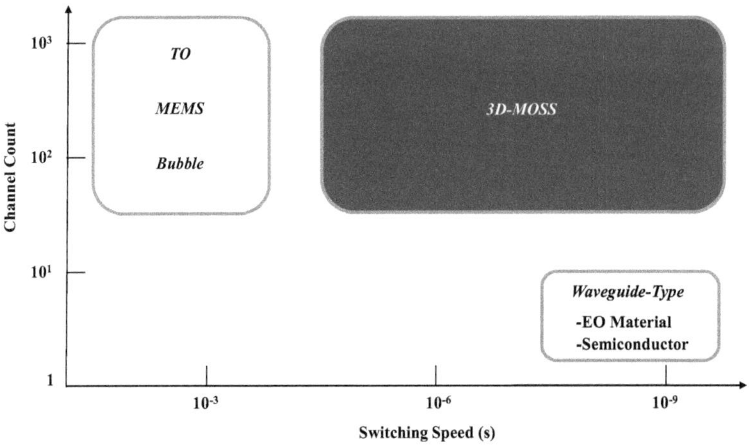

Figure 9.1 Expected performance of 3D-MOSS.

EO polymers, LN, PLZT, and III–V compounds, or semiconductors such as Si and III–V compounds, exhibit switching time of ns ~ ps order. Unfortunately, these optical switches have not accomplished large-scale integration.

Figure 9.1 shows expected performance of 3D-MOSS. In 3D-MOSS, since a large number of high-speed waveguide-type optical switches are integrated in a miniaturized 3D structure, "high-speed" switching and "massive" switching can be realized at the same time.

Figure 9.2 shows conceptual illustrations of 3D-MOSS implementation in OE systems. For optical switching in fiber communications, as depicted in Fig. 9.2a, a 3D-MOSS is mounted on an OE PCB, to which optical fibers are connected. For reconfigurable optical interconnects in supercomputers or high-performance servers in data centers, 3D-MOSSs are mounted together with 3D-stacked OE MCMs on OE PCBs, as depicted in Fig. 9.2b. Optical signals in optical interconnects are switched in 3D-MOSSs to enable dynamic routing between LSIs. In the case of multiprocessor systems, miniaturized X-BAR switch configurations are available on boards.

WDM switching systems can be constructed by using tunable wavelength filters or wavelength converters in 3D-MOSSs.

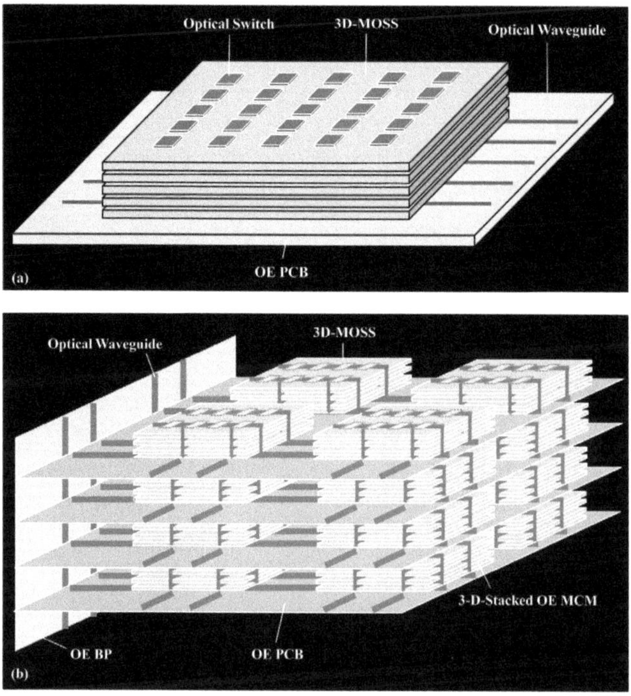

Figure 9.2 Conceptual illustrations of 3D-MOSS implementation in OE systems (a) for optical switching in fiber communications and (b) for reconfigurable optical interconnects.

9.2 Architecture of 3D-MOSS

9.2.1 3D-MOSS

Figure 9.3 shows a conventional 32 × 32 Banyan network consisting of five optical switch arrays. Each array contains sixteen 2 × 2 optical switches. Adjacent arrays are connected by optical waveguides in connection areas. In connection area 1, an optical waveguide experiences 15 waveguide crossing points, 7 in connection area 2, 3 in connection area 3, and 1 in connection area 4.

To construct a 3D-MOSS, the conventional Banyan network is divided into four sub-network blocks of BL 1, 2, 3, and 4, as shown in Fig. 9.3. Optical waveguides corresponding to inter- and intrablock

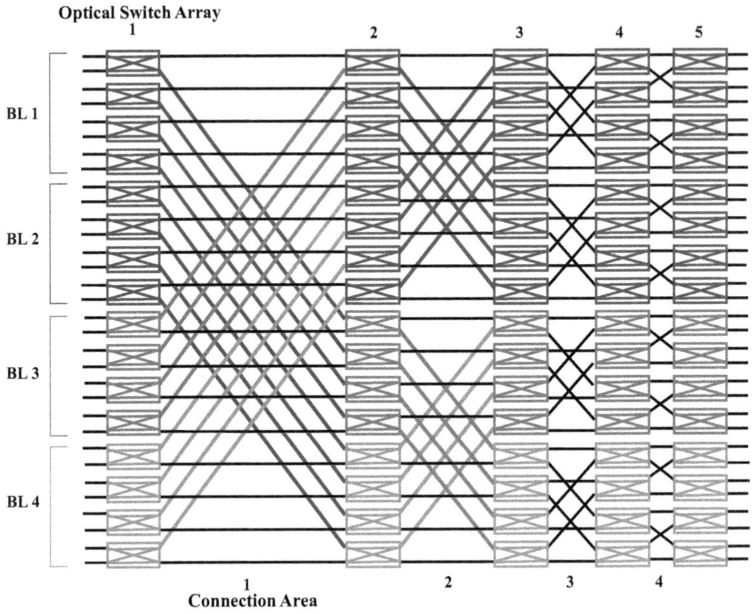

Figure 9.3 Conventional 32 × 32 Banyan network consisting of five optical switch arrays. Reprinted with permission from Yoshimura et al. [3].

connections are respectively represented by thick and thin lines. As shown in Fig. 9.4, the blocks are separated and stacked to each other to construct a 32 × 32 3D-MOSS with a four-layer structure of BL 1, 2, 3, and 4. The long optical waveguides in connection area 1 are replaced with short-distance optical Z-connections, connecting BL 1 to 3 and BL 2 to 4 and experiencing no waveguide crossing points. The long optical waveguides in connection area 2 are also replaced with optical Z-connections. Thus, the 3D-MOSS drastically reduces the waveguide crossing point count and the wiring distance.

It should be noted that the advantage of 3D-MOSS becomes remarkable when the channel count increases. For a 1024 × 1024 Banyan network, an optical waveguide in connection area 1 experiences 511 waveguide crossing points in the conventional planar configuration. The 3D-MOSS reduces the crossing point count to 0.

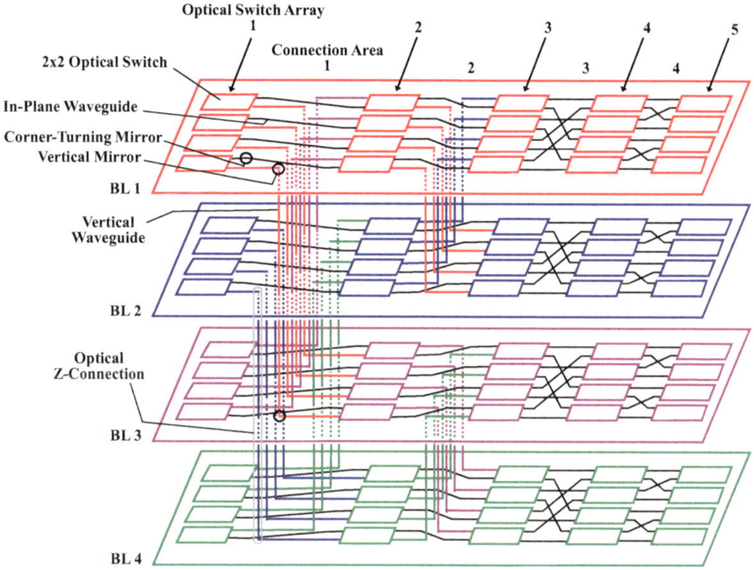

Figure 9.4 3D-MOSS for 32 × 32 Banyan network.

Connection areas consisting of in-plane waveguides and corner-turning mirrors are denoted by "in-plane connection areas," and those containing optical Z-connections by "vertical connection areas." In the present model, connection areas 3 and 4 are in-plane connection areas, and connection areas 1 and 2 are vertical connection areas.

A network configuration example is shown in Fig. 9.5 for BL 3. The 3D optical circuit backbone is assumed to consist of optical waveguides having a $4 \times 4\ \mu m^2$ core cross-section. The block size is 600 μm wide and 2250 μm long. The optical switch size and thickness are assumed to be $60 \times 390\ \mu m^2$ and 4 μm, respectively. The connection area between the optical switches is $60 \times 60\ \mu m^2$, indicating that $3600/(4 \times 4) = 225$ Z-connections can be placed in the area in the case that the cross-section of an optical Z-connection is $4 \times 4\ \mu m^2$. In the present model, there are only two or four optical Z-connections in the connection area.

Figure 9.6 shows a schematic illustration of a 3D-MOSS structure for a 32 × 32 Banyan network. On an OE PCB, OE-Films for BL

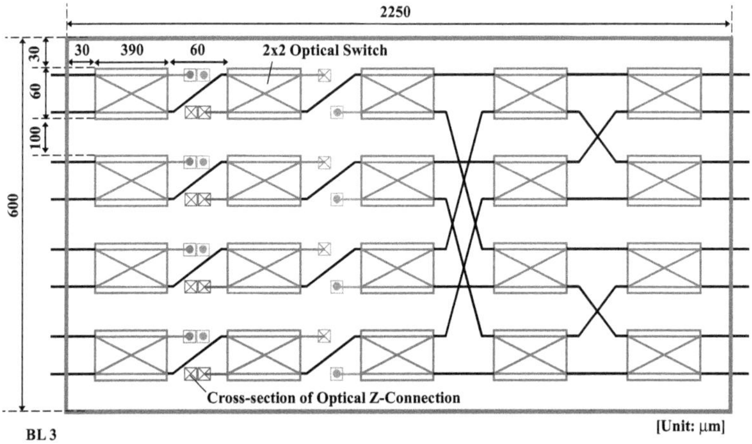

Figure 9.5 Network configuration in BL 3. Reprinted with permission from Yoshimura et al. [3].

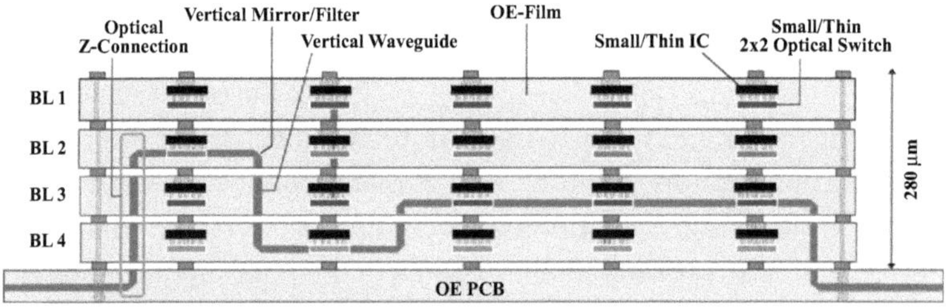

Figure 9.6 Schematic illustration of a 3D-MOSS structure for a 32 × 32 Banyan network.

1, 2, 3, and 4, in which small/thin optical switch dies and optical waveguides with vertical mirrors/filters are embedded, are stacked. In each block, an interface film, in which small/thin IC dies are embedded, is contained for driving the optical switches. The output terminals of the ICs are vertically connected to the electrodes of the optical switches by, for example, the metal diffusion technique. Connections between OE-Films and connections between OE-Films and an OE PCB are performed by optical Z-connections. The 3D-MOSS thickness is ~280 μm for the four-block structure consisting of four OE-Films and four interface films when the film thickness is 30 μm and the gap between the blocks is 10 μm.

In the model shown in Fig. 9.6, input optical signals introduced into BL 2 from the OE PCB through an optical Z-connection are switched in BL 2 to be led into BL 4. They are switched in BL 4 to be led into BL 3. The optical signals are switched in BL 3, and finally, are output into the OE PCB.

9.2.2 3D-MOSS with SOLNET Implementation

In 3D-MOSSs there are a large number of optical coupling points because the channel count is quite large. This forces us to align a lot of OE devices and layers precisely with enormous efforts, raising the system cost. Furthermore, vertical waveguide construction is necessary for interlayer connections. Implementation of SOLNETs, which enable us to construct self-organized 3D optical circuits as described in Chapter 7, is expected to achieve a drastic cost reduction in 3D-MOSSs.

Figure 9.7 shows the fabrication procedure of self-organized 3D optical circuits. First, small/thin OE dies are placed in optical waveguide films to make OE-Films. The OE-Films are stacked to construct a 3D structure. Next, by introducing excitation lights to the 3D structure from outside, coupling optical paths of SOLNETs are formed between the OE dies and the optical waveguides of the optical waveguide films, and at the same time, optical Z-connections are formed between the OE-Films to construct a self-aligned 3D optical circuit automatically. By the SOLNET implementation, even if some misalignments of the OE dies, optical waveguides and stacked films exist, self-aligned optical couplings are obtained. This reduces the fabrication cost, especially the cost for the OE device assembly and the optical Z-connection formation.

An example of the procedure for self-organization of 3D-MOSSs is shown in Fig. 9.8. Firstly, reflective targets of wavelength filters, which reflect write beams and pass signal beams, are deposited on core surfaces at the optical switch edges. Instead of wavelength filters, luminescent targets can also be used. These targets are deposited using ALD or MLD with processes mentioned in

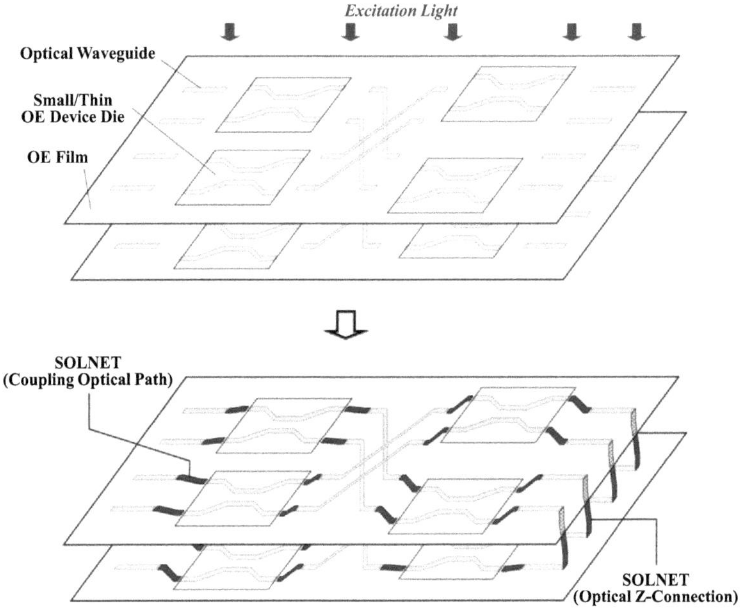

Figure 9.7 Fabrication procedure of self-organized 3D optical circuits. Reprinted with permission from Yoshimura et al. [1].

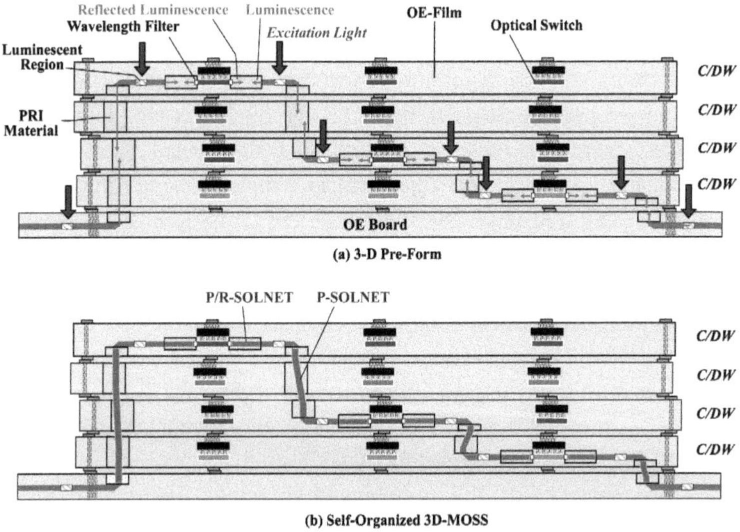

Figure 9.8 Procedure for self-organization of 3D-MOSSs. Reprinted with permission from Yoshimura et al. [1].

section 7.1.4. Next, OE-Films are fabricated by embedding optical switches into optical waveguide films using PL-Pack with SORT, described in Chapter 5. In optical waveguides, phosphor is doped to make luminescent regions for write beam generation. If the optical waveguides themselves have emissive characteristics, phosphor doping is not necessary. Then, as shown in Fig. 9.8a, a 3D pre-form is prepared by stacking OE-Films and inserting PRI materials in parts, where SOLNETs will be formed.

By illuminating the phosphor-doped regions with excitation light from outside, luminescence for write beams is generated from luminescent regions, and propagates in optical waveguides. As shown in Fig. 9.8b, for connections between optical switches and optical waveguides, write beams are introduced into PRI materials from optical waveguides, and are reflected by wavelength filters on optical switch edges to form P/R-SOLNETs. For the vertical waveguide construction in optical Z-connections, write beams are introduced into PRI materials from upper and lower optical waveguides through vertical mirrors to form P-SOLNETs. Thus, the self-organized 3D-MOSS is built.

9.3 Predicted Performance of 1024 × 1024 3D-MOSS

In the present section, to investigate a solution for the miniaturization of high-speed massive optical switching systems, performance of the 3D-MOSS with a 1024 × 1024 Banyan network architecture is evaluated theoretically. Basic concepts obtained in the Banyan network might be applicable to other network configurations.

9.3.1 Structural Model

Table 9.1 summarizes structural parameters for calculations in a 1024 × 1024 3D-MOSS model. The channel count N_c of the 3D-MOSS is 1024, and the layer count N_{L} is 1, 2^2, 2^4, 2^6, and 2^9. The 3D optical circuit backbone consists of optical waveguides distributed with a pitch P_{WG} of 20 μm. The core width w is 4 μm, core refractive index n_{Core} is 1.52, and cladding refractive index n_{Clad} is 1.50. In this

Table 9.1 Structural parameters in 1024 × 1024 3D-MOSS model

Channel count: N_C		1024
Layer count: N_L		1, 2^2, 2^4, 2^6, 2^9
OE-Film thickness/gap: T_F/T_G		46 μm/4 μm
Optical waveguide		
Width: w		4 μm
Refractive index	Core: n_{Core}	1.52
	Cladding: n_{Clad}	1.50
Pitch: P_{WG}		20 μm
2 × 2 WPD optical switch		
Length: L_{SW}, Width: W_{SW}		1190 μm, 100 μm
Pitch in a array: P_{SW}		200 μm
Step count: S_{SW}		10
Probe beam Wavelength		1.3 μm

case, two modes, that is, the ground mode of $m = 0$ and the first excited mode of $m = 1$, exist theoretically for a 1.3 μm probe beam, which is used for calculations of coupling efficiency (or insertion loss). It is known, however, that optical waveguides having a few modes actually behave as single-mode waveguides due to weak confinement for higher modes. So, the 3D optical circuit backbone in the 3D-MOSS is regarded as a single-mode network in practice. In the present work, calculations of coupling efficiency or insertion loss are carried out for ground-mode probe beams. OE-Film thickness T_F is 46 μm and the gap between the OE-Films T_G is 4 μm. So, the thickness per layer, $T_F + T_G$, is 50 μm.

For optical switches, 2 × 2 WPDs with a structure similar to that shown in Fig. 6.12 are assumed. They are located with a pitch P_{SW} of 200 μm in optical switch arrays. The optical switch array count, that is, step count S_{SW} is 10. One array contains $N_C/2$ optical switches with a length L_{SW} of 1190 μm and width W_{SW} of 100 μm. The length of the array equals individual optical switch length L_{SW}. Although in the model shown in Fig. 6.12 400 nm thick PLZT is used for the EO material, in the present model, 4 μm thick PLZT is assumed. In this case, switching voltage becomes 12 V to induce refractive index change equivalent to the change induced in the 400 nm thick PLZT in the model shown in Fig. 6.12.

Figure 9.9 shows top-view illustrations of 1024 × 1024 Banyan network models for a conventional planar structure and a 3D-MOSS

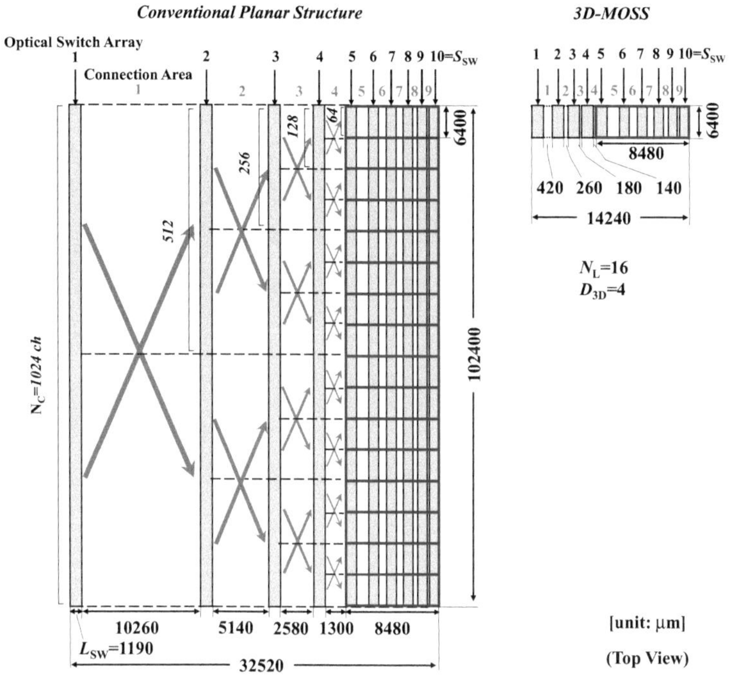

Figure 9.9 Top-view illustrations of 1024 × 1024 Banyan network models for a conventional planar structure and a 3D-MOSS structure. Reprinted with permission from Yoshimura et al. [1].

structure. Optical switch arrays are connected by optical waveguides in connection areas. In the 3D-MOSS, the network is divided into N_L blocks, which are stacked to make a N_L-layer 3D structure. The area of the network is drastically reduced in the 3D-MOSS with $N_L = 16$, compared with the conventional planar structure. The dimensions inserted in Fig. 9.9, as well as in Figs. 9.10–9.12, are calculated using the general expressions 9.2–9.6 described later.

In the N_L-layer 3D structure, the degree of 3D characteristics of the networks is expressed as follows:

$$D_{3D} = \log_2 N_L \tag{9.1}$$

D_{3D} indicates the number of vertical connection areas in a 3D-MOSS. This implies that connection areas 1 to D_{3D} are vertical connection areas, and connection areas $D_{3D} + 1$ to $S_{SW} - 1$ are

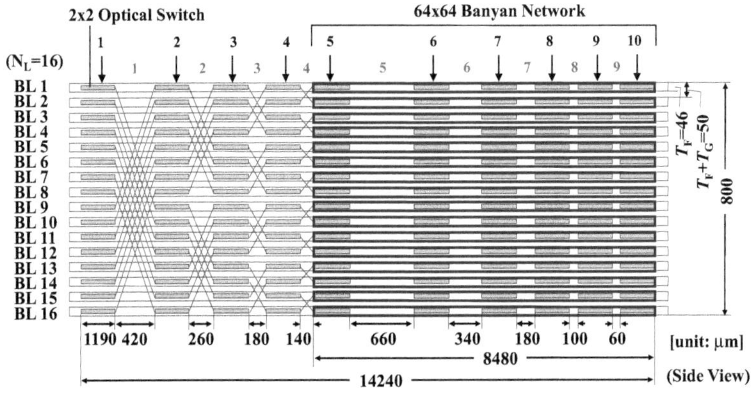

Figure 9.10 Side-view illustration of 1024 × 1024 3D-MOSS with N_L of 16. Reprinted with permission from Yoshimura et al. [1].

in-plane connection areas. In the case of $N_L = 16$, namely, $D_{3D} = 4$, connection areas 1–4 and 5–9 are respectively vertical and in-plane connection areas.

Figure 9.10 shows a side-view illustration of the 1024 × 1024 3D-MOSS with N_L of 16. Both vertical and in-plane optical connections exist in connection areas 1–4, and only in-plane optical connections in connection areas 5–9. The region that includes connection areas 5–9 corresponds to a 64 × 64 Banyan network with a conventional planar structure shown in Fig. 9.11.

As an example of an in-plane connection area, a detail of connection area 5 is shown in Fig. 9.12a. Connections between optical switch arrays are constructed by crank-shaped waveguides consisting of a pair of corner-turning mirrors. The interswitch connection line count is expressed by $N_C/2^5$ for connection area 5. Then, by adding P_{WG} for buffer spaces, the length of connection area 5 is given by $(N_C/2^5 + 1)P_{WG}$. The width is given by $((N_C/2)/2^{D_{3D}})P_{SW}$. Thus, the expressions for the connection area length $L(\mathrm{IP})_i$ and width $W(\mathrm{IP})$ of in-plane connection area i can be generalized as follows.

$$L(\mathrm{IP})_i = (N_C/2^i + 1)P_{WG} \tag{9.2}$$

$$W(\mathrm{IP}) = \left((N_C/2)/2^{D_{3D}}\right) P_{SW} \tag{9.3}$$

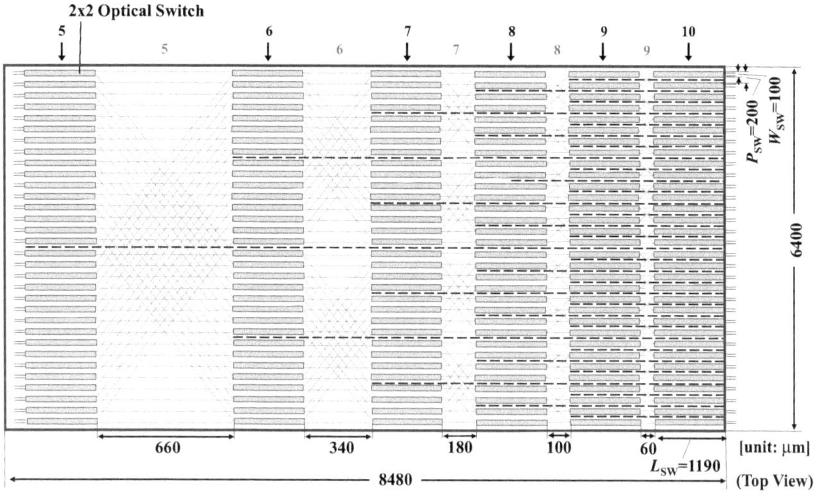

Figure 9.11 Top-view illustration of 64 × 64 Banyan network with in-pane optical connections. Reprinted with permission from Yoshimura et al. [1].

In the present model, $P_{SW} = 200\,\mu m$ and $P_{WG} = 20\,\mu m$. By substituting these parameters into Eqs. 9.2 and 9.3, the size of connection area 5 can be calculated to be 660 μm long and 6400 μm wide. In similar ways, the dimensions for $i = 6{\sim}9$ can be calculated as shown in Figs. 9.10 and 9.11. P_{WG} of 20 μm is wide enough for parallel wiring with a length of 1 cm to keep cross-talk less than −30 dB for the assumed 4 μm wide waveguides and the 1.3 μm probe beam wavelength.

As an example of a vertical connection area, a detail side view of connection area 1 and a top view of a part of the connection area in BL 12 are shown in Fig. 9.12b. Interswitch connection line count in a cross section is expressed as $N_L/2^{1-1}$ for connection area 1. Then, adding buffer spaces of $3P_{WG}$ for vertical mirrors at front and tail ends of the connection area, the length of connection area 1 is calculated to be $(N_L/2^{1-1} + 5)P_{WG}$. The width is the same as in the in-plane optical connection case. The height is $N_L(T_F + T_G)$. Thus, the expressions for connection area length $L(V)_i$, width $W(V)$, and height $H(V)$ of vertical connection area i can be generalized as follows:

$$L(V)_i = (N_L/2^{i-1} + 5)P_{WG} \tag{9.4}$$

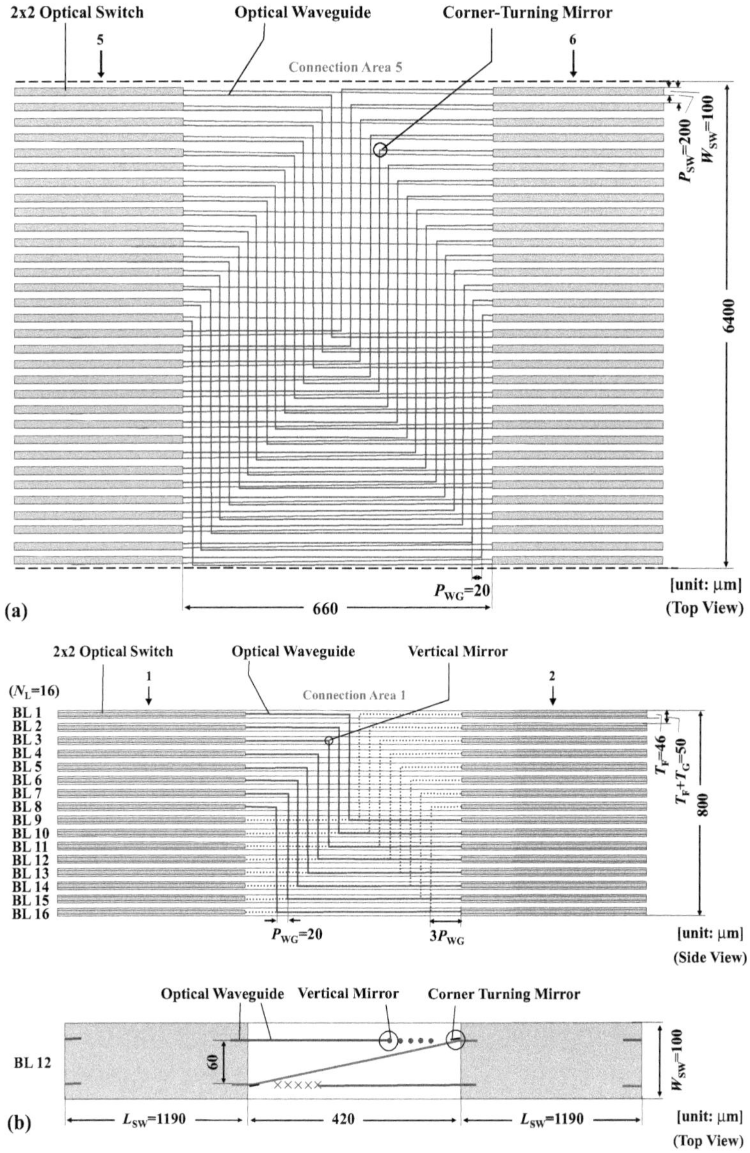

Figure 9.12 Details of assumed connection patterns. (a) Top view of in-plane optical connection of connection area 5, and (b) side view of vertical optical connection of connection area 1 and top view of a part of BL 12. Reprinted with permission from Yoshimura et al. [1].

$$W(\mathrm{V}) = \left((N_\mathrm{C}/2)/2^{D_\mathrm{3D}}\right) P_\mathrm{SW} \quad (9.5)$$

$$H(\mathrm{V}) = N_\mathrm{L}(T_\mathrm{F} + T_\mathrm{G}) \quad (9.6)$$

Parameter substitution into Eqs. 9.4–9.6 gives the size of connection area 1: 420 μm long, 6400 μm wide, and 800 μm high. In similar ways, the dimensions for $i = 2$–4 can be calculated as shown in Figs. 9.9 and 9.10.

Then, total 3D-MOSS length $L_\text{3D-MOSS}$, width $W_\text{3D-MOSS}$, and height $H_\text{3D-MOSS}$, including switch array length, are given by

$$L_\text{3D-MOSS} = L_\mathrm{SW} S_\mathrm{SW} + \left(\sum_{i=1}^{D_\mathrm{3D}} (N_\mathrm{L}/2^{i-1} + 5)\right) P_\mathrm{WG} + \left(\sum_{i=D_\mathrm{3D}+1}^{S_\mathrm{SW}-1} (N_\mathrm{C}/2^{i} + 1)\right) P_\mathrm{WG}, \quad (9.7)$$

$$W_\text{3D-MOSS} = \left((N_\mathrm{C}/2)/2^{D_\mathrm{3D}}\right) P_\mathrm{SW}, \quad (9.8)$$

$$H_\text{3D-MOSS} = N_\mathrm{L}(T_\mathrm{F} + T_\mathrm{G}). \quad (9.9)$$

The system size of 1024 × 1024 3D-MOSS can be obtained using Eqs. 9.7–9.9 as follows. For $D_\mathrm{3D} = 4$ ($N_\mathrm{L} = 16$), the length, width, and height are respectively 1.42 cm, 0.64 cm, and 0.08 cm. In the case of the conventional planar structure, which corresponds to the case of $D_\mathrm{3D} = 0$, the length, width, and height are respectively 3.25 cm, 10.2 cm, and 0.005 cm. These results indicate that 3D-MOSS is expected to realize a drastic reduction in occupation areas for optical switching systems.

9.3.2 Insertion Loss

In Fig. 9.13, major losses caused in 3D-MOSS are indicated by γ. $\gamma_\mathrm{P(IP)}$ is propagation loss of in-plane waveguides, $\gamma_\mathrm{P(V)}$ propagation loss of vertical waveguides, γ_CR loss induced at a right-angle waveguide–waveguide cross point, γ_CT loss of a pair of corner-turning mirrors in a crank-shaped waveguide, γ_Z loss of a pair of vertical mirrors in an optical Z-connection, γ_G loss at a gap between OE-Films, γ_SW loss within a 2 × 2 optical switch, and $\gamma_\text{W-S}$ waveguide–switch coupling loss.

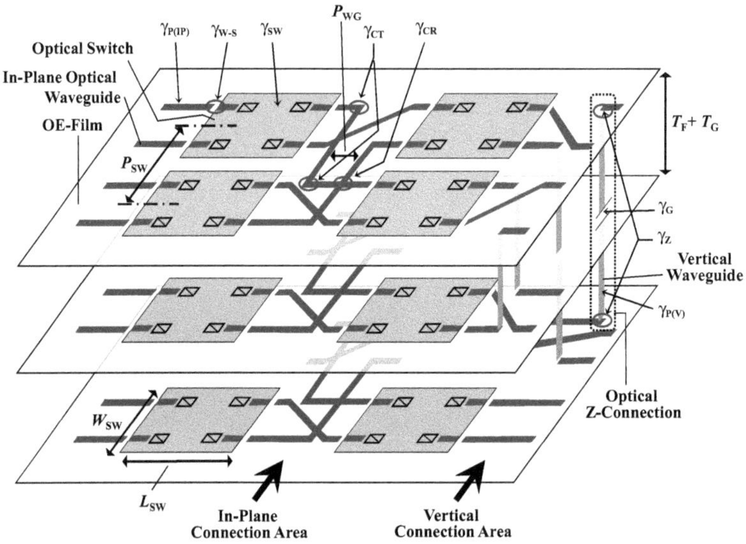

Figure 9.13 Schematic illustration of a part of 3D-MOSS. Major losses caused in 3D-MOSS are indicated by "γ." Reprinted with permission from Yoshimura et al. [2].

Estimated major losses in the present 3D-MOSS model are listed in Table 9.2. γ_P is assumed to be 0.5 dB/cm, which is typical in fluorinated polyimide. γ_{CR} was estimated by the BPM simulation to be 0.0088 dB. γ_Z of 0.97 dB was estimated by BPM/FDTD coupled simulation. Since a pair of corner-turning mirrors in a crank-shaped waveguide has the same structure as a pair of vertical mirrors in an optical Z-connection, γ_{CT} can be γ_Z of 0.97 dB. An example of the BPM/FDTD simulation for the optical Z-connection is shown in Fig. 9.14. A cross point of a vertical waveguide and a 4 μm gap between OE-Films is regarded as a kind of a right-angle waveguide–waveguide cross point. So, γ_G can be γ_{CR} of 0.0088 dB. The BPM simulation revealed that γ_G increases to 0.89 dB when a 25% misalignment exists between the OE-Films. Here, "25% misalignment" means that the misalignment is 25% of the waveguide core width. For a core width of 4 μm the 25% misalignment is 1 μm. Loss for a TB-SOLNET/P-SOLNET coupling with 25% misalignment was calculated to be 0.14 dB by the BPM

Table 9.2 Major losses in 3D-MOSS. Reprinted with permission from Yoshimura et al. [1]

Propagation in optical waveguide: γ_P		0.5 dB/cm
Waveguide–waveguide cross point: γ_{CR}		0.0088 dB
Pair of corner turning mirrors in crank-shaped waveguide: γ_{CT}		0.97 dB
Pair of vertical mirrors in optical Z-connection: γ_Z		0.97 dB
Gap between OE-Films: γ_G	with no misalignment	0.0088 dB
	with 25% misalignment	0.89 dB
	with 25% misalignment using SOLNET	0.14 dB
2 × 2 optical switch: γ_{SW}		1.7 dB
Waveguide–switch couplings: $\gamma_{W\text{-}S}$	with no misalignment	0 dB
	with 25% misalignment	0.83 dB
	with 25% misalignment using SOLNET	0.14 dB

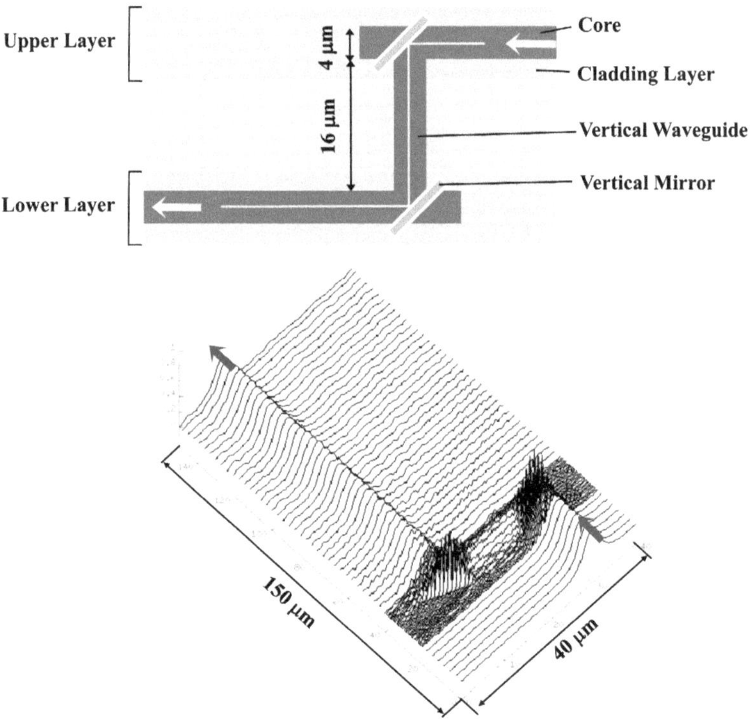

Figure 9.14 An example of the BPM/FDTD simulation for the optical Z-connection.

simulation. Then, γ_G with 25% misalignment can decrease to 0.14 dB by using the SOLNET. Here, the SOLNET length dependence of γ_G is neglected. γ_{SW} was estimated to be 1.7 dB by BPM simulation. $\gamma_{W\text{-}S}$ is basically 0 dB for no misalignments. The BPM calculation revealed that $\gamma_{W\text{-}S}$ is 0.83 dB for a 25% misalignment. When SOLNETs are applied to the couplings, as in the case of γ_G, $\gamma_{W\text{-}S}$ is 0.14 dB for a 25% misalignment. In the simulations described above, boundary reflection losses were neglected.

General expressions for the eight kinds of total losses in 3D-MOSSs are derived in terms of the losses listed in Table 9.2. The results are summarized in Table 9.3. In-plane optical connections include in-plane propagation, waveguide–waveguide cross points, pairs of corner-turning mirrors, optical switches, and waveguide–switch couplings. Vertical optical connections include vertical propagation, pairs of vertical mirrors, and gaps between OE-Films.

For in-plane propagation, the first and the second terms are caused from horizontal line length given by the second and the third terms of Eq. 9.7. The third term is caused from line length perpendicular to the horizontal direction. Since lines in vertical connection areas have no contribution to the line length perpendicular to the horizontal direction basically, the summation is carried out only for in-plane connection areas. For waveguide–waveguide cross points, an optical waveguide has a cross-point count of $N_C/2^i$ in in-plane connection area i, and no waveguide–waveguide cross points exist in vertical connection areas. Thus, summation for the total loss calculation is carried out for the in-plane connection areas. For pairs of corner-turning mirrors, the loss is determined by the number of pairs of corner turnings that equals in-plane connection area count since optical signals experience one pair of corner turnings in one in-plane connection area. Total losses within optical switches and at waveguide–switch coupling points are respectively induced by S_{SW} optical switches and $2S_{SW}$ waveguide–switch coupling points.

For vertical propagation, as can be seen from Fig. 9.10 and Fig. 9.12b, the total loss is determined by optical path length that corresponds to two times of the 3D-MOSS thickness. Since optical signals experience one optical Z-connection in one vertical

Table 9.3 General expressions for losses in 3D-MOSS. Reprinted with permission from Yoshimura et al. [1]

In-plane optical connections	
In-plane propagation: $\Gamma_{P(IP)}$	
$\left[\left(\sum_{i=1}^{D_{3D}} (N_L/2^{i-1}+5)\right) P_{WG} + \left(\sum_{i=D_{3D}+1}^{S_{SW}-1} (N_C/2^i+1)\right) P_{WG} + \left(\sum_{i=D_{3D}+1}^{S_{SW}-1} (N_C/2)/2^i\right) P_{SW}\right] \gamma_P$	
Waveguide–waveguide cross point: Γ_{CR}	Pair of corner turning mirrors: Γ_{CT}
$\left[\sum_{i=D_{3D}+1}^{S_{SW}-1} N_C/2^i\right] \gamma_{CR}$	$[S_{SW}-1-D_{3D}]\ \gamma_{CT}$
Optical switch: Γ_{SW}	Waveguide–switch coupling: Γ_{W-S}
$[S_{SW}]\ \gamma_{SW}$	$[2S_{SW}]\ \gamma_{W-S}$
Vertical optical connections	
Vertical propagation: $\Gamma_{P(V)}$	
$[2N_L(T_F+T_G)]\ \gamma_P$	
Pair of vertical mirrors: Γ_Z	Gap between OE-Films: Γ_G
$[2+D_{3D}]\ \gamma_Z$	$[2N_L]\ \gamma_G$ $\quad [2+D_{3D}]\ \gamma_G$ for SOLNET

connection area as shown in Fig. 9.12b, total loss caused by optical Z-connections, by adding two optical Z-connections for input and output, is calculated to be $[2 + D_{3D}]\gamma_Z$. Total loss due to gaps between OE-Films is determined by two times of the layer gap count contained in a 3D-MOSS. When the SOLNET is implemented in the vertical optical connections, total loss due to the gaps is replaced with a product of total Z-connection count times γ_G for the SOLNET. The expression for the total loss of the gaps between OE-Films for the SOLNET is obtained by assuming that each optical Z-connection is constructed by one SOLNET as depicted in Fig. 9.8. The factor 2 in the expression is for two optical Z-connections between the 3D-MOSS and the OE board.

By using the parameters in Table 9.1, the length, width, and height of 3D-MOSSs can be respectively calculated from Eqs. 9.7, 9.8, and 9.9. Insertion losses of 3D-MOSSs can be obtained by summation of eight losses listed in Table 9.3.

In Fig. 9.15, the occupation area, which is a product of the length and width of 3D-MOSSs, and the insertion loss of 3D-MOSSs are plotted as a function of D_{3D}. $D_{3D} = 0(N_L = 1)$ implies a conventional planar structure, and $D_{3D} = 9(N_L = 2^9)$ implies that all the connection areas are vertical optical connections.

The occupation area decreases monotonically with D_{3D}. The insertion loss, on the other hand, has a peak in the region of $D_{3D} = 4 \sim 6$ when no misalignments exist in the 3D-MOSS. Insertion loss at $D_{3D} = 0$ is 43 dB. With an increase in D_{3D}, the insertion loss decreases to 29 dB at $D_{3D} = 4$, then increases, reaching 40 dB at $D_{3D} = 9$. This result indicates that a certain optimum layer count exists for minimizing the insertion loss of 3D-MOSSs. The dashed curve represents the insertion loss for a 25% misalignment case, where the misalignments exist between adjacent layers and between in-plane waveguides and optical switch waveguides (note that the dashed curve is drawn with 1/24 reduction). The misalignments cause a monotonic increase in insertion loss with increasing D_{3D}. At $D_{3D} = 4$, the loss is 73 dB. By implementing SOLNETs into 3D-MOSSs with 25% misalignments, the insertion loss decreases, reaching a level comparable to that of 3D-MOSSs with no misalignments. The insertion loss at $D_{3D} = 4$ is 32 dB.

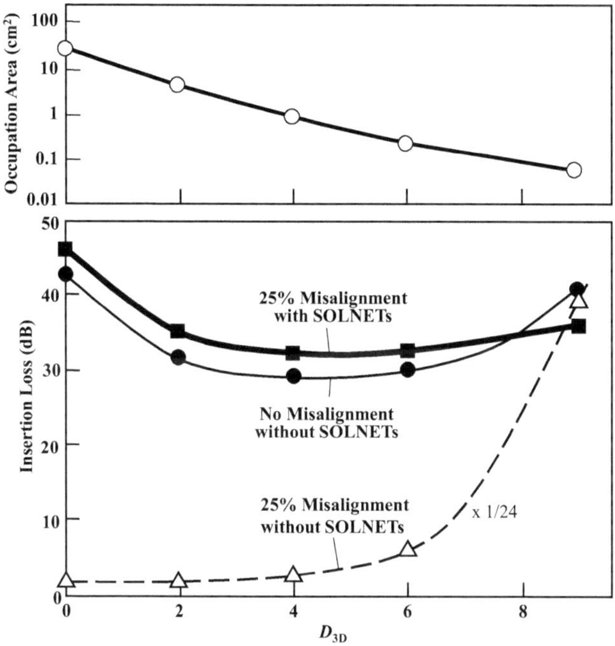

Figure 9.15 Calculated occupation area and the insertion loss of 3D-MOSSs plotted as a function of D_{3D}. Reprinted with permission from Yoshimura et al. [2].

Main losses in the 1024 × 1024 3D-MOSSs are separately shown in Fig. 9.16. For the case of no misalignments, dominant main losses are Γ_{SW} and Γ_{CT} in in-plane optical connections, and Γ_Z in vertical optical connections. With an increase in D_{3D}, Γ_{CT} decreases while Γ_Z increases, giving the minimum of the insertion loss at $D_{3D} = 4$ as shown in Fig. 9.15. When there are 25% misalignments, Γ_G, which increases with D_{3D}, becomes dominant, causing the large insertion loss shown in Fig. 9.15. By implementing SOLNETs, Γ_G is drastically reduced to reach loss levels of 3D-MOSSs with no misalignments, which contributes to drastic loss reduction in the SOLNET-implemented 3D-MOSSs shown in Fig. 9.15. Consequently, Γ_{SW}, Γ_{CT}, and Γ_Z appear as dominant main losses again in the SOLNET-implemented 3D-MOSSs. Thus, as can be seen from Figs. 9.16a and c, major losses at $D_{3D} = 4$ are Γ_{SW}, Γ_{CT},

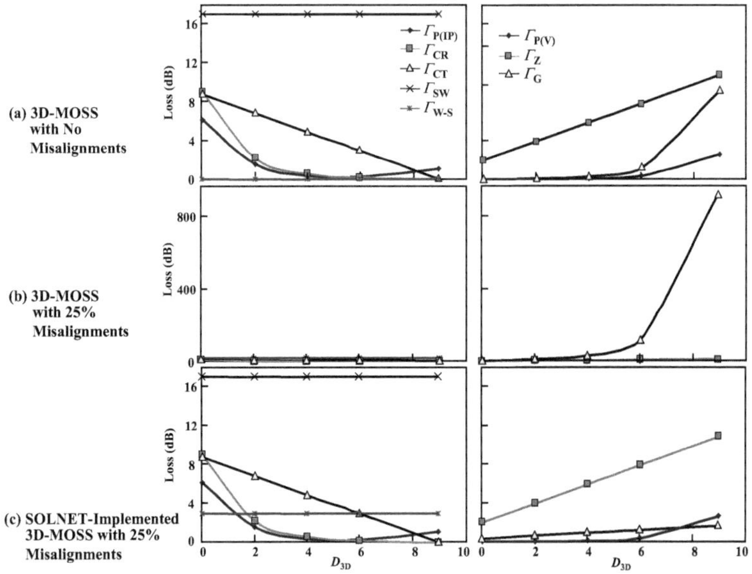

Figure 9.16 Calculated main losses in the 1024 × 1024 3D-MOSSs. Reprinted with permission from Yoshimura et al. [2].

and Γ_Z. These results suggest that main future projects to reduce the insertion loss of 3D-MOSSs should be concentrated into loss reduction of optical switches, corner-turning mirrors, and vertical mirrors.

It was reported that high-accuracy bonding equipment enables layer stacking with a positional accuracy of ~1 μm for 1 cm^2 chips [4]. Therefore, for 3D-MOSSs of 1.424 cm × 0.64 cm, it is consistent to assume positional misalignments of ~1 μm that corresponds to 25% misalignments for optical circuit backbones consisting of 4 μm wide optical waveguides.

The insertion loss of 32 dB for 1024 × 1024 SOLNET-implemented 3D-MOSSs is still high. To compensate for the loss, optical amplifiers embedded in the 3D-MOSSs might be necessary, which makes the systems complicated. For simplification of the systems, each loss listed in Table 9.2 should be decreased by optimizing individual structures and materials.

9.3.3 Electrical Characteristics

By regarding a WPD as a prism-shaped capacitor, switching speed and power consumption of the WPD optical switch were estimated based on the charge-discharge model. The WPD optical switch model used in the present calculation is the same as the model shown in Fig. 6.12, except that the PLZT thickness (namely, the electrode gap distance) is increased from 400 nm to 4 μm and the operation voltage is increased from 1.2 V to 12 V. The capacitance C, switching speed τ, and power consumption Power that are given by Eqs. 6.7, 6.8, and 6.10 for one prism-shaped capacitor are respectively calculated to be 0.57 pF, 57 ps for driver on-resistivity $R_{ON} = 100\ \Omega$, and 41 mW for a data rate F of 5×10^8 1/s.

When the number of prism-shaped capacitors in a WPD optical switch is N_{Prism}, τ for a WPD optical switch is given by

$$\tau = N_{Prism}\ R_{ON} C. \tag{9.10}$$

Because the total WPD optical switch count in a 3D-MOSS is $(N_c/2)S_{SW}$, the total count of the prism-shaped capacitors contained in a 3D-MOSS is given by $N_{Prism}(N_c/2)S_{SW}$. Then, total capacitance arising from the prism-shaped capacitors is given by $C\,N_{Prism}(N_c/2)S_{SW}$. As a result, Power in a 3D-MOSS is expressed as follows:

$$\text{Power} = F\ C V^2\ N_{Prism}\ (N_c/2)S_{SW} \tag{9.11}$$

By substituting $C = 0.57$ pF, $V = 12$ V, $N_{Prism} = 8$, $N_c = 1024$, and $S_{SW} = 10$ into Eqs. 9.10 and 9.11, $\tau = 454$ ps for $R_{ON} = 100\ \Omega$, and Power = 33 W for $F = 1 \times 10^7$ 1/s are obtained for the 1024 × 1024 3D-MOSS. The power consumption corresponds to power consumption density of 36 W/cm^2 since the area of the 1024 × 1024 3D-MOSS model is 1.42 × 0.64 cm^2. This value is comparable to acceptable heat generation of 30 W/cm^2 reported by K. Takahashi [4] for stacked LSI of 1 cm^2 area.

Expected performance of the 1024 × 1024 3D-MOSS, which uses optical switches of WPDs with a 4 μm thick PLZT and operation voltage of 12 V, is summarized in Table 9.4 together with the system size.

Figure 9.17 shows the switching rate dependence of the power consumption density in 1024 × 1024 3D-MOSSs, where WPD optical

Table 9.4 Expected performance of 1024 × 1024 3D-MOSS (optical switch: WPD with 4 μm thick PLZT, operation voltage: 12 V)

Size		Length	Width	Height
Conventional planar structure		3.25 cm	10.2 cm	0.005 cm
3D-MOSS ($D_{3D} = 4$)		1.42 cm	0.64 cm	0.08 cm
Insertion loss				
3D-MOSS ($D_{3D} = 4$)	without misalignment	29 dB		
	with 25% misalignment	73 dB		
	with 25% misalignment using SOLNET	32 dB		
Conventional planar structure		43 dB		
Electrical characteristics				
Operation voltage: V		12 V		
Switching speed: τ		454 ps (driver on-resistivity: 100 Ω)		
Power consumption density		36 W/cm^2 (switching rate: 1×10^7/s)		

switches with a 4 μm thick PLZT, 400 nm thick PLZT, or 400 nm thick Polymer MQD are used. The power consumption density increases with the switching rate. Possible switching rate is determined by a balance of heat generation and releasing rates in 3D-MOSSs. For a given heat releasing rate, the lower the heat generation rate becomes, the higher the available switching rate becomes as far as the switching speed of the WPD optical switch is sufficiently fast.

The power consumption density versus switching rate relationship varies depending on EO materials. In a 4 μm thick PLZT, the switching rate is limited below 1×10^7 1/s for a case that the limit of the heat releasing capability is 30 W/cm^2. When the PLZT thickness is decreased to 400 nm, the operation voltage decreases to 1.2 V (see Table 6.2) and the allowable switching rate increases to 1×10^8 1/s. Drastic switching rate improvement is expected in 400 nm thick polymer MQDs (operation voltage: 4 V, see Table 6.2), which might have a potentiality of an EO coefficient of ~1000 pm/V, a dielectric constant of $\sim 4\varepsilon_0$ and a refractive

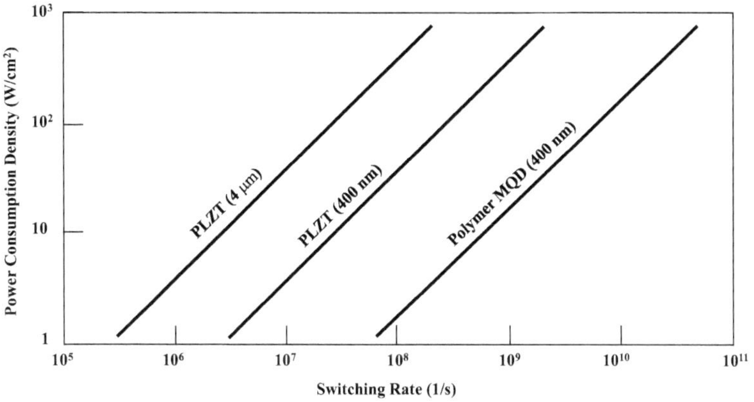

Figure 9.17 Calculated dependence of the power consumption density on the switching rate in 1024 × 1024 3D-MOSSs, where WPD optical switches with 4 μm thick PLZT, 400 nm thick PLZT, or 400 nm thick Polymer MQD are used.

index of ~2, resulting in a possibility of switching rate close to 10^{10} 1/s.

These results of performance simulation suggest that the 1024 × 1024 3D-MOSS is viable for switching systems in integrated optical interconnects within computers as well as for massive optical packet switching systems from viewpoints of the size, insertion loss, switching speed, and thermal management.

9.3.4 Impact of HIC Waveguide Implementation into 3D-MOSS

When HIC waveguides having a 400 × 400 nm^2 core cross section are used for 3D optical circuit backbones in the 3D-MOSS, the PLZT thickness, namely, the electrode gap in WPD optical switches can be reduced from 4 μm to 400 nm, and consequently, the switching voltage can be reduced from 12 V to 1.2 V. Thus, HIC optical circuit backbones are expected to achieve low-voltage switching, which is favorable for high-speed and low-power operation, as described in section 9.3.3.

References

1. T. Yoshimura, M. Ojima, Y. Arai, and K. Asama, "Three-dimensional self-organized micro optoelectronic systems for board-level reconfigurable optical interconnects–performance modeling and simulation–," *IEEE J. Sel. Top. Quantum Electron.* **9**, 492–511 (2003).
2. T. Yoshimura, Y. Arai, H. Kurokawa, and K. Asama, "Predicted insertion loss reductions achieved by implementing three-dimensional micro optical network in chip-scale optical interconnects," *IEEE Photonics Technol. Lett.* **16**, 647–649 (2004).
3. T. Yoshimura, S. Tsukada, S. Kawakami, M. Ninomiya, Y. Arai, H. Kurokawa, and K. Asama, "Three-dimensional micro-optical switching system architecture using slab-waveguide-based micro-optical switches," *Opt. Eng.* **42**, 439–446 (2003).
4. K. Takahashi, "Research and development on ultra-high-density 3dimensional LSI-chip-stack packaging technologies," *The 3rd Annual Meeting on Electronics System Integration Technologies Digest* (edited and published by the Electronic System Integration Technology Research Department, Association of Super-Advanced Electronics Technologies (ASET)), 43–94 (2002).

Chapter 10

Film-Based Integrated Solar Energy Conversion Systems

In the present chapter, integrated solar films, which are film-based integrated solar energy conversion systems involving waveguide-type thin-film solar cells and artificial photosynthesis cells, are proposed [1–12]. They are one of the applications of the S-FOLM, and their structures and fabrication processes are similar to those of the 3D integrated optical interconnects.

The advantages of the integrated solar films are summarized in Table 10.1 [10]. Size and weight reduction due to their film structures, and cost reduction due to the reduced semiconductor material and indium tin oxide (ITO) consumption are possible. By using light beam collecting films, wide-angle light beam collection is possible. Integration of optical waveguides and small/thin OE dies enables us to demultiplex collected light beams and deliver them to appropriate semiconductor dies with matched spectral responses, improving the conversion efficiency. Integration of batteries and electronic dies adds charge storage and current regulation functions to solar cells. Flexibility is also their important feature. These advantages produce novel applications as listed in Table 10.1.

Self-Organized 3D Integrated Optical Interconnects with All-Photolithographic Heterogeneous Integration
Tetsuzo Yoshimura

ISBN 978-981-4877-04-6 (Hardcover), 978-1-003-04305-8 (eBook)
www.jennystanford.com

Table 10.1 Advantages and applications of the integrated solar films

Advantages

• Size, weight, and cost reduction
• Wide-angle light beam collection capability
• Integration of optical waveguides, small/thin OE and electronic dies, batteries, etc.
  ○ Demultiplexing to deliver appropriate wavelength light beams to semiconductors with matched spectral responses
  ○ Charge storage, current regulation
• Flexibility

Applications
• Curtains, sunshade/umbrella, sheet, buildings, walls, fences, roads, cars, robots, laptop computers, smart phones, hand-carrying equipment, cloths, etc.

Some examples of novel applications are illustrated in Fig. 10.1. Integrated solar films will enable curtains, sunshades/umbrellas, and sheets with electric power generation capabilities [10]. They will also be put on buildings, walls, fences, roads, cars, robots, etc. The solar cells have an effect to suppress heat generation arising from solar beam exposure. Therefore, they act as the "green curtains of plants" to cool down the temperature of surrounding areas.

In addition, film-based integrated systems will provide ubiquitous functions depicted in Fig. 10.2, including wearable photovoltaics, wearable computers, system glasses with thin-film displays consisting of organic light-emitting diodes (OLEDs) and thin-film transistors (TFTs), thin-film speakers, and so on [9, 10, 12].

10.1 Integrated Solar Films

In conventional solar cells, semiconductors such as Si are placed all over the solar energy conversion modules, so it is foreseen that the cost of the systems will rise because of shortage of semiconductor materials. In systems with Fresnel lens/mirror concentrators, semiconductor materials can be saved because solar

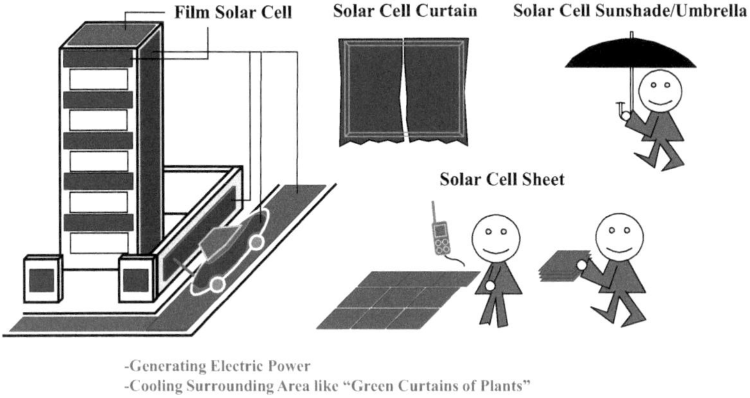

Figure 10.1 Expected applications of the integrated solar films.

beams are focused on small semiconductor cells. However, the system size becomes large, and furthermore, they need solar-beam tracking mechanisms to keep the focusing point of the solar beams on the semiconductor cells.

To reduce semiconductor consumption and to provide wide-angle light beam collecting capability to solar cells and, at the same time, to make systems flexible and compact, we proposed integrated solar films with tapered vertical waveguides [1, 3, 7–12] shown in Fig. 10.3. In the distributed-conversion-type structure shown in Fig. 10.3a, thin-film solar cells made from small/thin semiconductor dies are embedded in the film by the heterogeneous integration process such as PL-Pack with SORT, described in Chapter 5. The incident light beams received by tapered vertical waveguides in a light beam collecting film are guided to thin-film solar cells, whose areas are smaller than the areas of the light beam collecting aperture of the tapered vertical waveguide. This enables reduction of semiconductor consumption, namely, cost for solar energy conversion. In the example shown in Fig. 10.3a, since the ratio of the solar cell size to the aperture size of the tapered vertical waveguide is ~1:3, the semiconductor consumption can be reduced to ~1/9 of that in the conventional solar cells.

In the distributed-conversion-type structure shown in Fig. 10.3b, incident light beams collected by the tapered vertical waveguides

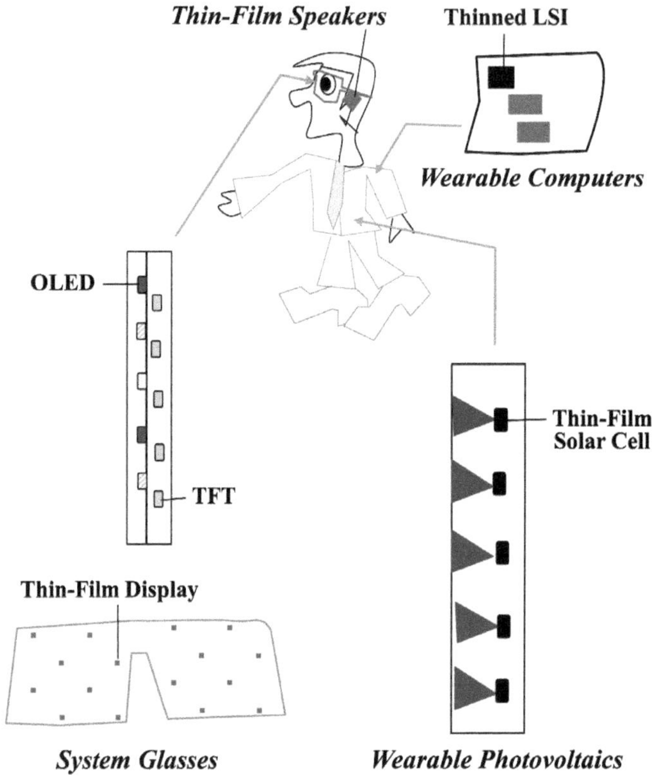

Figure 10.2 Integrated film systems with ubiquitous functions.

are guided to the bottom parts of the waveguides, where cone-shaped vertical mirrors are placed, to be introduced into the embedded waveguide-type thin-film solar cells via the vertical mirrors.

In the centralized-conversion-type structure shown in Fig. 10.3c, a film with tapered vertical waveguides and an optical waveguide film with planar waveguides having vertical mirrors are stacked to build a light beam collecting film. Incident light beams collected by the tapered vertical waveguides are introduced into the planar waveguides via the vertical mirrors, and are guided to thin-film solar cells or waveguide-type thin-film solar cells. By embedding wavelength filters into the optical waveguide film, light beams can

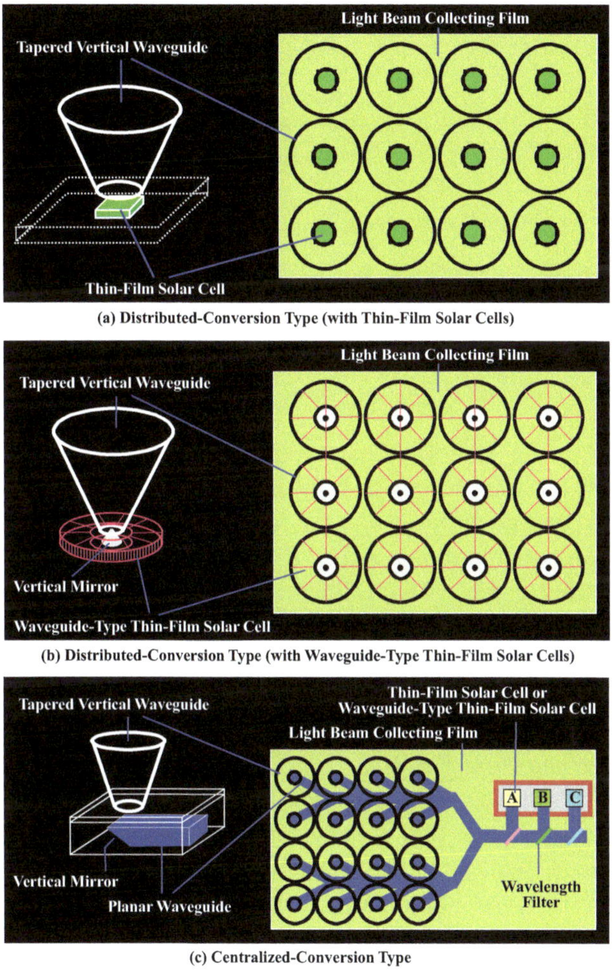

Figure 10.3 Integrated solar films with tapered vertical waveguides. (a) Distributed-conversion type structure with thin-film solar cells, (b) distributed-conversion type structure with waveguide-type thin-film solar cells, and (c) centralized-conversion type structure with thin-film solar cells or waveguide-type thin-film solar cells.

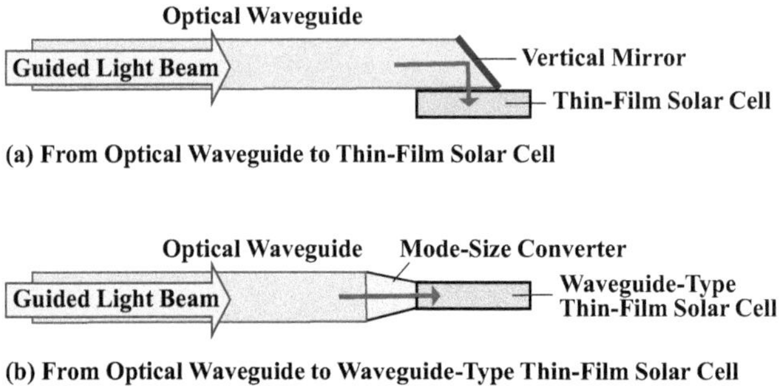

Figure 10.4 Guided beam insertion (a) into thin-film solar cells via vertical mirrors and (b) into waveguide-type thin-film solar cells via mode-size converters.

branch to the solar cells having matched spectral responses, for example, UV-blue beams to cells with GaN, visible beams to cells with GaAs, and infrared beams to cells with InP. Guided beam insertion into thin-film solar cells is performed via vertical mirrors, as depicted in Fig. 10.4a. Guided beam insertion into waveguide-type thin-film solar cells is performed via mode-size converters, as depicted in Fig. 10.4b.

The centralized-conversion-type integrated solar film configurations enable drastic reduction in semiconductor consumption. In the example shown in Fig. 10.3c, since the ratio of the solar cell size to the aperture size of the vertical mirror is ~1:3 and light beams from 16 tapered vertical waveguides come together into one planar waveguide, semiconductor consumption is reduced to $\sim(1/9) \times (1/16) = \sim 1/144$ of that in conventional solar cells. The reduction of the internal resistivity of the solar cells is also expected.

An additional feature of integrated solar films is improved appearance. As shown in Fig. 10.5, coloring of solar cells is possible by back lights. In conventional solar cells, since semiconductors are placed on the whole substrate, the cells look black. In some dye-sensitized solar cells, colored appearance is available by suppressing light absorption in particular visible wavelength regions. This, however, causes reduction in energy conversion efficiency. In the case of integrated solar films, back light can pass through the light

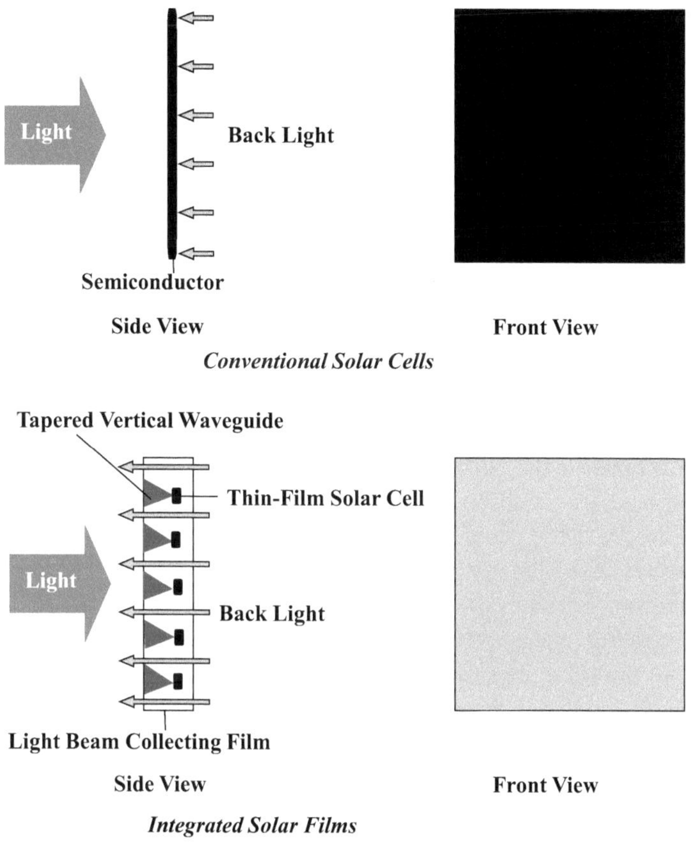

Figure 10.5 Coloring of solar cells by back lights.

beam collecting film. By adjusting the color of the back lighting, favorite coloration of the film might be achievable.

10.2 Waveguide-Type Thin-Film Solar Cells

Thin-film solar cells and waveguide-type thin-film solar cells are key devices in integrated solar films. Thin-film solar cells, in which thin-film semiconductors are used, are preferable to the reduction of semiconductor materials and internal resistivity in solar cells. The thin-film structures, however, sometimes cause insufficient absorption for normally incident light beams. Waveguide-type thin-film solar cells [1, 2, 4, 7–12], in which light beams propagate in thin-film semiconductors with the guided light configuration, solve the problem, enhancing light beam absorption to raise the energy conversion efficiency. In the present section, the concept of the waveguide-type thin-film solar cell is described for cases of sensitized solar cells.

Sensitized solar cells have received much attention as cost-effective next-generation solar energy conversion devices [13]. In conventional sensitized solar cells, porous semiconductors are used to increase the number of adsorbed dye molecules for sensitization on semiconductor surfaces, which enhances light absorption. In such structures, however, the internal resistivity of the cells becomes large because the porous structures decrease the transport channel width and the mobility of carriers.

To avoid the problem, it is preferable to use thin-film semiconductors with high crystalline quality in the cells. In such structures, however, normally incident light beams pass through very thin sensitizing layers, resulting in small light absorption. Waveguide-type thin-film sensitized solar cells, in which the light beam absorption is enhanced by the guided light configuration, solve the problem.

Figure 10.6a shows a waveguide-type thin-film sensitized solar cell consisting of an n-type thin-film semiconductor, a sensitizing layer, and a p-type thin-film semiconductor [4, 7–11]. High-crystalline-quality thin-film semiconductors with high carrier mobility decrease the internal resistivity to improve the photovoltaic performance. Light beams are guided in the cell. In such a guided-light configuration, light beams pass through a lot of sensitizing agents in the sensitizing layer to enhance photocurrents. Thus, an improvement in photovoltaic performance is expected. In the waveguide-type sensitized solar cell shown in Fig. 10.6b, one of the thin-film semiconductors is replaced with an electrolyte.

In preliminary experiments, a photocurrent enhancement by a factor of $4 \sim 10$ was observed in the guided-light configuration compared to the normally incident light configuration [4, 7–9]. In addition, the guided-light configuration enables us to replace

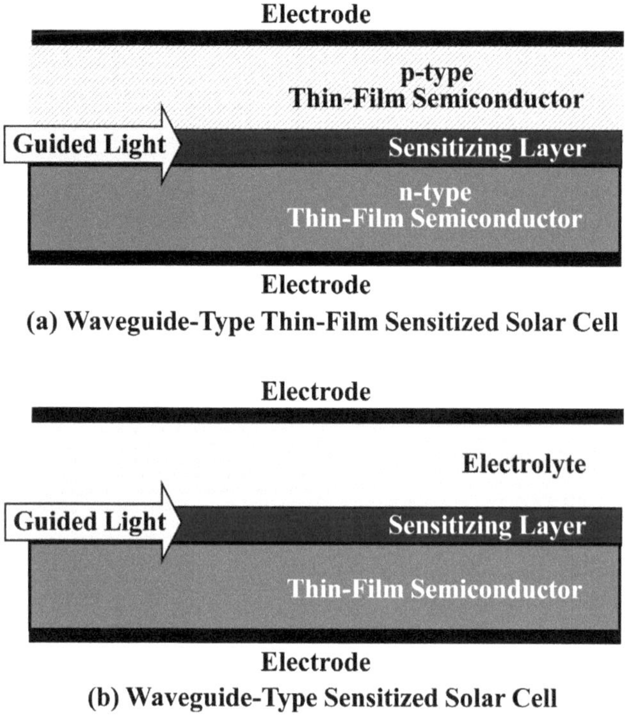

Figure 10.6 (a) Waveguide-type thin-film sensitized solar cell with a structure of n-type semiconductor/sensitizing layer/p-type semiconductor, and (b) waveguide-type sensitized solar cell with a structure of semiconductor/sensitizing layer/electrolyte.

expensive and electrically resistive transparent electrodes with low-cost and low-resistivity metal electrodes.

The sensitizing layer can be polymer MQDs and molecular MQDs that are grown on the thin-film semiconductor surfaces by MLD [2, 4–9]. Figure 10.7a shows an example of polymer MQD sensitization. Four quantum dots, QD1 $\sim$ 4, having different lengths are contained in a single polymer wire to form a polymer MQD. Excited electrons in the quantum dots are injected into the semiconductor. Absorption bands of individual quantum dots are narrow because of their zero-dimensional characteristics, and the peak positions are determined by the dot lengths. The superposition of the absorption bands of

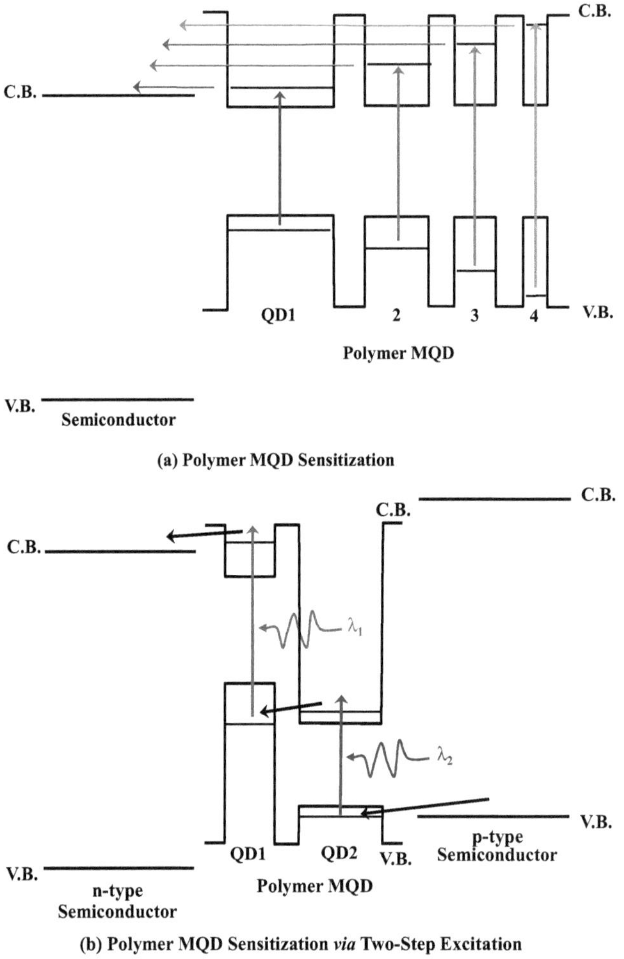

Figure 10.7 (a) Polymer MQD sensitization and (b) polymer MQD sensitization *via* two-step electron excitation.

the individual quantum dots covers a wide range of wavelengths, enabling electron injection into the semiconductor by light beams in a wide range of wavelengths.

For the polymer MQD sensitization via two-step electron excitation [6, 8, 9], as shown in Fig. 10.7b, the "z-scheme" process in the photosynthesis of plants is utilized. An electron in QD1 excited by a photon with a wavelength of λ_1 is injected into the n-type semiconductor. Another electron in QD2 excited by a photon with a wavelength of λ_2 is transferred to the valence band of QD1. The hole left in QD2 is compensated by an electron from the p-type semiconductor. Thus, electrons travel cyclically as [p-type semiconductor] $\rightarrow$ [V.B./QD2] $\rightarrow$ [C.V./QD2] $\rightarrow$ [V.B./QD1] $\rightarrow$ [C.V./QD1] $\rightarrow$ [n-type semiconductor] $\rightarrow$ [p-type semiconductor] $\rightarrow$. This sensitization mechanism suppresses the energy loss arising from excess photon energy and simultaneously increases the energy difference between the Fermi level of the n-type semiconductor and the valence band of the polymer MQDs, which increases the voltage generated from the solar cell.

Figure 10.8a shows an example of molecular MQD sensitization. Five molecules, M1 $\sim$ M5, having different energy gaps form a molecular MQD for a light-harvesting antenna. Excited electrons in the molecules are injected into the semiconductor through the antenna with a gradual energy slope [8, 9, 14].

For molecular MQD sensitization via two-step electron excitation [8, 9], as shown in Fig. 10.8b, an electron excited by a photon with a wavelength of λ_1 in molecule M1 is injected into the n-type semiconductor. An electron excited by a photon with a wavelength of λ_2 in molecule M2 is transferred to HOMO of M1. The hole left in HOMO of M2 is compensated by an electron from the p-type semiconductor. Thus, electrons travel cyclically as [p-type semiconductor] $\rightarrow$ [HOMO/M2] $\rightarrow$ [LUMO/M2] $\rightarrow$ [HOMO/M1] $\rightarrow$ [LUMO/M1] $\rightarrow$ [n-type semiconductor] $\rightarrow$ [p-type semiconductor] $\rightarrow$.

10.3 Key Fabrication Processes for Integrated Solar Films

Figure 10.9 shows a whole structure of the integrated solar film consisting of tapered vertical waveguides and planar waveguides with vertical mirrors shown in Fig. 10.3c. On the optical waveguide film, electrodes for electrical power transmission are formed. To fabricate such structures, the following key processes are necessary.

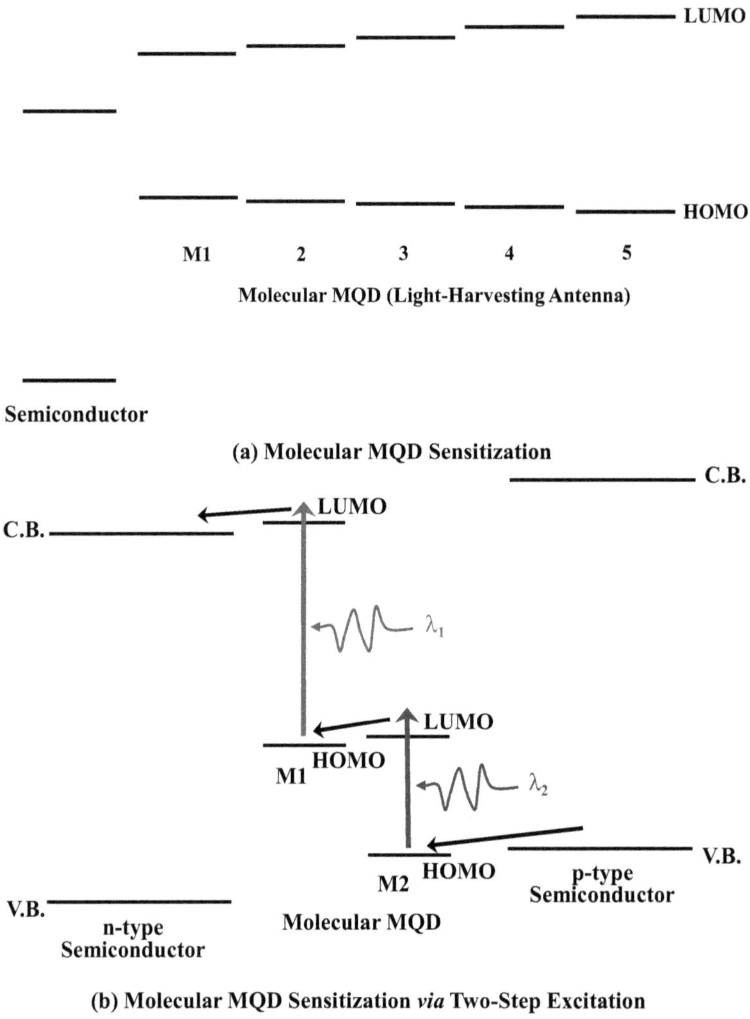

Figure 10.8 (a) Molecular MQD sensitization with a light-harvesting antenna and (b) molecular MQD sensitization *via* two-step electron excitation.

- Embedding of thin-film solar cells or waveguide-type thin-film solar cells
- Formation of planar waveguides with vertical mirrors
- Formation of tapered vertical waveguides

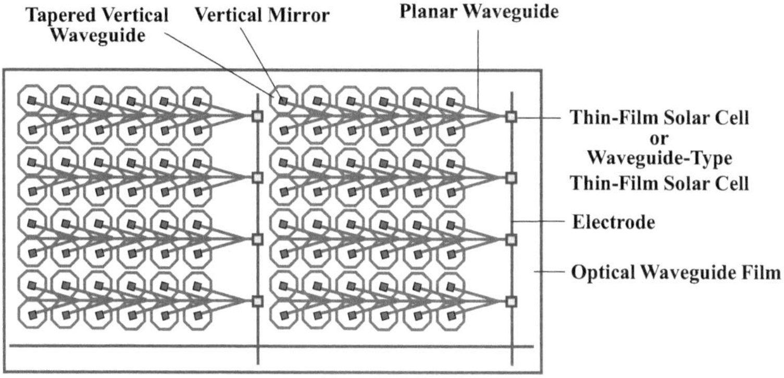

Figure 10.9 Whole structure of the integrated solar film consisting of tapered vertical waveguides and planar waveguides with vertical mirrors.

- Optical couplings between planar waveguides and waveguide-type thin-film solar cells.

For embedding of solar cells and formation of planar waveguides with vertical mirrors, PL-Pack with SORT described in Chapter 5 and the built-in mask method described in Chapter 4 can directly be applied, respectively.

For formation of tapered vertical waveguides, the built-in mask method can be applied. An example of the fabrication process is shown in Fig. 10.10. By exposing photopolymers to tilted UV lights, cured regions having beveled walls are formed. The tilted UV exposure is performed from various directions. By development, a tapered vertical waveguide film is made on the built-in mask. After a supporting substrate is attached to the top of the film, the removable layer is removed to pick up the tapered vertical waveguide film. By stacking the film on an optical waveguide film, a light beam collecting film is obtained

Figure 10.11 shows photographs of an array of straight vertical waveguides and an array of tapered vertical waveguides fabricated by the built-in mask method. The light beam collecting efficiency of the tapered vertical waveguide is shown in Fig. 10.12. The efficiency was evaluated by introducing white light of a halogen lamp from the front side and measuring the power of the light that transmits through the built-in mask window. Compared to the efficiency for

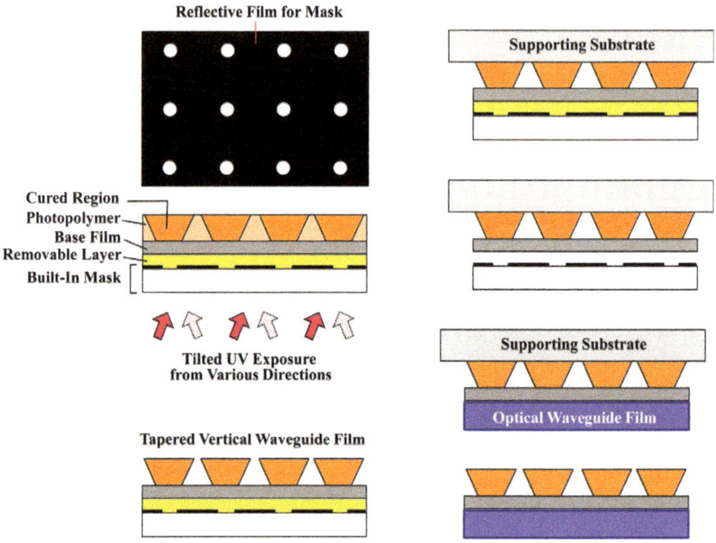

Figure 10.10 Fabrication process of a light beam collecting film consisting of tapered vertical waveguides and an optical waveguide film with planar waveguides.

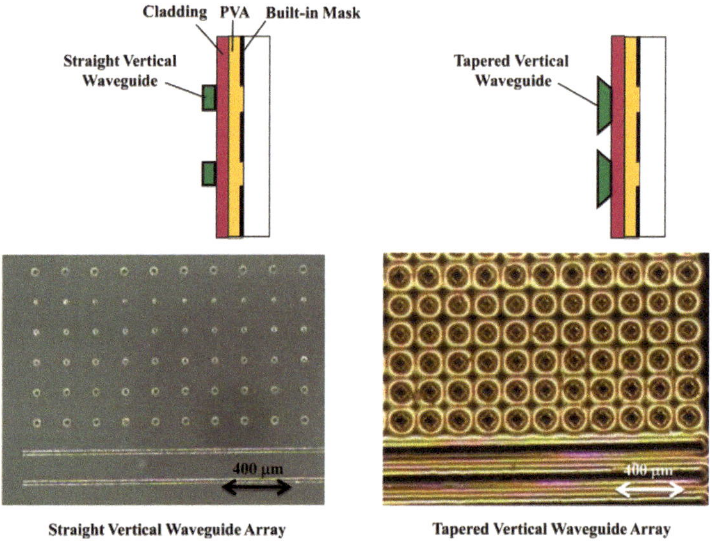

Figure 10.11 Photographs of an array of straight vertical waveguides and an array of tapered vertical waveguides fabricated by the built-in mask method. The waveguide arrays were fabricated by Chie Yoshino.

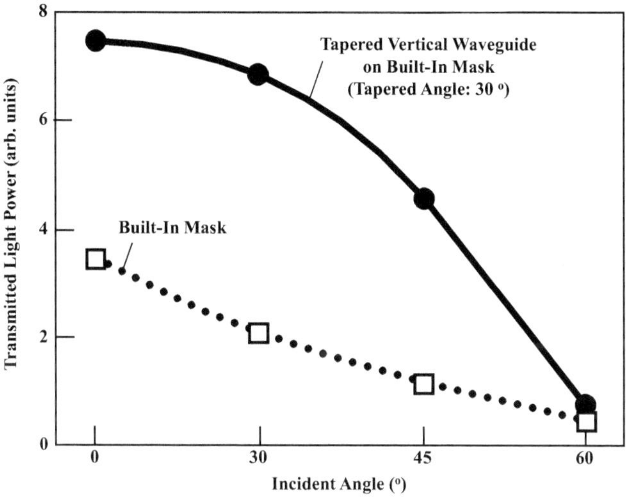

Figure 10.12 Light beam collecting efficiency of the tapered vertical waveguide. The measurements were performed by Chie Yoshino.

the case that only the built-in-mask is placed, the efficiency increases by 2 $\sim$ 3 times by placing tapered vertical waveguides with a taper angle of 30° in front of the built-in mask, confirming the light collecting capability of the tapered vertical waveguides. The efficiency might be improved by optimizing the structure and the fabrication process.

For the waveguide fabrication, in addition to the built-in mask method, imprinting, stamping, and so on, can also be used.

In integrated solar films consisting of waveguide-type thin-film solar cells, two types of light beam couplings are involved. One is the coupling from free spaces into microscale planar waveguides in optical waveguide films, and the other is from the microscale planar waveguides into nanoscale waveguides in waveguide-type thin-film solar cells.

The SOLNETs are promising for these couplings, as illustrated in Fig. 10.13. For "free spaces to microscale planar waveguides" couplings, an array of luminescent targets (or wavelength filters) is deposited at the vertical mirror sites on the optical waveguide film. After a PRI material layer is placed on the film surface, the layer is exposed to blanket write beams through a mask with a window

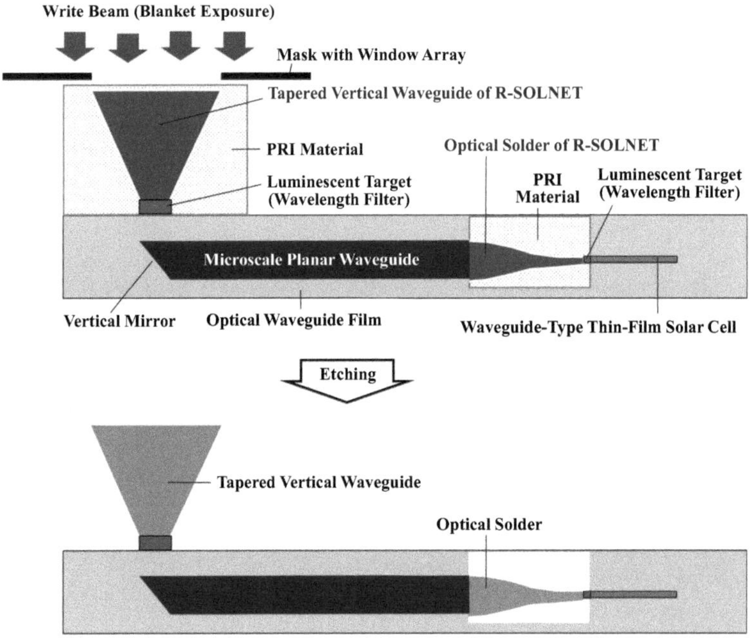

Figure 10.13 Implementation of the SOLNETs into integrated solar films.

array corresponding the light beam collecting apertures to fabricate an array of tapered vertical waveguides of R-SOLNETs self-aligned to the vertical mirrors of the microscale planar waveguides. For "microscale planar waveguides to nanoscale waveguides" couplings, luminescent targets (or wavelength filters) are deposited on the nanoscale waveguide edges of the waveguide-type thin-film solar cells. After the PRI material is inserted between the microscale planar waveguides and the nanoscale waveguides, optical solder of R-SOLNETs is formed by introducing write beams from the microscale planar waveguides to the PRI material. By etching the surrounding regions, optical couplings of high-index-contrast optical solder are obtained. If necessary, the optical solder is covered with low-refractive-index cladding materials.

In addition, preliminary simulation results for light beam collecting films are briefly described. Figure 10.14 shows a basic simulation model for light beam collection from a free space into

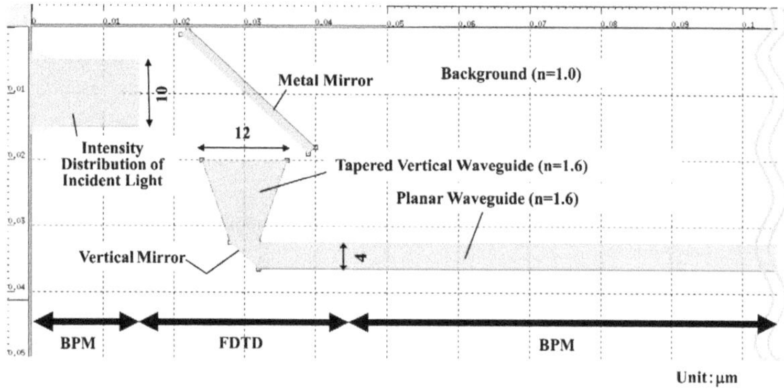

Figure 10.14 Basic simulation model for light beam collection from a free space into a planar waveguide via a tapered vertical waveguide and a vertical mirror.

a planar waveguide via a tapered vertical waveguide and a vertical mirror. The simulation is performed by the 2D FDTD method and the BPM. For incident light beams, light beams guided in the fundamental mode of a 10 μm wide optical waveguide are used. The light beams are reflected by a metal mirror to be introduced into the tapered vertical waveguide. Incident angles of the light beams are adjusted by the metal mirror. The light beam wavelength is 600 nm. Refractive indexes of optical waveguides are 1.6, and the refractive index of the background is 1.0 assuming air. The planar waveguide width is 4 μm. The length of the model from the left boundary to the right boundary is 200 μm. Light intensity at the output (the right boundary of the model) is calculated to estimate the light beam collecting efficiency.

Figure 10.15 shows models and propagating light beam intensity for light beam collecting films involving tapered vertical waveguides. In Model UH, the refractive index of the upper part in the tapered vertical waveguide is higher than that of the lower part. In Model LH, the refractive index of the lower part is higher than that of the upper part. The tapered angle of the tapered vertical waveguide is $\theta_T = 17^\circ$. While the light beam collecting efficiency is 61% in Model UH, the efficiency is 77% in Model LH. This can be explained as follows. Because the light beam spot size becomes small with the light beam propagation toward the lower part of the tapered

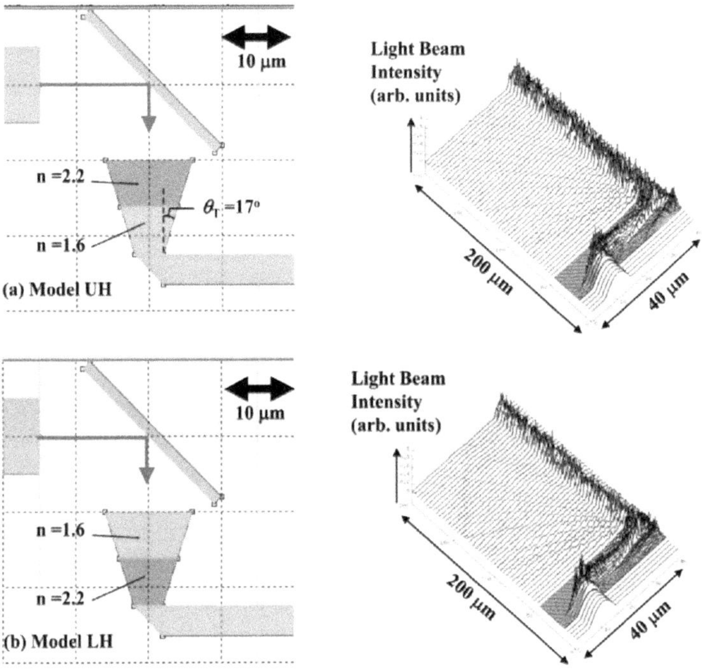

Figure 10.15 Models and the propagating light intensity for light beam collecting films involving tapered vertical waveguides. (a) Upper part of the tapered vertical waveguide has high refractive index (Model UH). (b) Lower part of the vertical waveguide has high refractive index (Model LH).

vertical waveguide, stronger light beam confinement is necessary in the lower part. By putting the high refractive index material in the lower part, strong light beam confinement is achieved effectively, resulting in the enhancement of the light beam collecting efficiency.

Since the light beam collecting aperture of the tapered vertical waveguides is 12 μm and that of the vertical mirror is 4 μm, light beam collection with a factor of 3 is achieved in the present two-dimensional model. For the three-dimensional model, light beam collection with a factor of $3 \times 3 = 9$ is expected.

It should be noted that the incident light beams have a profile, where the intensity has its maximum at the center of the spot as shown in Fig. 10.14. Due to this inhomogeneous intensity distribution, the effect of the tapered vertical waveguides might be underestimated. When wider incident light beams are used for simulations, the light beam collecting efficiency enhancement might become more remarkable. In order to increase the efficiency further, the tapered angles and refractive index distribution profiles in the tapered vertical waveguides should be optimized.

10.4 Multilayer Waveguide-Type Light Beam Collecting Films

Figure 10.16 shows a schematic structure of the integrated solar film with a multilayer waveguide-type light beam collecting film. Many films having planar waveguides with vertical mirrors are stacked. The incident light beams are introduced through the vertical mirrors into the planar waveguides to be guided to solar cells. In the present

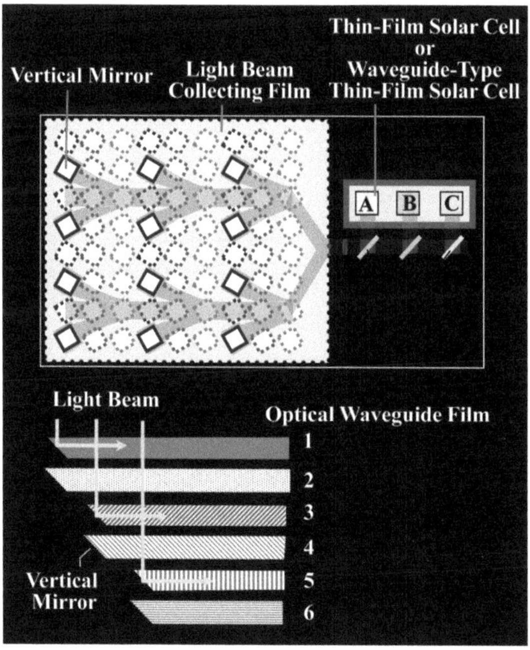

Figure 10.16 Schematic structure of the integrated solar film with a multi-layer waveguide-type light beam collecting film.

section, predicted light beam collecting ability of the multilayer waveguide-type light beam collecting films is described [3].

10.4.1 Simulation Procedure

The simulation procedure for collection of light beams from a free space into a planar waveguide via a tapered vertical waveguide and a vertical mirror is the same as the procedure for the simulations described in Fig. 10.14 in section 10.3. Incident angles of the light beams with a wavelength of 600 nm are adjusted by the metal mirror. Anti-reflecting (AR) coating with a thickness of 100 nm and a refractive index of 1.3 is applied to the top and the bottom of the optical waveguide.

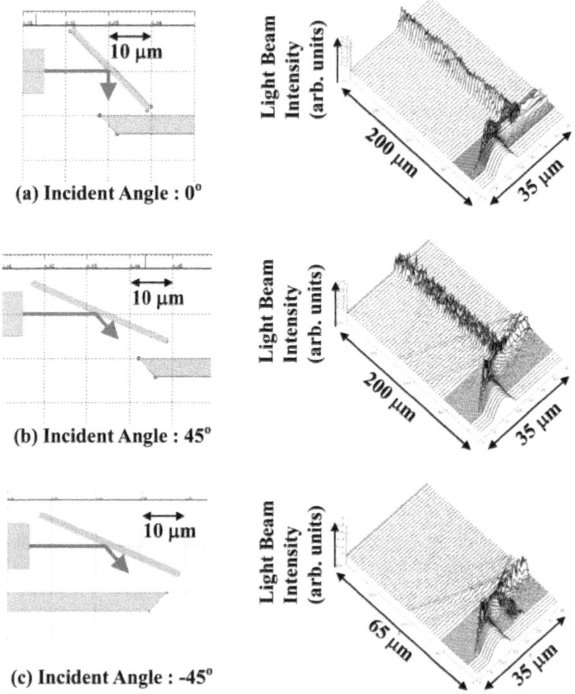

Figure 10.17 Models with 45° mirrors and propagating light beam intensity for a single light beam collecting film without the tapered vertical waveguide. Incident angles are 0°, 45°, and −45°. Reprinted with permission from Shioya and Yoshimura [3].

10.4.2 Light Beam Collection by Light Beam Collecting Films

Figure 10.17 shows models with 45° mirrors and propagating light beam intensity for single light beam collecting films without the tapered vertical waveguides. The angle of the vertical mirror is 45°. Incident angles are 0°, 45°, and −45°. Here, in the model of −45°, the waveguide direction is flipped horizontally, which effectively reproduces a configuration for −45° incidence. For an incident angle of 0°, although considerable leakage is observed at the 45° mirror, a part of the incident light beam is introduced into the optical waveguide through the 45° mirror, and then guided in the optical waveguide smoothly, indicating that the fundamental mode is dominant. The light beam collecting efficiency is calculated to be 62%. For an incident angle of 45°, a lot of leakage of the light beams passing through the 45° mirror are observed. At the same time, considerable higher modes are excited in the guided beams. For an incident angle of −45° almost all the light beams pass through the 45° mirror.

Figure 10.18 shows models with 30° mirrors and the propagating light beam intensity. For an incident angle of 0°, the light beams guided in the optical waveguide contain considerable higher modes. For an incident angle of 45°, on the other hand, the guided beams propagate smoothly with small amount of higher modes. Figure 10.19 shows models with 60° mirrors and the propagating light beam intensity. The guided beams contain a lot of higher modes for both incident angles of 0° and 45°.

The light beam collecting efficiency is summarized in Fig. 10.20 for incident angles of 0° and 45°. For an incident angle of 0°, the 45° mirror exhibits the largest efficiency of 62%. For an incident angle of 45°, the 30° mirror exhibits the largest efficiency of 47%. The 60° mirror does not exhibit large efficiency for both incident angles of 45° and 0°. These results indicate that optical waveguides with 45° mirrors and optical waveguides with 30° mirrors are respectively effective for incident light beams with a vertical direction and for those with tilted directions.

A model for the multilayer waveguide-type light beam collecting film is shown in Fig. 10.21. It consists of stacked waveguide films [7].

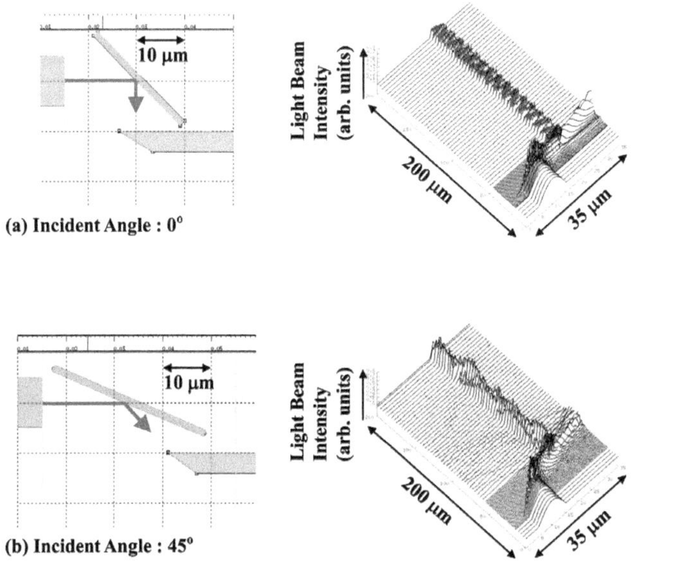

Figure 10.18 Models with 30° mirrors and the propagating light beam intensity. Reprinted with permission from Shioya and Yoshimura [3].

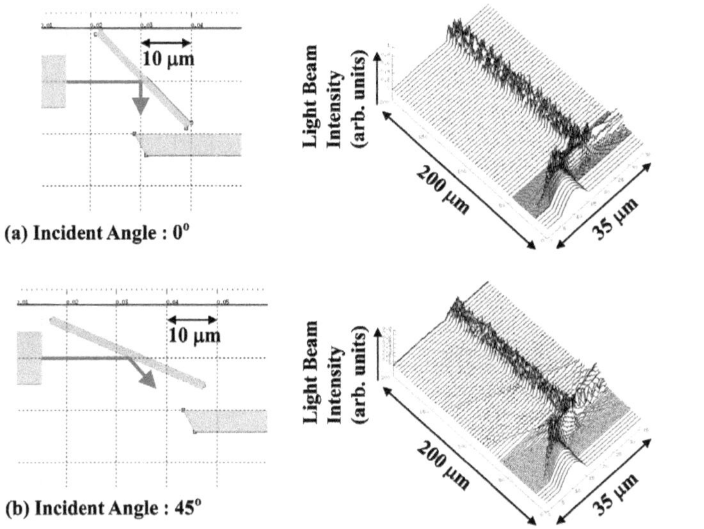

Figure 10.19 Models with 60° mirrors and the propagating light beam intensity. Reprinted with permission from Shioya and Yoshimura [3].

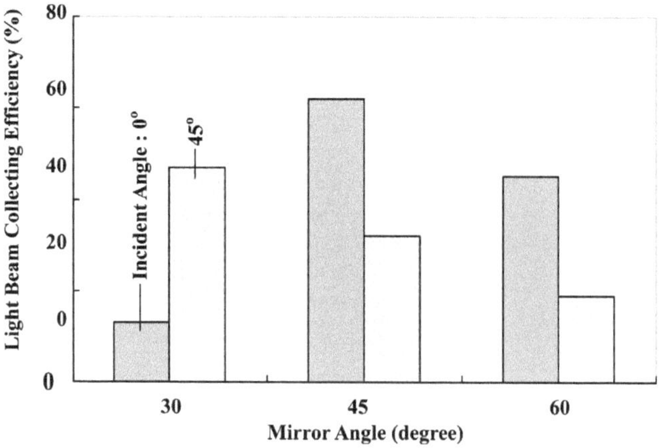

Figure 10.20 Light beam collecting efficiency in models with 30°, 45°, and 60° mirrors for incident angles of 0° and 45°. Reprinted with permission from Shioya and Yoshimura [3].

In this example, as shown in the side view, 18 optical waveguide films are stacked. Numbers in the top view indicate vertical mirror apertures, namely, the light beam collecting apertures, of the films denoted by the numbers. An example of optical circuits in a waveguide film is illustrated in Fig. 10.22. Light beams trapped by the vertical mirrors are collected through branching waveguides, which merge into main waveguides at joints. By arranging the optical circuits in the films regularly, the surface can be occupied by the vertical mirror apertures in high density.

Figure 10.23 shows a model of a three-layer waveguide structure with 45° mirrors and the propagating light beam intensity for an incident angle of 0°. The distance between optical waveguides is 1 μm. The light beam wavelength is 600 nm. It is found that the incident light beams, which go through the first layer, are received by the second layer and the third layer to increase the light beam collecting efficiency. As Table 10.2 shows, light beam collecting efficiency increases from 62% to 87% by increasing the layer count from 1 to 3.

As mentioned in section 10.4.2, optical waveguides with 45° mirrors and optical waveguides with 30° mirrors are respectively preferable to collect light beams with vertical incident angles and

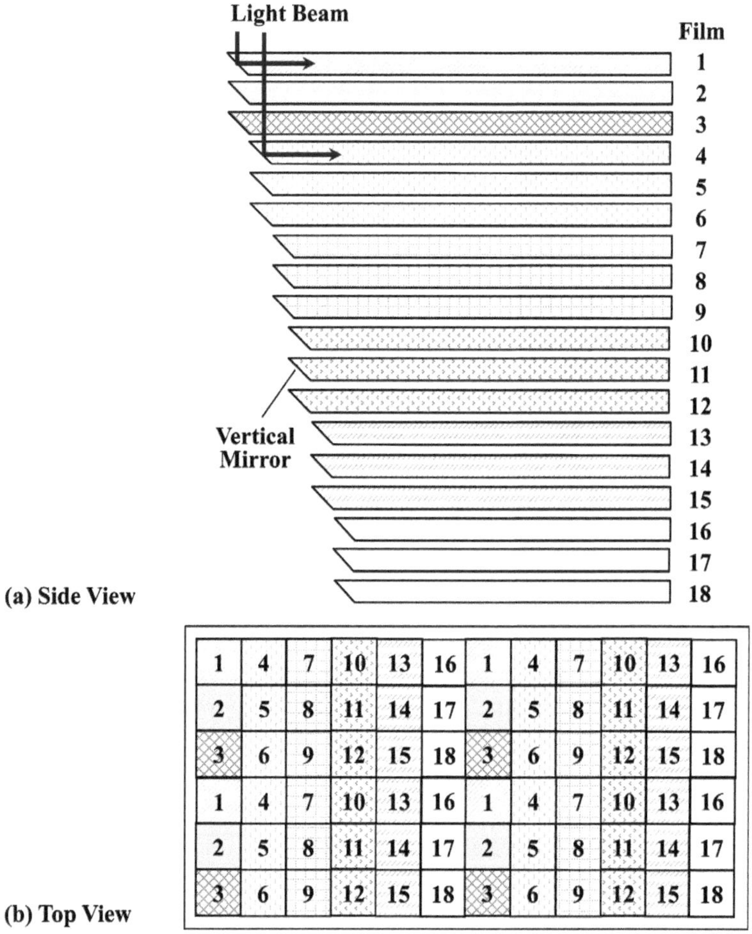

Figure 10.21 Model for the multilayer waveguide-type light beam collecting film. Reprinted with permission from Shioya and Yoshimura [3].

tilted incident angles. The results suggest that light beam collecting films consisting of multilayer optical waveguides with 45° and 30° mirrors, denoted by 45°/30°/30°, are expected to exhibit high light beam collecting efficiency. An example of the three-layer structure is illustrated in Fig. 10.24. The first layer is an optical waveguide with 45° mirror, and the second layer and the third layer are optical waveguides with 30° mirrors. The two optical waveguides with 30°

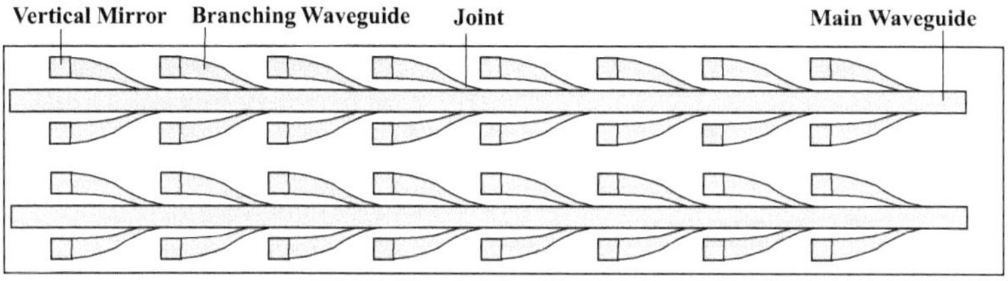

Figure 10.22 An example of optical circuits in a waveguide film.

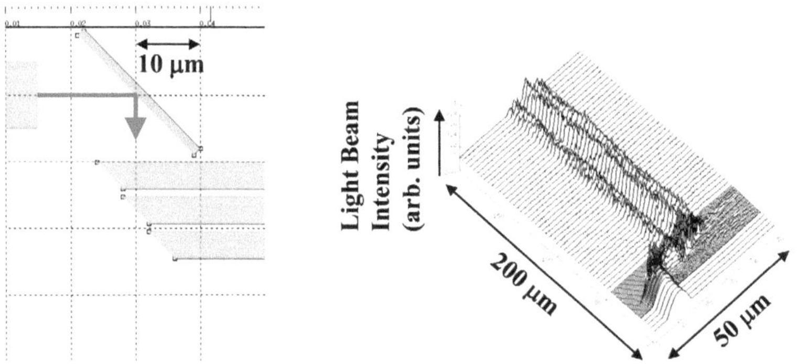

Figure 10.23 Model of a three-layer waveguide structure with 45° mirrors and the propagating light beam intensity for an incident angle of 0°. Reprinted with permission from Shioya and Yoshimura [3].

Table 10.2 Light beam collecting efficiency for layer counts of one, two, and three. Reprinted with permission from Shioya and Yoshimura [3]

Layer count	Light beam collecting efficiency (%)
1	62
2	83
3	87

mirrors are arranged in opposite directions. The first layer is for incident light beams with vertical directions. The second layer is for beams from the upper left, and the third layer is for beams from the upper right. This structure is expected to provide wide-angle light beam collecting capability without solar beam tracking mechanisms.

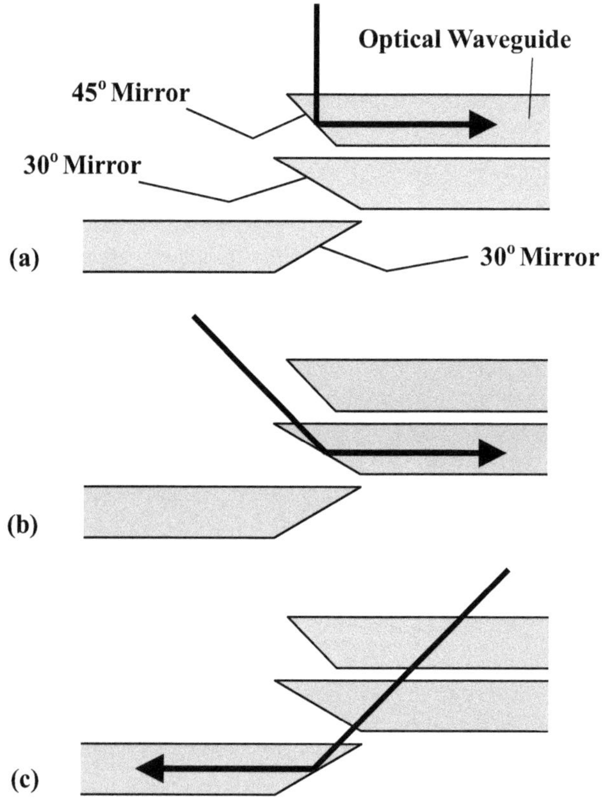

Figure 10.24 An example of the three-layer structure with 45° and 30° mirrors, denoted by 45°/30°/30°. (a) Incident light beams with vertical directions, (b) incident light beams from the upper-left, and (c) incident light beams from the upper-right. Reprinted with permission from Shioya and Yoshimura [3].

Figure 10.25 shows models of the three-layer structure of 45°/30°/30° and the propagating light beam intensity for incident angles of 0°, 45°, and −45°. The distance between optical waveguides is 1 μm. The wavelength of incident light beams is 600 nm. It is found that the light beams are mainly collected by the first layer for the vertical incidence, by the second layer for the tilted incidence from the left, and by the third layer for the tilted incidence from the right.

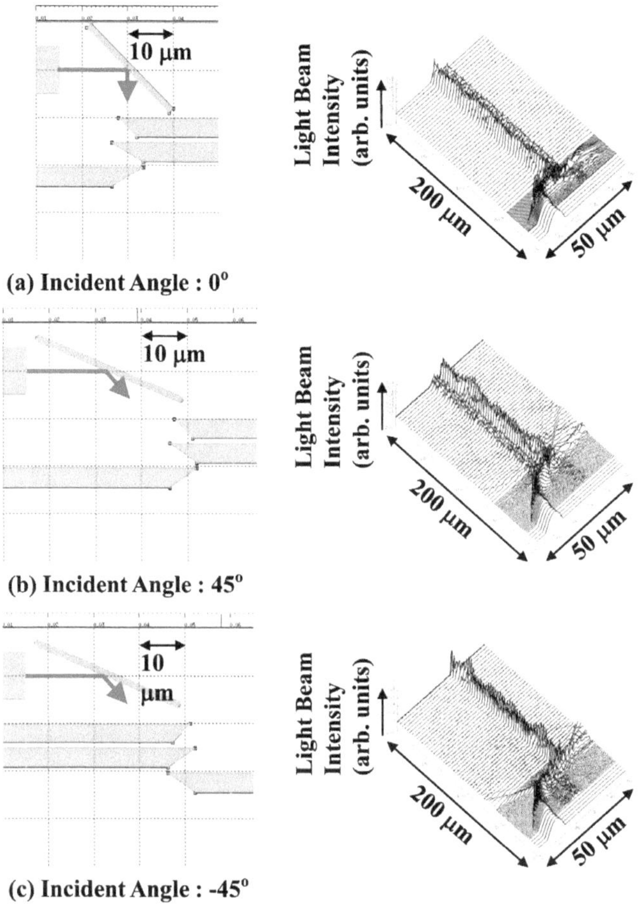

Figure 10.25 Models of the three-layer structure of 45°/30°/30° and the propagating light beam intensity for incident angles of 0°, 45°, and −45°. Reprinted with permission from Shioya and Yoshimura [3].

For comparison, light beam collecting characteristics of a light beam collecting film consisting of a three-layer structure with only 45° mirrors, denoted by 45°/45°/45°, are examined. The structure shown in Fig. 10.26 is the same as the model shown in Fig. 10.25 except for the mirror angles in the second and third layers. While the collected light beams propagate in the optical waveguides smoothly for the vertical incidence, for the tilted incidence the collected

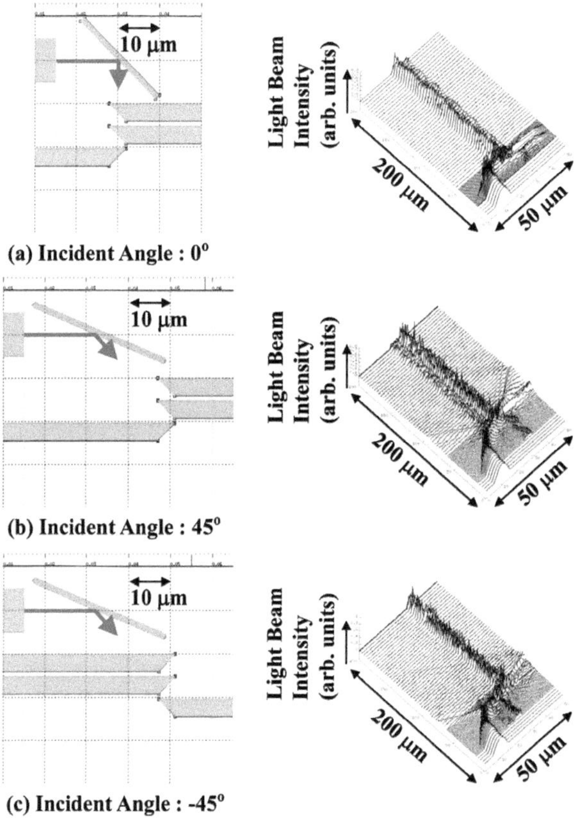

Figure 10.26 Models of the three-layer structure with only 45° mirrors, denoted by 45°/45°/45°, and the propagating light intensity for incident angles of 0°, 45°, and −45°. Reprinted with permission from Shioya and Yoshimura [3].

light beams are reduced and propagate in considerably higher modes.

Figure 10.27 shows dependence of light beam collecting efficiency on incident angles for the three-layer structures of 45°/30°/30° and 45°/45°/45°. 45°/30°/30° exhibits higher efficiency than 45°/45°/45° in whole incident angle regions. Especially, for incident angle of 45°, the efficiency is 1.5 times larger in 45°/30°/30° than in 45°/45°/45°. This indicates that the light beam collecting film consisting of multilayer optical waveguides

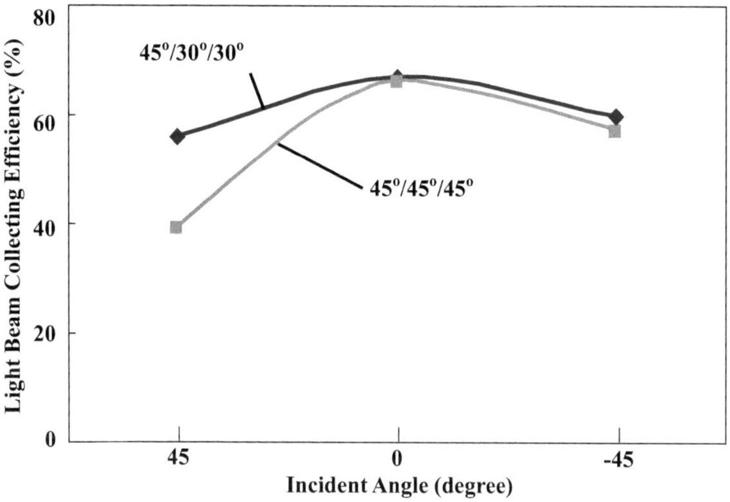

Figure 10.27 Dependences of light beam collecting efficiency on incident angles for the three-layer structures of 45°/30°/30° and 45°/45°/45°. Reprinted with permission from Shioya and Yoshimura [3].

with mirrors having different angles provides wide-angle light beam collecting capability.

It should be noted that incident angle dependence of the efficiency shown in Fig. 10.27 is asymmetry. Namely, when incident angle is −45°, the efficiency of 45°/45°/45° is close to that of 45°/30°/30°. This might happen accidentally by prism effect of the 45° mirror part of the upper waveguide layers.

The model shown in Fig. 10.25 have 4 μm × 4 μm aperture in top views. This corresponds to the aperture occupation for a layer count of one in Table 10.2. So, when 3 × (4 μm × 4 μm) aperture, which corresponds to a layer count of three in Table 10.2, is built by stacking two more film units of 45°/30°/30°, the light beam collecting efficiency close to 80% is expected for incident angles of 45°, 0°, and −45°.

In addition, the light beam collecting efficiency for an incident angle of 45° in 45°/30°/30° was calculated to be 60%, 57%, and 65% for wavelengths of 400 nm, 600 nm and 800 nm, respectively. Thus, it is found that the wavelength dependence of the efficiency is relatively small, indicating that the proposed model could operate in wavelength regions from 400 nm to 800 nm.

10.4.3 Overall Consideration for Light Beam Collecting Efficiency

Overall light beam collecting efficiency η_{Overall} in the light beam collecting films, whose basic structure is shown in Fig. 10.22, is given as follows:

$$\eta_{\text{Overall}} = \eta_{\text{Mirror}} \times \eta_{\text{Waveguide}} \times \eta_{\text{Die}} \tag{10.1}$$

Here, η_{Mirror}, $\eta_{\text{Waveguide}}$, and η_{Die} are respectively the light beam trapping efficiency by a vertical mirror, the total light beam propagation efficiency through the branching and main waveguides, and the coupling efficiency from an optical waveguide into a small/thin semiconductor die for solar cells.

The reflectance of the vertical mirrors might be around 80%, roughly speaking. In this case η_{Mirror} is ~80%. $\eta_{\text{Waveguide}}$ can be calculated by the BPM. The model is similar to that shown in Fig. 10.22. The branching angle is 10°. The branching waveguides are tapered from a width of 4 μm at the vertical mirror to 0.1 μm at the joint, which is effective in increasing the light beam propagation efficiency. Figure 10.28a shows the dependence of $\eta_{\text{Waveguide}}$ on the number of joints when light beams are introduced into all the vertical mirrors. The efficiency decreases with increasing number of joints. When the number of joints is 8, $\eta_{\text{Waveguide}}$ is ~ 50%. Reflection loss at the waveguide–semiconductor die interfaces is around 16% when a polymer with a refractive index of 1.52 and ZnO with a refractive index of 2.1 are respectively used for the waveguides and the semiconductor dies. The coupling efficiency might increase near 100% by inserting appropriate AR coating at the interfaces, enabling η_{Die} of 100%. By substituting these estimated values for η_{Mirror}, $\eta_{\text{Waveguide}}$, and η_{Die} into Eq. 10.1, overall light beam collecting efficiency is estimated to be ~40% for eight-step merging, that is, for number of joints of 8.

In a light beam collecting film with the eight-step merging, light beams from 16 vertical mirrors are collected into one small/thin semiconductor die. Therefore, the factor of semiconductor con-

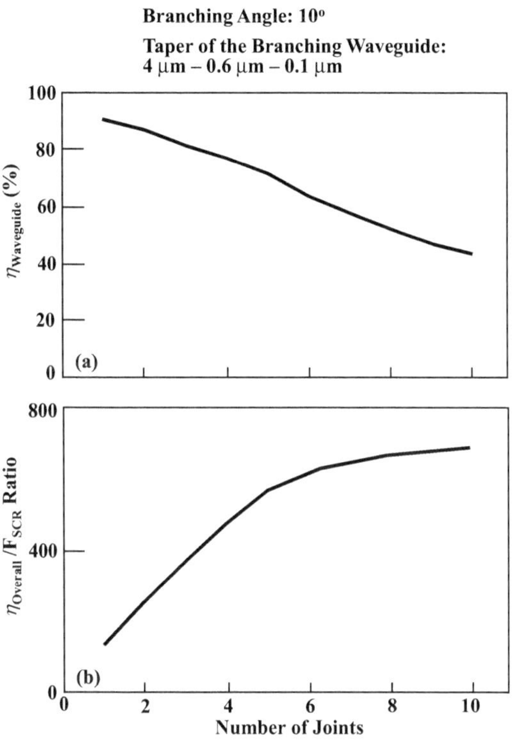

Figure 10.28 (a) Dependence of $\eta_{Waveguide}$ on the number of joints when light beams are introduced into all the vertical mirrors. (b) Dependence of $\eta_{Overall}/F_{SCR}$ ratio on the number of joints.

sumption reduction, denoted by F_{SCR}, is 1/16. When a light beam collecting film, in which a film with tapered vertical waveguides and a film with planar waveguides having vertical mirrors are stacked as shown in Fig. 10.3c, is used, semiconductor consumption reduction of 1/9 is obtained at the tapered waveguide part. Therefore, the eight-step merging gives us F_{SCR} of $(1/9) \times (1/16) = 1/144$. In the conventional solar cell, F_{SCR} is 1 because the semiconductor is placed all over the module.

$\eta_{Overall}/F_{SCR}$ ratio is defined as the figure of merit of the resource-saving ability. This ratio is proportional to the amount of energy conversion achieved by semiconductors of unit area. In the conventional solar cell, since $\eta_{Overall}$ and F_{SCR} are respectively nearly equal to 100% and 1, the $\eta_{\text{Overall}}/F_{\text{SCR}}$ ratio is $100/1 = 100$. In Fig. 10.29b, dependence of $\eta_{\text{Overall}}/F_{\text{SCR}}$ ratio on the number of joints is shown for the light beam collecting film with planar waveguides. The ratio increases with the number of joints, followed by saturation. In the case of the eight-step merging, the ratio is $40/(1/16) = 640$, which is 6.4 times larger than the ratio in the conventional solar cell. For the light beam collecting film, in which a film with tapered vertical waveguides and a film with planar waveguides having vertical mirrors are stacked, the $\eta_{\text{Overall}}/F_{\text{SCR}}$ ratio is $40/(1/144) = 5760$, which is 58 times larger than the ratio in the conventional solar cell. Thus, it is concluded that the proposed integrated solar films are expected to realize drastic material-saving.

In the multilayer waveguide-type light beam collecting film, in order to achieve aperture occupation of 100%, 39 layers are necessary for one angle of the vertical mirrors when the branching angle and the joint counts are respectively 10° and 8 in the merging waveguide. This 39-layer structure forms a module unit. To construct a multilayer structure consisting of optical waveguides with 45° and 30° mirrors, one module unit for 45° mirrors and four module units for 30° mirrors are necessary to be stacked in order to collect light beams with vertical incident angles and tilted incident angles (from the left, right, front, and back sides). When the waveguide core thickness is 4 μm and the distance between waveguides is 1 μm, the module unit thickness is around 195 μm. Then, total thickness of the stack of five module units is 975 μm.

In the present design, we assumed 4 μm waveguide cores. In practice, larger cores with several tens micrometer sizes can be used. On the contrary, for the light beam collecting film, in which films with tapered vertical waveguides and films with planar waveguides having vertical mirrors are combined, namely, in the case that a number of tapered vertical waveguide films and planar waveguide films with 45° mirrors and 30° mirrors are stacked as illustrated in Fig. 10.29, the waveguide core sizes should be reduced to ~1 μm or less in order to prevent the total thickness of the light beam collecting film from increasing abnormally.

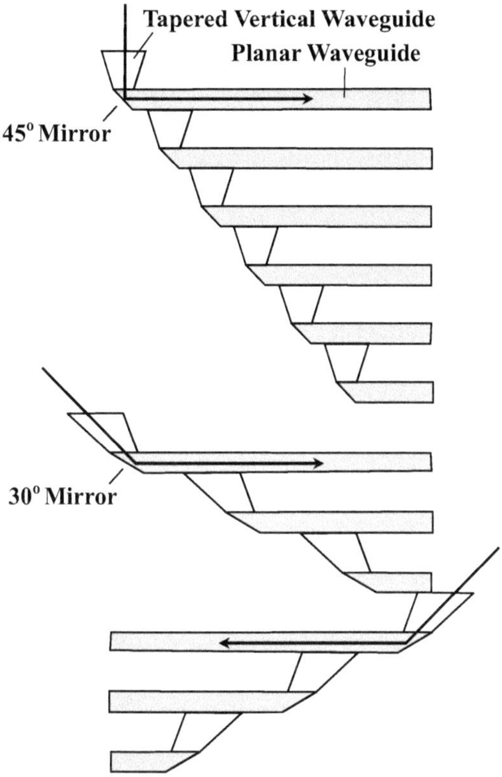

Figure 10.29 Light beam collecting film, in which films with tapered vertical waveguides and films with planar waveguides having vertical mirrors are combined.

The estimated efficiency of the light beam collecting films is lower than that of Fresnel lens/mirror concentrators. So, at present the light beam collecting films seem not suitable for large-scale solar energy conversion centers. However, they have an outstanding feature of flexible light-weight films, enabling us to carry and put them as sheets everywhere to provide possible applications illustrated in Fig. 10.1. By precise investigation on the merging structures in the horizontal waveguides, further improvement of the collecting efficiency is expected.

10.5 Thin-Film Artificial Photosynthesis Cells

To date, artificial photosynthesis of hydrogen/oxygen has been achieved [15], and photosynthesis of formate utilizing semiconductor/complex hybrid photocatalysts has also been reported [16].

The photosynthesis process for H_2/O_2 using two kinds of semiconductors was developed by Sayama et al. using water, in which the semiconductor powders are dispersed [15]. In this case, however, there is a drawback that relative positions of the semiconductors cannot be controlled, preventing the synthesizing efficiency from being optimized.

In order to improve the photosynthesis performance, we proposed photosynthesis cells utilizing thin-film semiconductors with controlled relative positions [1, 9–12, 17, 18]. Several examples for positional configurations of thin-film semiconductors in thin-film artificial photosynthesis cells are shown in Fig. 10.30. The H_2/O_2 generation is achieved by the z-scheme process in the photosynthesis of plants [15]. The structure with divided thin films and that with thin films having windows promote H^+ diffusion in water.

Possible fabrication processes for thin-film photosynthesis cells are shown in Fig. 10.31. In process I, thin-film semiconductors 1 and 2 are respectively grown on growth substrates 1 and 2. The thin films are attached to each other with spacers between them. Then

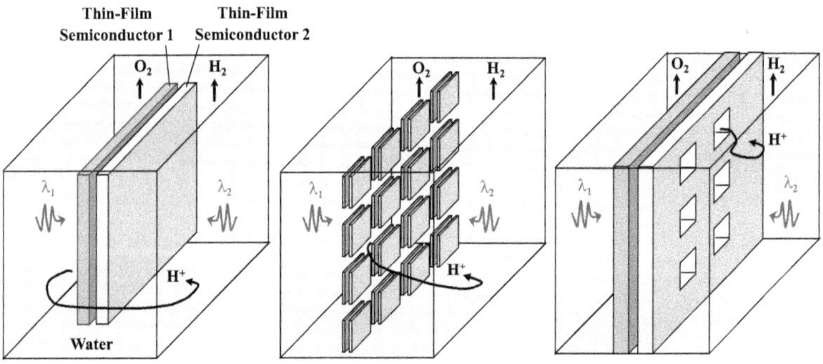

Figure 10.30 Examples for positional configurations of thin-film semiconductors in thin-film artificial photosynthesis cells.

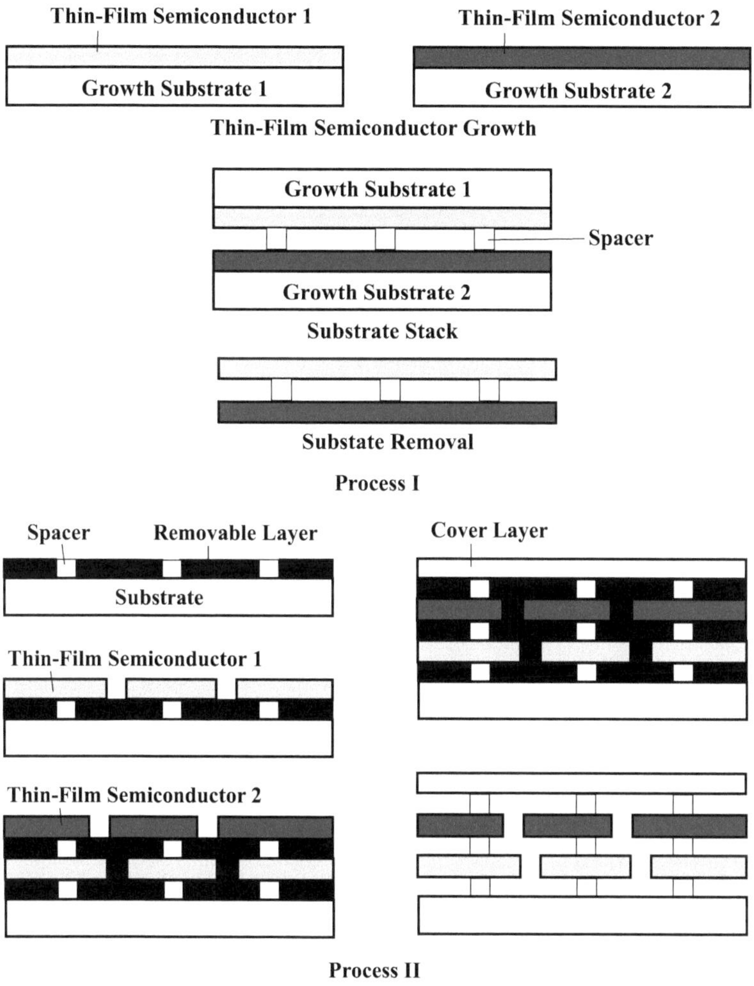

Figure 10.31 Possible fabrication processes for thin-film photosynthesis cells.

the growth substrates are removed to obtain stand-alone coupled thin films. In process II, patterned thin-film semiconductor 1 is deposited on a removable layer of a substrate with spacers. Then, another removable layer and spacers are formed on it, and patterned thin-film semiconductor 2 is deposited on them. After one more removable layer with spacers and a cover layer is formed on them,

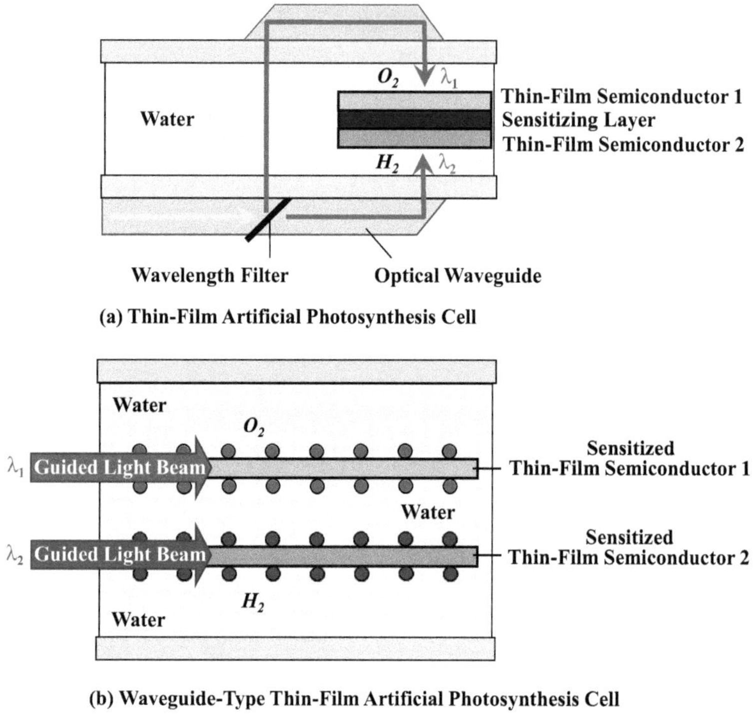

Figure 10.32 Concept of the photosynthesis cells with thin-film semiconductors. (a) Thin-film artificial photosynthesis cell and (b) waveguide-type thin-film artificial photosynthesis cell.

all the removable layers are removed by etching to obtain a structure with thin-film semiconductors located in a designated configuration three-dimensionally.

Figure 10.32a shows an example of a thin-film artificial photosynthesis cell, in which 3D optical circuits with embedded wavelength filters are implemented [1, 9–12, 17, 18]. Light beams branch into two lightwave paths for wavelengths of λ_1 and λ_2, and the two light beams are respectively introduced onto two thin-film semiconductors in water, generating H_2/O_2 through the z-scheme process. Figure 10.32b shows the waveguide-type thin-film artificial photosynthesis cell. In this example the thin-film semiconductors are sensitized by agents. Guided light beams in thin-film semiconductors pass through a lot of sensitizing

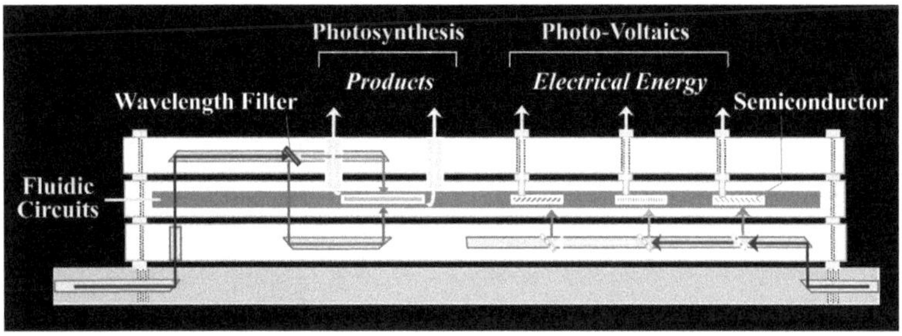

Figure 10.33 Concept of the integrated photonic/electronic/chemical system (IPECS).

agents. The guided light configuration enhances the light absorption of the thin-film semiconductors to generate H_2/O_2 gases efficiently.

In the photosynthesis of H_2, high-power-density light beam exposure is required to achieve high conversion efficiency because the H_2O oxidization is induced by the four-electron process. The waveguide-based system is preferable to increase the light beam power density. Therefore, the waveguide-type thin-film artificial photosynthesis cell is suitable to improve the conversion efficiency in the photosynthesis.

When the molecular MQD sensitization with two-step electron excitation shown in Fig. 10.8b is applied to the water splitting, H_2 would be generated from the n-type semiconductor/water interface (left-hand side in Fig. 10.8b), where co-catalyst is deposited if necessary, and O_2 from the p-type semiconductor/water interface (right-hand side in Fig. 10.8b). Meanwhile, it is known that light-harvesting antennas (Fig. 10.8a) raise density of charge carriers for the H_2O oxidization. The sensitization utilizing light-harvesting antennas grown by MLD might enhance the photosynthesis efficiency further [24].

A concept of the integrated photonic/electronic/chemical system (IPECS) is shown in Fig. 10.33 [1, 9–12, 17, 18]. IPECS contains fluidic circuits for supplying liquids such as electrolyte, water, and various solvents to the system. Embedded wavelength filters in IPECS enable efficient artificial photosynthesis as well as photovoltaics. Since the structure of the IPECS is similar to that of the 3D-stacked OE MCM shown in Fig. 3.2b, similar fabrication processes can be applied to the IPECS construction.

References

1. T. Yoshimura, "Proposed applications of 3D optical interconnect technologies to integrated chemical systems," *2007 Digest of the LEOS Summer Topical Meetings (Bio-Inspired Sensors and Application/Imprinting on Photonic Integrated Circuits)*, Portland, Oregon, 129–130 (2007).
2. T. Yoshimura, A. Oshima, D. Kim, and Y. Morita, "Quantum dot formation in polymer wires by three-molecule molecular layer deposition (MLD) and applications to electro-optic/photovoltaic devices," *ECS Trans.* **25**(4), "Atomic Layer Deposition Applications 5," from the Vienna, Austria Meeting, 15–25 (2009).
3. R. Shioya and T. Yoshimura, "Design of solar beam collectors consisting of multi-layer optical waveguide films for integrated solar energy conversion systems," *J. Renewable Sustainable Energy* **1**, 033106-1-15 (2009).
4. T. Yoshimura, H. Watanabe, and C. Yoshino, "Liquid-phase molecular layer deposition (LP-MLD): potential applications to multi-dye sensitization and cancer therapy," *J. Electrochem. Soc.* **158**, 51–55 (2011).
5. T. Yoshimura, R. Ebihara, and A. Oshima, "Polymer wires with quantum dots grown by molecular layer deposition of three source molecules for sensitized photovoltaics," *J. Vac. Sci. Technol. A* **29**, 051510 (2011).
6. T. Yoshimura and S. Ishii, "Effect of quantum dot length on the degree of electron localization in polymer wires grown by molecular layer deposition," *J. Vac. Sci. Technol. A* **31**, 031501 (2013).
7. T. Yoshimura, "Nanostructured materials formed by molecular layer deposition for enhanced solar energy utilization with optical waveguides," in *Energy Efficiency and Renewable Energy through Nanotechnology, Green Energy and Technology*, Springer-Verlag London, 351–390 (2011).
8. T. Yoshimura, "Thin-film molecular nanophotonics," Chapter 8 in *Photonics, Vol. 2: Nanophotonic Structures and Materials*, Wiley, 261–310 (2015).
9. T. Yoshimura, *Thin-Film Organic Photonics: Molecular Layer Deposition and Applications*, CRC/Taylor & Francis, Boca Raton, Florida (2011).
10. T. Yoshimura, *Optical Electronics: Self-Organized Integration and Applications*, Pan Stanford Publishing, Singapore (2012).

11. T. Yoshimura, *Self-Organized Lightwave Networks: Self-Aligned Coupling Optical Waveguides*, CRC/Taylor & Francis, Boca Raton, Florida (2018).
12. T. Yoshimura, *Molecular Nano Systems: Applications to Optoelectronic Computers and Solar Energy Conversion*, Corona Publishing Co., Tokyo (2007) [in Japanese].
13. B. O'Regan and M. Gratzel, "A low-cost, high-efficiency solar cell based on dye-sensitized colloidal TiO_2 films," *Nature* **353**, 737–740 (1991).
14. T. Yoshimura, S. Bai, H. Tateno, and C. Yoshino, "*In situ* photocurrent spectra measurements during growth of three-dye-stacked structures by the liquid-phase molecular layer deposition," *J. Appl. Phys.* **122**, 015309 (2017).
15. K. Sayama and H. Arakawa, "Dream of artificial synthesis," in *Newton: Next-Generation Technology*," Newton Press, Tokyo, 102–109 (2002) [in Japanese].
16. S. Sato, T. Arai, T. Morikawa, K. Uemura, T. M. Suzuki, H. Tanaka, and T. Kajino, "Selective CO_2 conversion to formate conjugated with H_2O oxidation utilizing semiconductor/complex hybrid photocatalysts," *J. Am. Chem. Soc.* **133**, 15240–15243 (2011).
17. T. Yoshimura, "Integrated photonic/electronic/chemical systems, solar energy conversion systems, light collectors, and optical waveguides," Japanese Patent, Tokukai 2009-4717 (2009) [in Japanese].
18. T. Yoshimura and K. Asama, "Integrated chemical systems and integrated light energy conversion systems," Japanese Patent, Tokukai 2007-107085 (2007) [in Japanese].

Chapter 11

Embodiments Disclosed in Patents

Inventions related to self-organized 3D integrated optical interconnects described in previous chapters have been disclosed in US patents [1–15]. In the present chapter, some embodiments in patents are presented for integrated OE MCMs, 3D optical interconnects, micro optical link modules, and active optical sheets/boards/connectors.

11.1 Integrated OE MCMs

Figure 11.1a shows an embodiment of integrated OE MCM 10. The OE MCM interconnects LSI chips 1a–1d by optical links and electrical traces for inter-chip connections. It also interconnects inputs/outputs within individual chips for intra-chip connections. The OE MCM comprises base substrate 12 and active layer 20. The active layer comprises optical waveguides 24a–24h, OE devices such as optical switches/light modulators 26a–26c and PDs 28a–28c, electrical traces 30, and electrical connection pads 32 for LSI chips. LSI chips are attached to the active layer and are electrically coupled to the connection pads by connection technologies like flip-chip bonding. Waveguides and OE devices are embedded in the active layer.

Self-Organized 3D Integrated Optical Interconnects with All-Photolithographic Heterogeneous Integration
Tetsuzo Yoshimura

ISBN 978-981-4877-04-6 (Hardcover), 978-1-003-04305-8 (eBook)
www.jennystanford.com

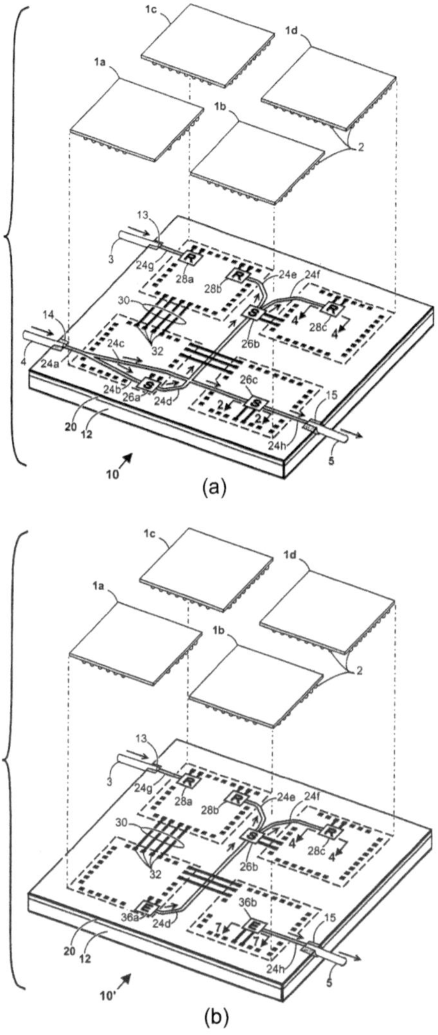

Figure 11.1 Integrated OE MCM for inter-chip and intra-chip connections. E-O conversion is performed by (a) optical switches/light modulators and (b) light-emitting devices such as LEDs, LDs, VCSELs.

Signals are conveyed electrically by the traces and optically by the waveguides. When the signals are conveyed optically, the optical switches/light modulators and PDs respectively perform the E-O and O-E conversions.

For example, lightwaves are brought to the OE MCM by optical fiber 4 and are coupled to waveguide 24a. Here, the lightwaves are continuous waves or pulse trains. The coupling between the fiber and the waveguide is accomplished by standard V-groove connector 14. The coupling is also possible using a vertical mirror or a grating in the waveguide. The lightwaves propagate in the waveguide and branch to be guided in two waveguides 24b and 24c using a Y-branching waveguide. The lightwaves in waveguide 24b are routed to optical switch/light modulator 26a, which selectively routes the lightwaves into waveguide 24d depending upon the electrical signals for the optical switch/light modulator. The electrical signals are provided to the optical switch/light modulator by two connection pads, which in turn are coupled to the circuitry in LSI chip la through solder bumps. In this manner, electrical output signals from the LSI chip are converted to optical signals.

From optical switch/light modulator 26a, waveguide 24d is routed to another optical switch/light modulator 26b. The optical switch/light modulator has one optical input, which is coupled to waveguide 24d, and one electrical input, which is coupled to the circuitry in LSI chip 1d. The optical switch/light modulator also has two optical outputs, which are coupled to waveguides 24e and 24f, respectively. Depending upon the electrical input to the optical switch/light modulator, it either routes the optical signals, which are generated in optical switch/light modulator 26a, to one of its optical outputs. The optical signals on waveguides 24e and 24f are provided to PD 28b and PD 28c, respectively. They convert the optical signals to electrical signals, and provide the signals to LSI chips 1c and 1d, respectively, through corresponding connection pads 32. In the case that the switchable routing is not necessary, the output of waveguide 24d can be directly connected to PD 28b or PD 28c, or to both PDs.

In a similar manner, the lightwaves in waveguide 24c are routed to optical switch/light modulator 26c, which is controlled by an electrical output signal from LSI chip 1b. Optical signals generated in the optical switch/light modulator are routed into waveguide 24h, which is coupled with fiber 5, to be transported off of OE MCM 10.

PDs are also used to receive optical signals from outside of the OE MCM. An example is shown with PD 28a, which receives optical signals from fiber 3 through waveguide 24g. The electrical signals from the PD are provided to LSI chip 1c through connection pads.

Although it is not shown in the figure, waveguides can be routed from one terminal of an LSI chip to another terminal of the same LSI chip, thereby providing intra-chip optical interconnects.

The number of waveguides, OE devices, electrical traces, electrical connection pads, fibers, and LSI chips shown in Fig. 11.1a have been kept to a low number for the sake of visual simplicity. OE MCMs have many more of these components than shown. The widths of the waveguides are on the order of 0.5 μm to 50 μm, and the dimensions of LSI chips are on the order of 1 cm. Typical widths, lengths, and thicknesses of OE devices are respectively on the order of 1–50 μm, 1—5,000 μm, and 0.5–10 μm. Although Fig. 11.1a shows that each OE device is provided with two electrical connections from a corresponding LSI chip, it might be appreciated that the active layer comprises a ground plane (or power plane) within it, and that an OE device has one of its electrical terminals connected to this plane and its other electrical terminal coupled to output or input of a corresponding LSI. Instead of fibers 3–5, arrayed fibers, film waveguides, or imaging guides can be used.

Figure 11.1b shows another embodiment of integrated OE MCM 10′. Common reference numbers have been used to designate common elements of OE MCM 10′ and OE MCM 10. In OE MCM 10′, light emitting devices 36a and 36b are used in place of optical switches/light modulators 26a and 26b of OE MCM 10. Since light-emitting devices do not need lightwaves from external light sources, OE MCM 10′ does not require fiber 4. The light-emitting devices could be LEDs, LDs, VCSELs, and so on.

A structural example of the optical switch/light modulator is depicted in Fig. 11.2 for the case of 26c. A top view of optical switch/light modulator 26c in relation to waveguides 24c, 24h, and 24i, and a partial cross-sectional view of active layer 20 are depicted in Figs. 11.2a and b, respectively. The active layer comprises a patterned waveguide core layer for waveguides 24c, 24h, and 24i. The waveguide layer is formed above undercladding layer 21, and is covered over by overcladding layer 23. The overcladding layer extends over the sides of the waveguides. The materials of the cladding layer and the core layer can be, for example,

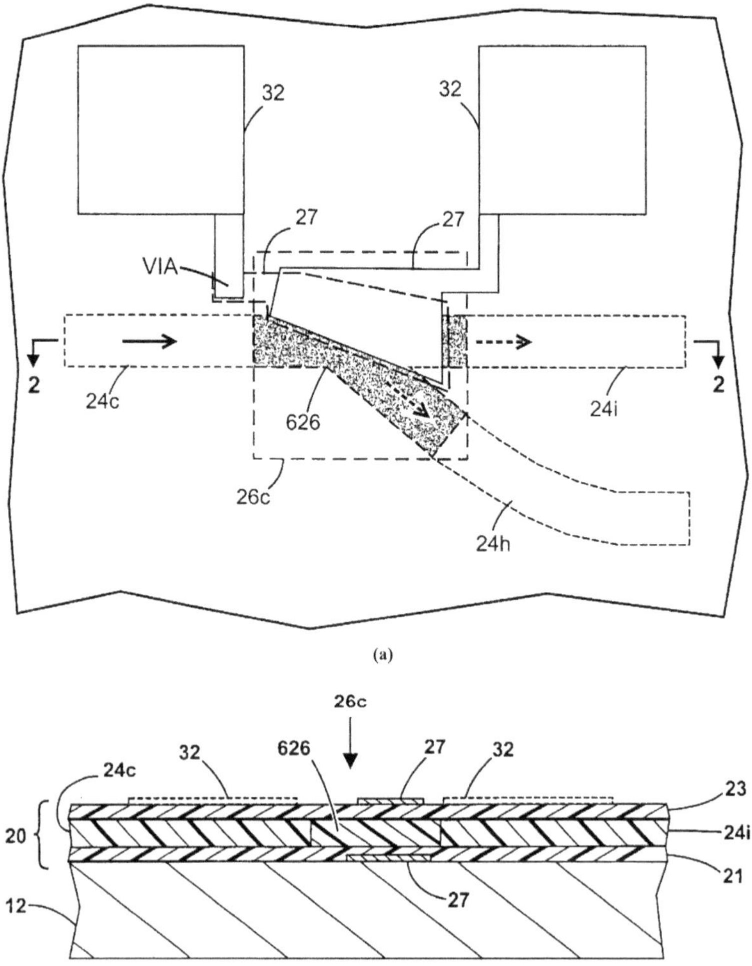

Figure 11.2 Structural example of optical switch/light modulator. (a) Top view and (b) cross-sectional view.

fluorinated polyimide OPI-N1005 and OPI-N3405 (Hitachi Chemical Co.), respectively. If base substrate 12 can function as a suitable cladding layer, undercladding layer 21 can be omitted.

The optical switch/light modulator can be MZMs, MZMs combined with directional couplers, MRs, WPD optical switches, ITR optical switches, digital optical switches, grating-type optical switches, electro-absorption (EA) light modulators, semiconductor optical gate switches, etc. In the example shown in Fig. 11.2, an ITR optical switch is used. It comprises EO material 626, whose refractive index changes by applying an electric field across it. The EO material, which is positioned between the undercladding layer and the overcladding layer, is formed in a Y-shaped structure. The electric field is applied by two opposing electrodes 27. Short electrical traces connect the electrodes to respective connection pads 32. When no voltage is applied across the electrodes, lightwaves travel from waveguide 24c to waveguide 24i. When a voltage is applied between the electrodes to reduce the refractive index of the EO material, total internal reflection occurs at the boundary between the area with electrodes and that without the electrodes. As a result, the lightwaves are switched to travel from waveguide 24c to waveguide 24h. The EO material can be PLZT, LN, III–V semiconductor, polymer MQDs, molecular crystals, etc.

A structural example of PD 28c is depicted in Fig. 11.3. Small/thin semiconductor die 628 with two electrodes 27 formed at the top surface is embedded in the active layer by the ELO. The die is independently prepared and is then placed on the top of undercladding layer 21 on base substrate 12, on which waveguides have already formed. When optical signals are introduced from waveguide 24f to the die, they are guided in the die to induce a voltage and/or a current across the electrodes, generating electrical signals for the input of LSI chips. The PD can have interdigitated electrode configurations. The PD can also have sandwich-type electrode configurations. The die is a thin-film semiconductor with a p–n or p–i–n junction. The p-type doped region is electrically coupled to one electrode 27 and the n-type doped region electrically coupled to the other electrode 27. Intrinsic semiconductors with two ohmic contacts to them, bipolar phototransistors, and field-effect phototransistors can also be used.

Figures 11.4a and b show another structural example. PD 28c′ is embedded into the active layer by the ELO. Load resistor 29′, which is required to generate voltage signals from a PD, is a serpentine NiCr film formed on the active layer. Optical signals are introduced from the waveguide to the PD using a vertical mirror. One of the

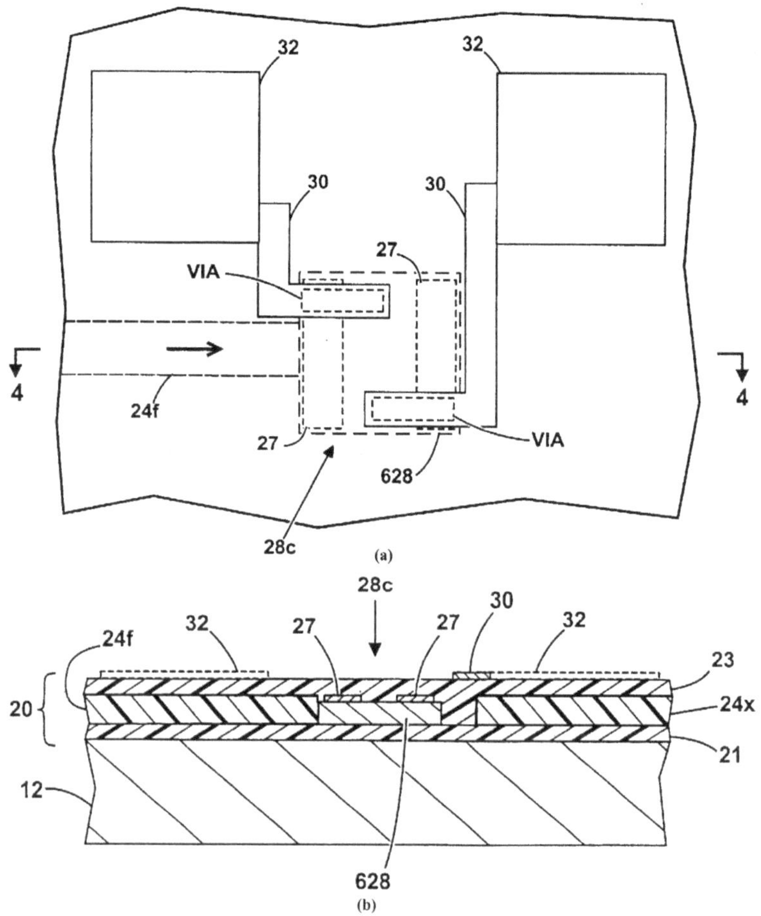

Figure 11.3 Structural example of PD. (a) Top view and (b) cross-sectional view.

electrodes is led to the bottom side of the structure using via. Although this example is for vertical-type PDs, the arrangements and considerations can be applied to lateral-type PDs as well. The load resistor can be integrated into the ELO PD. Alternately, it might be possible to integrate preamplifiers into the ELO PD.

Figures 11.5a and b respectively show a top view in relation to waveguide 24h and a partial cross-sectional view of light-emitting device 36b′. The device comprises VCSEL 638 formed in small/thin

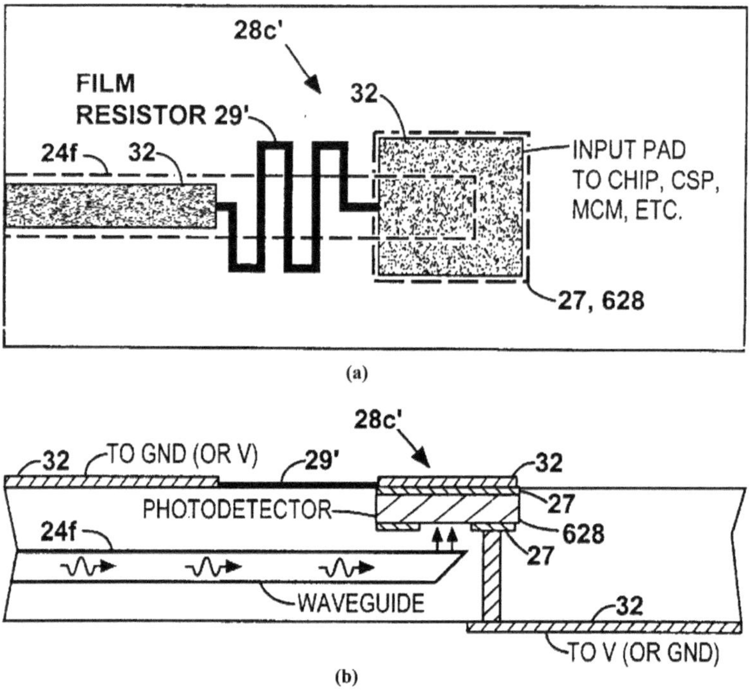

Figure 11.4 Another structural example of PD. (a) Top view and (b) cross-sectional view.

semiconductor die 636. The die lies below core layer 24h, and a vertical mirror 639 is positioned in front of the VCSEL to reflect the emitted lightwaves into waveguide 24h. The waveguide end located over the VCSEL is beveled at 45° angle with respect to the surface-normal direction of the VCSEL surface. The beveling can be accomplished by tilted laser abrasion, tilted RIE, or tilted lithographic exposures with the built-in mask method. The vertical mirror is built upon the beveled surface by depositing a reflective metal or reflective dielectric multilayers over this area.

As shown in Fig. 11.5b, small/thin semiconductor die 636 is embedded in a material layer 25b, which underlies undercladding layer 21. To ensure that the die is attached to base substrate 12, if necessary, adhesive layer 25a is formed over the base substrate prior to forming the material layer. These layers are made of polymeric

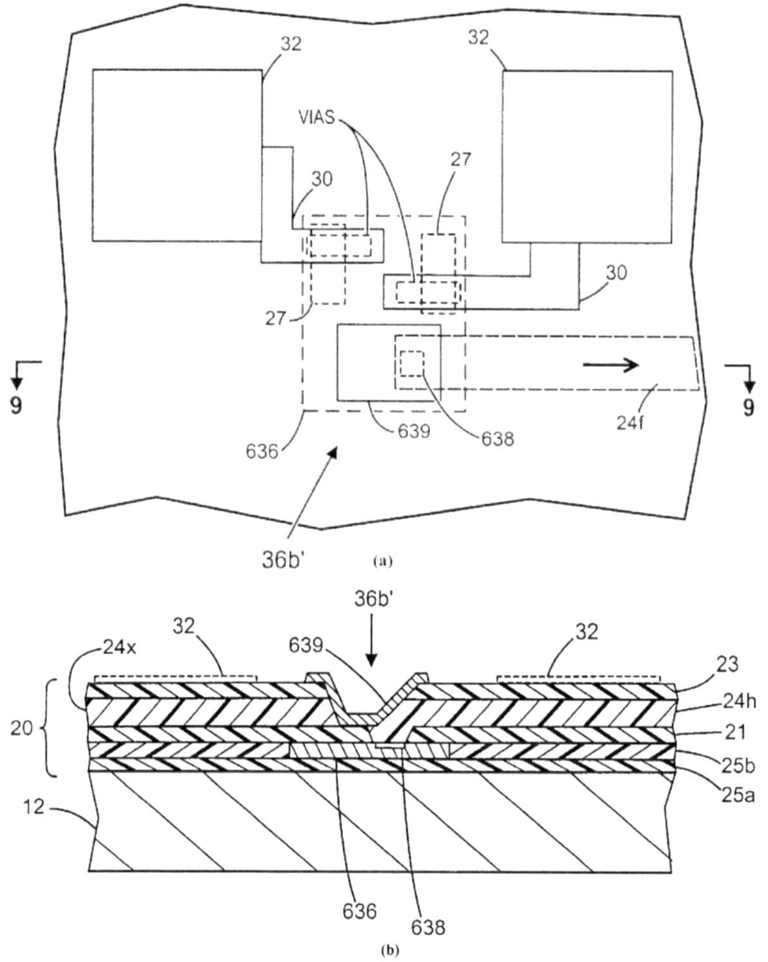

Figure 11.5 Light-emitting device of VCSEL formed in a small/thin semiconductor die. (a) Top view and (b) cross-sectional view.

materials such as polyimide, or materials of the cladding/core layer. The electrodes 27 of the light-emitting device are sometimes located on the bottom surface of the device, or on the bottom and top surfaces. In these case, the bottom electrodes can be diffusion-bonded to electrical traces formed on the surface of the base substrate (or on the cured adhesive layer). High temperature underfill material is then preferably dispensed to fill the air pockets underneath the component that has been diffusion bonded. The material of the material layer can also fill the air gap. A vertical-type PD can be embedded using a similar process to construct a structure similar to that shown in Fig. 11.5.

11.2 3D Optical Interconnects

11.2.1 Horizontal Layer Attachment

As described in Chapter 3, various types of optical interconnect structures can be built by S-FOLMs. The first example is depicted in Fig. 11.6. The structure is constructed by the horizontal attachment of OE MCM films 20 to OE PCB 100. Here, the OE MCM film is active layer 20 that is obtained by removing OE MCM base substrate 12 from OE MCM 10 or 10′ shown in Fig. 11.1. The OE PCB houses four OE MCM films, each of which houses four LSI chips. The OE PCB comprises active layer 120 having a plurality of optical waveguides 124a–124h. The waveguides are made from undercladding layer 121, patterned core layer 124, and overcladding layer 123 (see Fig. 11.7). In the waveguides a plurality of vertical beam splitters 154 as well as vertical optical couplers 156 and 160 are formed for transmitting optical signals from the OE PCB to the OE MCM films and vice versa. In order to show the routing of the waveguides, two of the four OE MCM films are not shown in Fig. 11.6. The distributions of waveguides, vertical beam splitters, and vertical optical couplers underneath the upper-left and upper-right OE MCM films are the same as those underneath the lower-left and lower-right OE MCM films.

Light-emitting device 136, whose structure is the same as that illustrated in Fig. 11.5, is embedded in the OE PCB for a light source. The output of the light-emitting device is coupled to waveguide 124a, which is routed to the upper-left OE MCM film, and is further extended to the upper-right OE MCM film. Waveguide124a passes through horizontal beam splitter 164, which divides off a portion of the guided lightwave to waveguide 124b. Waveguide 124b is routed to the lower-left OE MCM film, and further over to the lower-right

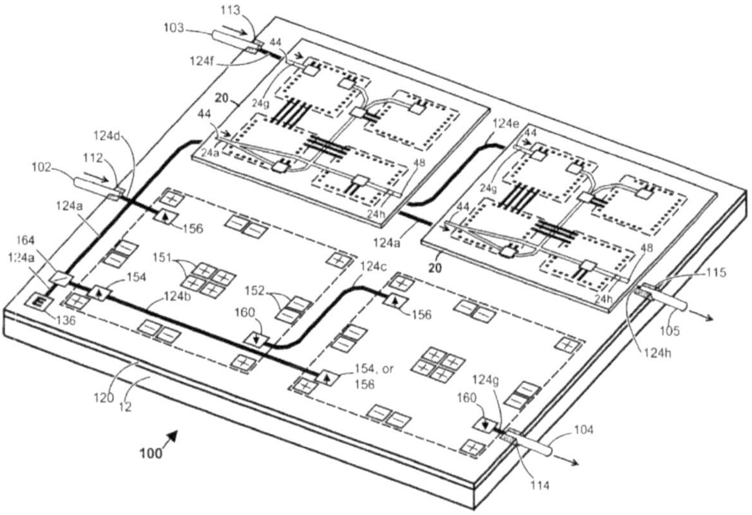

Figure 11.6 3D Optical Interconnects constructed by the horizontal attachment of OE MCM films onto OE PCB.

OE MCM film. The waveguide passes through vertical beam splitters 154, which direct a portion of the guided lightwave upward toward the OE MCM films to couple it into waveguides 24a in the lower-left and lower-right OE MCM films via receiving vertical optical couplers 44.

Waveguide 124c conveys a lightwave between the lower-left OE MCM film and the lower-right OE MCM film. The lightwave guided in waveguide 24h in the lower-left OE MCM film is transmitted vertically toward the surface of the OE PCB via a vertical optical coupler 48. The lightwave is introduced into waveguide 124c via vertical optical coupler 160 and is conveyed to vertical optical coupler 156, which is positioned underneath the lower-right OE MCM film. The lightwave is introduced into waveguide 24g via vertical optical coupler 156 and receiving vertical optical coupler 44. Waveguide 124e is configured similarly to waveguide 124c to convey a lightwave from the upper-left OE MCM film to the upper-right OE MCM film.

Waveguide 124d is coupled to optical fiber 102 by conventional V-groove connector 112 to receive optical signals from outside.

Waveguide 124d is coupled to vertical optical coupler 156 at its another end to introduce optical signals into waveguide 24g of the lower-left OE MCM film via receiving vertical optical coupler 44. Optical waveguide 124f is similarly coupled to optical fiber 103 and vertical optical coupler 156 to introduce optical signals into waveguide 24g of the upper-left OE MCM film via receiving vertical optical coupler 44. Waveguide 124g is coupled to optical fiber 104 and receiving vertical optical coupler 160. The optical signals from the lower-right OE MCM film are conveyed to the optical fiber to be sent outside. Waveguide 124h is similarly coupled to receiving vertical optical coupler 160 and optical fiber 105.

Power supply voltages are supplied from the OE PCB to the OE MCM films by conductive pads 151 and 152. The OE PCB has electrical traces for conveying electrical signals between OE MCM films. In some cases, these electrical traces might be formed in dielectrically isolated conductive layers with vias being formed to respective connection pads.

To convey optical signals to and from OE MCM films or the OE PCB, arrayed waveguide films can be used instead of optical fibers. Such a waveguide film comprises a flexible polymer sheet having plurality of optical waveguides, and usually having vertical optical couplers at the end of the waveguides. One edge of the waveguide film adheres to an area of active layer 120 in such a way that its vertical optical couplers in its edge are aligned with corresponding vertical optical couplers in active layer 120. Optical fiber arrays or imaging guides are also useful. In some cases, optical power is supplied from outside by an external light source instead of light-emitting device 136 placed on the OE PCB.

Figure 11.7 shows a cross-sectional view of OE PCB 100 and upper-left OE MCM film 20 in a vertical coupling region, where optical couplings are performed by vertical beam splitter 154 and receiving vertical optical coupler 44. The base substrate 12 has been removed from OE MCM 10 or 10′ to have OE MCM film 20. Generally speaking, it is desirable to use materials with high thermal stability (e.g., can withstand high processing temperatures like polyimide, epoxy) for layers 21, 23, and 24 in the OE MCM film. The OE PCB comprises conductive layer 151 and second conductive layer 152 for providing the ground level or voltages from electrical power

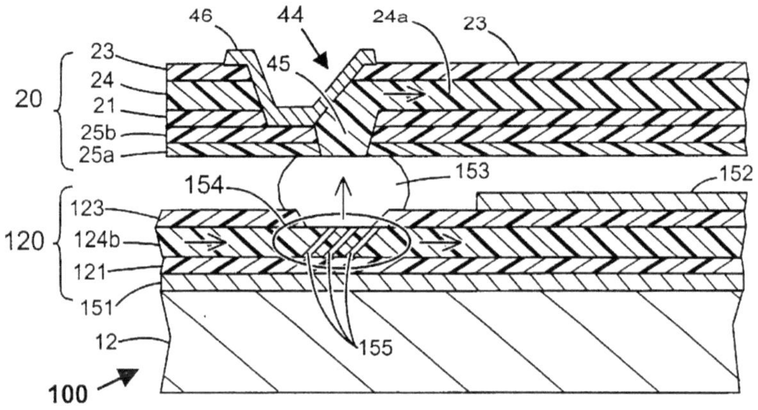

Figure 11.7 Cross-sectional view of OE PCB and upper-left OE MCM film in a vertical coupling region shown in Fig. 11.6. Optical couplings are performed by a vertical beam splitter and a receiving vertical optical coupler.

supplies, undercladding layer 121, core layer (waveguide) 124b, and overcladding layer 123. Base substrate 12 of OE PCB 100 comprises aluminum, silicon, quartz, glass, polymer, and so on. If necessary, the OE PCB base substrate can be removed by, for example, etching or laser ablation. For the etching, a protective etch-stop layer disposed between the OE PCB base substrate and OE PCB active layer 120 improves the process reliability. By using a silicon wafer with an aluminum top layer as the OE PCB base substrate, the lateral etching of the aluminum layer from the sides of the wafer can be achieved to separate the OE PCB active layer from the silicon wafer.

For receiving vertical optical coupler 44 in the OE MCM film, a vertical mirror is formed on the end portion of the core and cladding layer. The fabrication process is similar to that for the vertical mirror, which couples the output of a VCSEL to a waveguide illustrated in Fig. 11.5. That is, after a beveled surface is constructed, a layer of metal or reflective material is deposited over the beveled surface. Optical via 45 is prepared through layers 25a, 25b, and undercladding layer 21 by forming an aperture in these layers by laser ablation or RIE, and then filling them with the core material. In some cases, the optical vias are not necessary. The anti-reflection

(AR) coating applied to the film surfaces is effective to reduce reflection of lightwaves.

For vertical beam splitter 154 in the OE PCB, tilted grating 155 with a periodic refractive index distribution is formed in waveguide 124b. The spacing in the periodic structure is usually on the order of one wavelength of the lightwave. To achieve the vertical reflection, the grating surface should be tilted with an angle of $\sim 45^\circ$ with respect to the lightwave path. When an incident lightwave passes through the tilted grating, a portion of it is reflected to the vertical direction while other portion is transmitted through the grating. The ratio of the reflected and transmitted portions is dependent upon the refractive index difference in the periodic structure, and upon the number of periods in the grating. To fabricate beveled trenches for the tilted grating, the built-in mask method, MNE, or MNA with the tilted exposure can be used (see sections 4.1–4.5). MLD and ALD are suitable to fill the beveled trenches with dielectric materials due to their ultrathin/conformal thin-film growth capability as described in section 12.2. Instead of using a grating, a semitransparent mirror or a multilayer dielectric filter can be used to perform the same function of vertically reflecting a fraction of the lightwave upwards. The tilted multilayer dielectric filter can be deposited on a beveled waveguide wall by MLD or ALD.

The lightwave reflected vertically from tilted grating 155 passes through optical glue 153 to OE MCM film 20. The optical glue has a refractive index, which is relatively close to that of core materials 124 and 24. The optical glue improves transmission efficiency between the OE PCB and the OE MCM film by reducing the lightwave reflection. When PRI materials are used for the optical glue, a self-aligned vertical waveguide of SOLNETs between the OE PCB and the OE MCM film can form in the optical glue by providing write beams through the waveguides in a similar way as that illustrated in Figs. 8.1 and 8.3. In addition to the optical glue, bonding sheets or underfill materials can be used to improve the optical coupling performance. Bonding sheets and underfill materials with PRI characteristics enable us to form self-aligned vertical waveguide of SOLNETs

Figure 11.8 shows a cross-sectional view of OE PCB 100 and upper-right OE MCM film 20 in a vertical coupling region, where

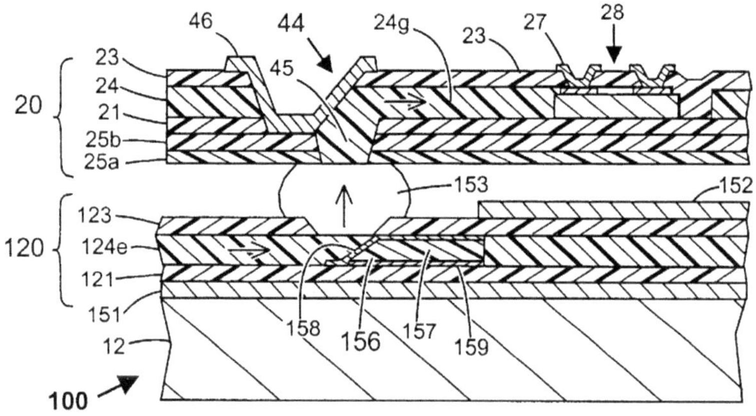

Figure 11.8 Cross-sectional view of OE PCB and upper-right OE MCM in a vertical coupling region shown in Fig. 11.6. Optical couplings are performed by a vertical optical coupler and a receiving vertical optical coupler.

waveguide 124e meets vertical optical coupler 156. The structure is almost the same as that depicted in Fig. 11.7 except for the structure of the vertical optical coupler in the OE PCB. The vertical optical coupler comprises a reflective metal layer 158 formed on a beveled edge of the waveguide. Detailed examples are disclosed by Yoshimura et al. [5]. The beveled structure is formed by the built-in mask method, MNE, and MNA with tilted exposure, tilted RIE, or stamping. Vertical optical coupler 48 and receiving vertical optical coupler 160 have similar structure as 44 and 156, respectively.

Figure 11.9a shows a cross-sectional view of chip layer 350 of OE-Film (*C*) attached onto active layer 320. Two thinned LSI chips 351 and 352 are encapsulated in a dielectric film to form the chip layer. The active layer contains VCSEL 336 and PD 328. The chip layer is laminated to the active layer by anisotropic conductive film 302. Although the active layer structure has only been shown schematically for simplicity, the basic layer structure is the same as that shown in previous embodiments. The relative sizes of the active components to the LSI chips and pads are not to scale. The VCSEL and PD are drawn larger than in constructed configurations. The VCSEL is controlled by chip 351. Optical signals emitted from the VCSEL are transmitted to waveguide 324a in the active layer

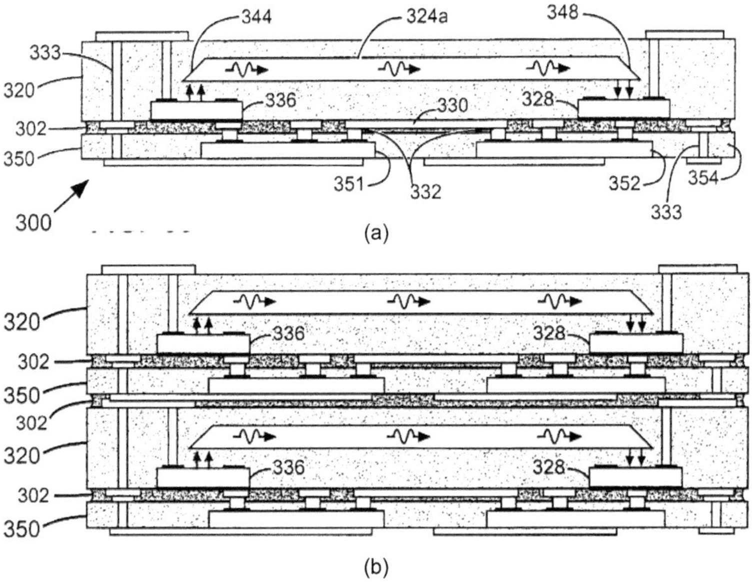

Figure 11.9 (a) Cross-sectional view of chip layer of OE-Film (C) attached to active layer, and (b) multilayer structure of chip layer/active layer/chip layer/active layer.

via vertical optical coupler 344. The waveguide conveys the signals to the PD via vertical optical coupler 348. Electrical outputs of the PD are coupled to circuitry of chip 352. The chip layer and the active layer have electrical traces 330, contact pads 332, and electrical vias 333 that convey electrical signals. The contact pads are coupled to one another through spheres of a conductive material that is dispersed in the anisotropic conductive film to construct Z-connections.

Instead of the anisotropic conductive film, an adhesive bonding sheet can be used for joining the two layers. Holes are made through the sheet in the locations of the contact pads, and a conductive bonding material is disposed on the pads. The two layers are then laminated together by the adhesive bonding sheet with heat and pressure, and at this time, Z-connections are constructed. Z-connections can also be constructed by the solder joint or the metal diffusion joint, which is especially useful for making fine

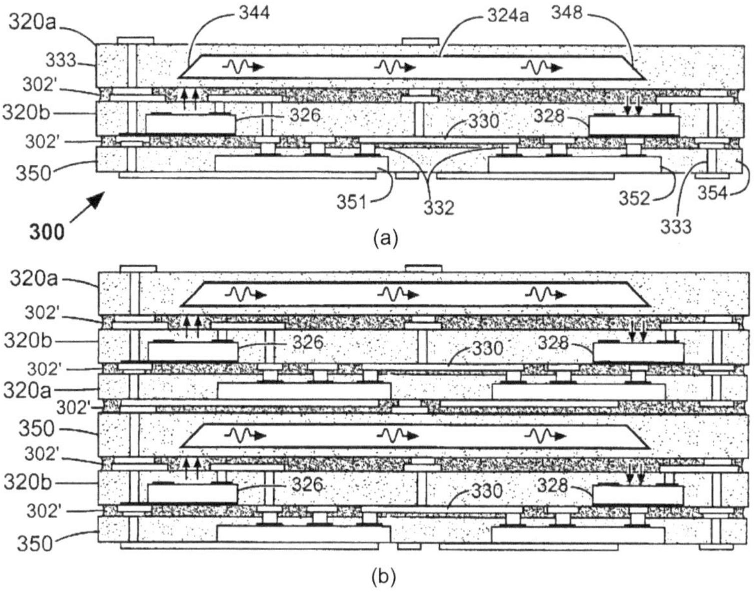

Figure 11.10 (a) Cross-sectional view of chip layer of OE-Film (*C*) attached to active layer, and (b) multilayer structure of chip layer/active layer/chip layer/active layer. The active layer is separated into two layers, namely, a waveguide layer and an active-component layer.

Z-connections. It is possible to use an underfill material instead of the bonding sheet.

Figure 11.9b shows a multilayer structure of chip layer/active layer/chip layer/active layer. The layers are separated from the base substrates used to construct them. The vias in the stacked layers convey ground and power supply voltages to all of the layers, and also covey electrical signals from one layer to another layer. Although optical Z-connections are not shown in the figure, they can be readily realized by vertical optical couplers.

In some cases, the waveguides and the other active components in active layer 320 can be embedded in separate layers, as shown in Figs. 11.10a and b. The active layer has been divided into waveguide layer 320a and active-component layer 320b. These layers are joined together by the bonding sheet, the underfill material, the solder, or metal diffusion described above referring to Fig. 11.9.

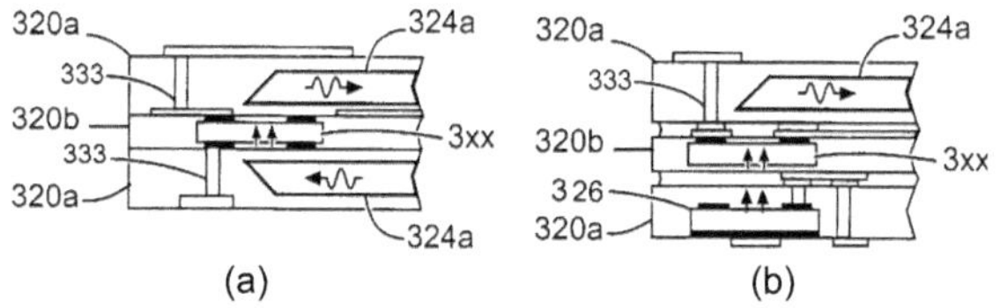

Figure 11.11 Structures, where light modulators are incorporated. (a) The layers are built upon one another by the build-up process, and (b) the layers are separately manufactured and joined together with the Z-connection technology.

Figure 11.11 shows structures, where light modulators are incorporated in the multilayer structures. Figure 11.11a presents a case in which the layers are built upon one another by the build-up process. Two waveguides 324a in two separate waveguide layers 320a are optically coupled by an optical Z-connection through vertical optical couplers. Light modulator 3xx, which is embedded in separate layer 320b, is placed in the optical path between the two vertical optical couplers. The light modulator comprises a body, which is a die of an EO material or an electro-absorption (EA) material, and two electrodes made on the body surfaces. Figure 11.11b presents a case in which the layers are separately manufactured and joined together with the Z-connection technology. A layer containing waveguide 324a, a layer containing a light modulator 3xx, and a layer containing a VCSEL 326 are stacked. A lightwave emitted from the VCSEL is modulated by the light modulator to generate optical signals, and the signals are guided in the waveguide. By inserting PRI underfills or PRI bonding sheets between the individual layers, vertical waveguides of SOLNETs can be formed through several stacked layers and underfill/bonding sheet layers.

In Figs. 11.9–11.11, chips 351 and 352 can be a VCSEL driver and a PD amplifier. Chip layers can be stacked on each other. For example, a first chip layer for processors stacked upon a second chip layer for drivers and amplifiers. The waveguides can also be multilayer structures if the connection density is high. The multiple-layer structures can be fabricated by the build-up process or by the Z-connection process. An additional benefit of structures

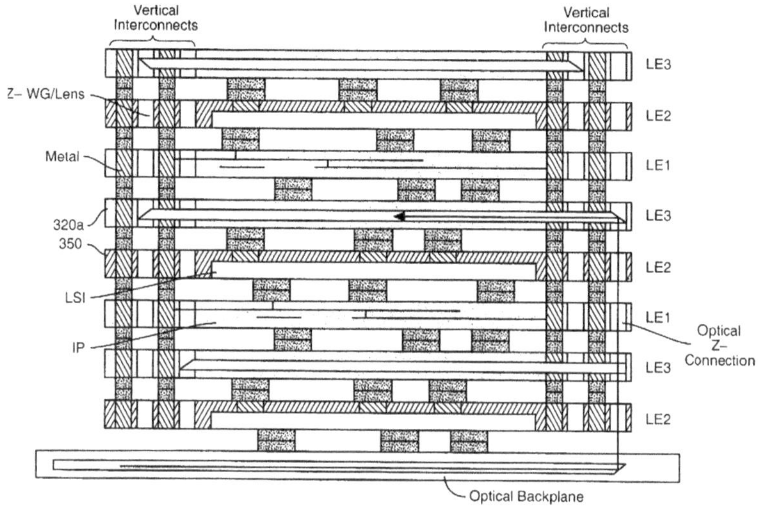

Figure 11.12 Multilayer structure, in which waveguide layers, active layers containing OE components, chip layers, and interposer (IP) layers are made of PRI materials to form vertical waveguides or micro-lenses of SOLNETs for the optical Z-connections.

constructed by the Z-connection process is that unoccupied gaps between the layers 320a, 320b, and 350 are used as channels for a coolant gas or liquid to flow through the structures. One can use driver-integrated VCSELs/light modulators and amplifier-integrated PDs. In such cases, the chips for drivers and amplifiers can be removed while additional pads for the ground/power supply are needed.

Figure 11.12 shows a multilayer structure, in which waveguide layers 320, active layers containing OE devices, chip layers 350, and interposer (IP) layers are made of PRI materials to form vertical waveguides or microlenses of SOLNETs for optical Z-connections. In the figure, the active layers are not drawn for the sake of clarity. The layers are separately manufactured and joined together by the Z-connection process. Bonding sheets made of the PRI materials laminate the layers together.

In the example shown in Figs. 11.13 and 11.14, the optical Z-connections are formed in separate optical Z-connection units

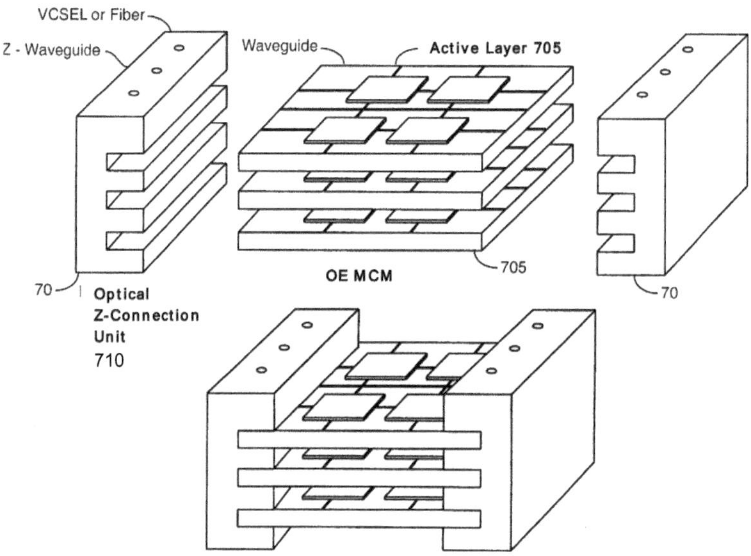

Figure 11.13 Example of optical Z-connection formation using separate optical Z-connection units.

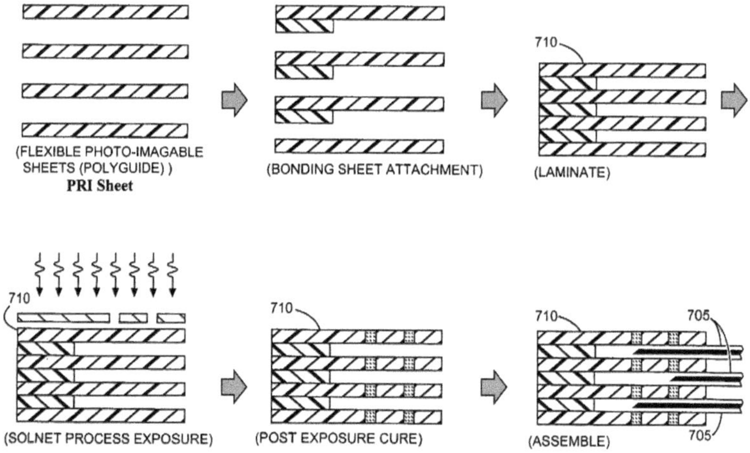

Figure 11.14 Vertical waveguide formation using the sequence of SOLNET fabrication steps.

710, which are coupled to a set of active layers 705. The active layers may take the form shown in Fig. 11.10a. The units 710 have a plurality of vertical waveguides formed through their z direction using the sequence of SOLNET fabrication steps shown in Fig. 11.14. Starting with a plurality of PRI sheets (e.g., polyguide), short sections of bonding sheets are attached to the left edges of the PRI sheets. The PRI sheets are bonded together, and then the SOLNET process is applied to the PRI sheets using writing beams through a photomask to form vertical waveguides. The sheets are cured and are assembled to the active layers 705. The vertical waveguides optically connect to waveguides at the edges of the active layers 705.

11.2.2 Vertical Layer Attachment

In the examples illustrated in Figs. 11.15 and 11.16, the optical interconnect structure is constructed by the vertical attachment of OE MCM 10″ to OE PCB 210. Figure 11.15 shows an overall view of system configuration 200 in a region where the front-most OE MCM abuts against the OE PCB. Figure 11.16 shows examples of the details.

As illustrated in Fig. 11.15, a plurality of OE MCMs 10″ are mounted vertically to OE PCB 210. The OE MCMs are fabricated in the same manner as OE MCMs 10 and 10′ illustrated in Fig. 11.1, except that most of the waveguides are routed to the edge of the OE MCMs which abut the OE PCB. A plurality of waveguides 224a–224g are formed in the OE PCB. These waveguides are used to covey optical signals between the OE MCMs.

When light modulators are used to transmit optical signals instead of light-emitting devices, external light sources are required. An optical power source board 280, which interfaces to the OE PCB in the same manner as OE MCMs, can be used for the external light supplier. The light beam power is conveyed by waveguides 224a–224g. The optical power source board comprises a plurality of LDs whose outputs are routed to the edge of the board. The outputs of two or more LDs can be combined to increase the power level and/or to add lightwaves with different wavelengths on the same waveguide, achieving MUX for WDM. DEMUX is performed by

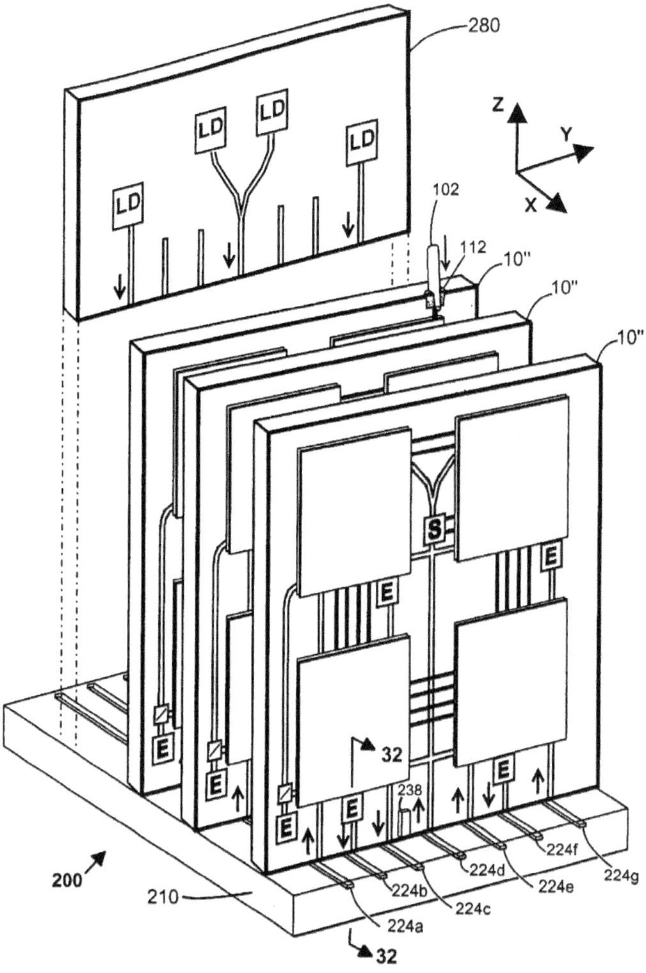

Figure 11.15 3D Optical Interconnects constructed by the vertical attachment of OE MCM onto OE PCB.

grating filters, multilayer dielectric filters, or MR filters on the OE MCM or OE PCB.

In the structural example shown in Fig. 11.16a, OE PCB 210 comprises base substrate 212, active layer 120″, and waveguides formed in the active layer. OE MCM 10″ comprises base substrate 12, active layer 20, and waveguides formed in the active layer. The

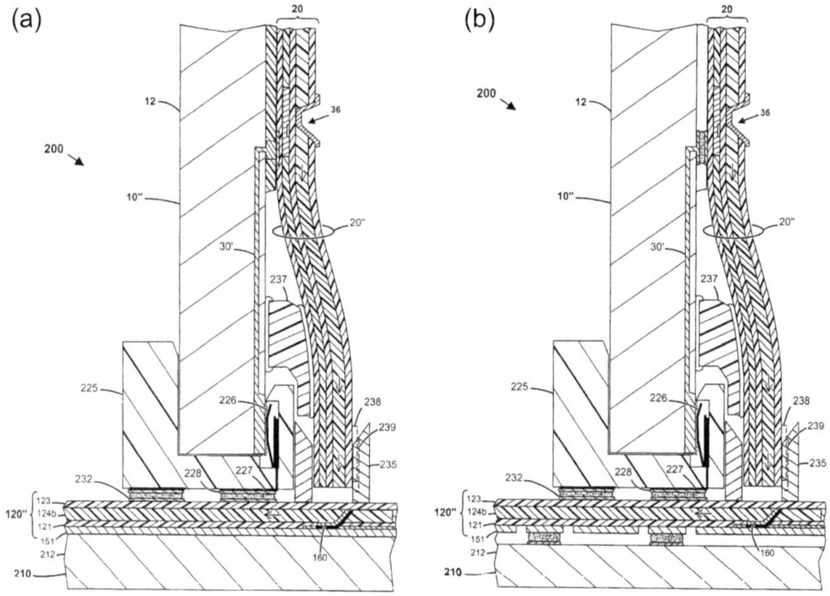

Figure 11.16 Details of the 3D Optical Interconnects. (a) Active layers are directly constructed on respective base substrates. (b) Active layers are constructed separately and then bonded to respective base substrates.

active layers are directly constructed on respective base substrates. Vertical optical couplers 160 are located within the active layer of the OE PCB to couple lightwaves between the waveguides of the OE PCB and the waveguides of the OE MCM or the optical power source board. For the WDM optical interconnects vertical beam splitters, which have a similar structure as that of vertical beam splitter 154 shown in Fig. 11.7, or multilayer dielectric filters are used instead of the vertical optical couplers.

Active layer 20″ of the OE MCM is separated from the base substrate near the edge, where the OE MCM abuts the OE PCB. To construct such a structure, a sacrificial patch of a material disposed on the base substrate near the edge is used. After the active layer is built up on the base substrate, the sacrificial patch is etched away laterally from the edge to make the active layer be separated selectively from the base substrate along the edge.

This configuration provides several benefits. First, it enables the base substrate of the OE MCM to be firmly held by receptacle 225 without damaging the waveguides. Second, it enables electrical traces 30′ on the top surface of the OE MCM base substrate to be routed to the edge of the substrate and mechanically coupled to conventional electrical connector 226 in the receptacle. Each electrical signal on connector 226 is routed to corresponding connection pad 232 of the OE PCB by respective wire 227 and connection pad 228. These pads are soldered together, which conveys the electrical signals, ground, and power, and which also serves to attach the receptacle to the EO PCB. As a third benefit, the separation of the OE MCM active layer from the base substrate enables the waveguides in the layer to be more accurately aligned to vertical optical couplers 160 in the OE PCB active layer. Oftentimes the OE MCM base substrate has a few microns of warpage, which could cause large misalignments between the waveguides and the vertical optical couplers. Second receptacle 235 adhered to the top surface of the OE PCB enables the waveguides to be more accurately positioned with respect to the vertical optical couplers. When the OE MCM is inserted into receptacle 225, the flexibility of the OE MCM active layer enables it to be guided into the second receptacle 235, and therefore into very accurate alignments in the x-axis direction with the vertical optical couplers (see Fig. 11.15 for the coordinate system definition). To achieve accurate alignments of the waveguides to the vertical optical couplers in the y direction, receptacle 235 may comprise a slot 239, whose length is parallel to the z-direction, and the OE MCM active layer may have key 238 disposed on the top surface, which fits snugly within the slot. For the z-direction positioning, a small amount of optical glue (or liquid refractive material) may be disposed in the bottom of receptacle 235 and dried to a tacky state, and the end of the OE MCM active layer may then contact the tacky layer and be adhered thereto. If necessary, anti-reflection coatings may be applied on the edge of the OE MCM active layer and the surface of the OE PCB active layer.

It may be appreciated that conventional optical connectors can be used to couple the waveguides between the OE MCM and the OE PCB. A first waveguide connector is attached to the edge or at

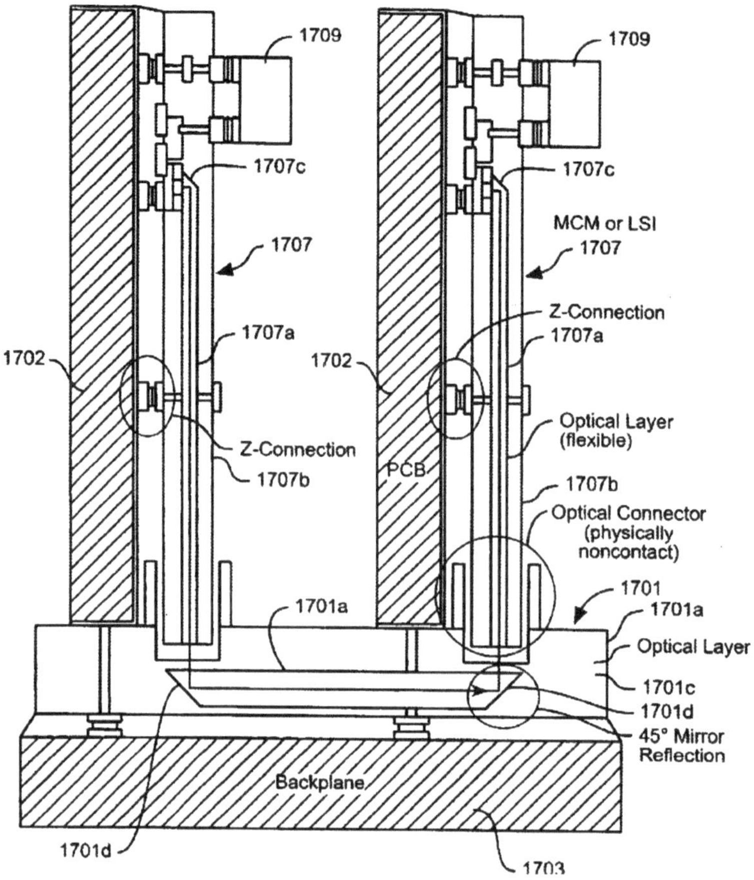

Figure 11.17 3D Optical Interconnects constructed by the vertical attachment of OE PCB onto OE BP.

the surface of the OE MCM active layer, whereas a second surface-normal connector is attached to the OE PCB surface proximate to the vertical optical coupler to connect the first connector and the second connector.

In the structural example shown in Fig. 11.16b, the OE MCM active layer and the OE PCB active layer are constructed separately and then bonded to respective base substrates. This approach enables us to use different technologies for fabricating the electrical interconnections and the optical interconnections. It also makes it easy to separate the OE MCM active layer from the base substrate near the edge where the active layer abuts to the OE PCB.

Figure 11.17 shows an optical interconnect structure constructed by the vertical attachment of OE PCB to OE BP. Waveguide assembly 1701 of the OE BP comprises core layer 1701a with a pair of vertical mirror surfaces 1701d in cladding material 1701c. Each waveguide assembly 1707 of the OE PCB comprises core layer 1707a optically coupled to core layer 1701a of the OE BP through one of mirror surfaces 1701d. Each core layer 1707a is in a cladding material 1707b. Each core layer 1707a also terminates in a vertical mirror surface 1707c. MCMs or LSIs 1709 are coupled to the PCB 1702 through the cladding material 1707b.

11.3 Micro Optical Link Modules

One of the practical methods to improve the optical link performance in computers is optical co-packaging, which extends bandwidths. The micro optical link module based on the S-FOLM is applicable to optical co-packaging because it enables us to directly extract optical signals from LSIs or MCMs and send the signals to other LSIs, MCMs, or system units.

Figure 11.18 shows an example of the micro optical link module. LSI chips are mounted on an OE-Film (*DW*) containing waveguides, VCSELs, and PDs. The OE-Film is attached onto an MCM. An optical fiber array, an image guide, or a waveguide array is connected to

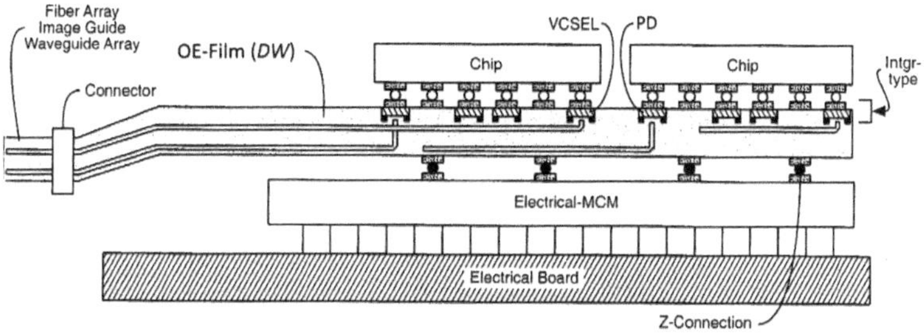

Figure 11.18 Micro optical link module, where an OE-Film (DW) containing waveguides, VCSELs and PDs is attached onto an electrical MCM.

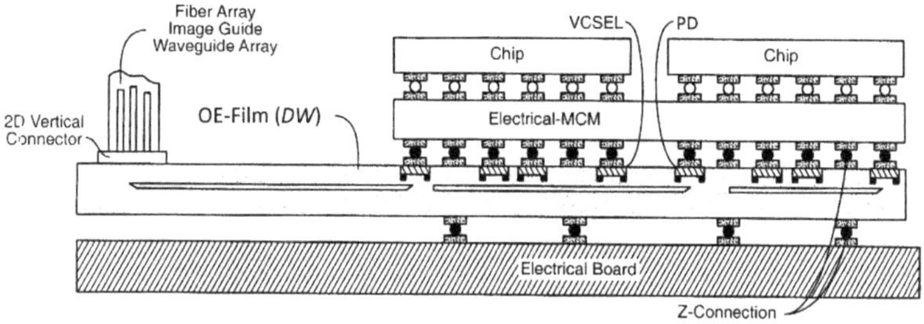

Figure 11.19 Micro optical link module, where an OE-Film (*DW*) containing waveguides, VCSELs and PDs is placed under an electrical MCM.

the waveguides of the OE-Film, enabling optical links between LSIs, MCMs, or system units. The VCSELs, in response to electrical outputs of the LSI chips, emit optical signals, which are transmitted through the optical links. Conversely, optical signals, which come to micro optical link module from outside, are coupled to the waveguides of the OE-Film. The optical signals are converted into electrical signals by the PDs to be input to the LSI chips. In Fig. 11.19 an OE-Film (*DW*) is placed under an MCM and a 2D vertical connector is attached onto the OE-Film surface. In these embodiments, interface films containing driver/amplifier chips can be stacked on the OE-Film (*DW*).

In the S-FOLM-based optical co-packaging, the distances between electrical input/output terminals and E-O/O-E conversion parts are smaller than those distances in conventional optical links. This improves interconnection performance. Furthermore, while extra space is needed for conventional optical link modules, no extra space is needed in S-FOLMs.

Figure 11.20 shows the micro optical link module for optical co-packaging utilizing high-speed and low-heat-generation OE-ADLES. The light modulators are directly driven by the outputs of the LSI chips to generate optical signals from external lights. By increasing the external light power, the optical signals at PDs become sufficiently strong so that no amplifiers are necessary. The OE-ADLES is preferred for optical co-packaging because it permits the elimination of drivers and amplifiers to simplify the optical links.

In Fig. 11.21, details of a micro optical link module for optical co-packaging is shown. Each of the 16 optical connectors contain 12

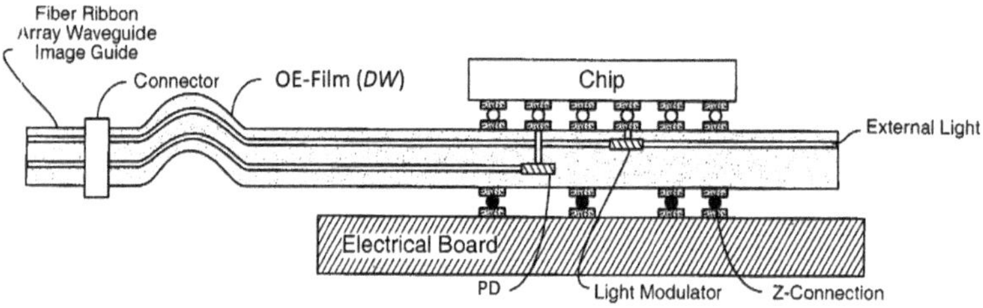

Figure 11.20 Micro optical link module utilizing the high-speed and low-heat-generation OE-ADLES.

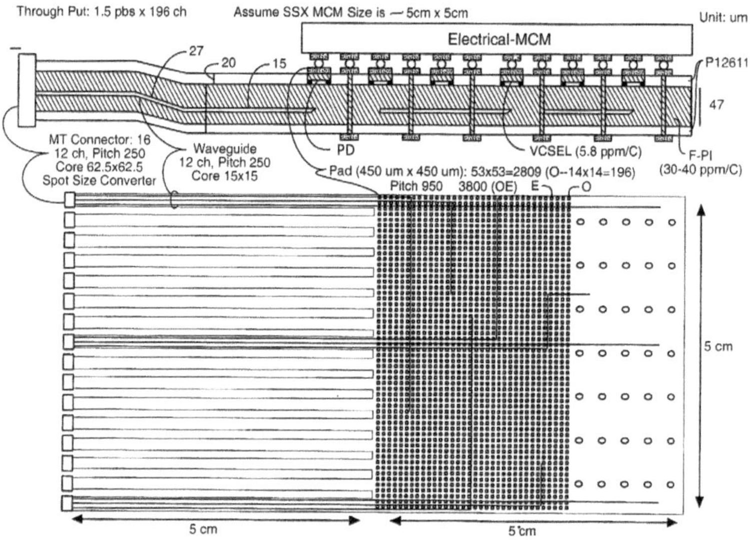

Figure 11.21 Details of a micro optical link module for the optical co-packaging.

waveguides of the S-FOLM, so that there are 196 channels for the communication of optical signals. Spot size converters are inserted to couple S-FOLM waveguides to commercially available optical connectors.

Figure 11.22 shows details of a portion, where a VCSEL is optically coupled to waveguides by a metallized vertical mirror in the S-FOLM. A portion of the Au contact layer of the VCSEL is electrically connected to the mirror metalization, facilitating a convenient electrical connection to the VCSEL. The optical emitting

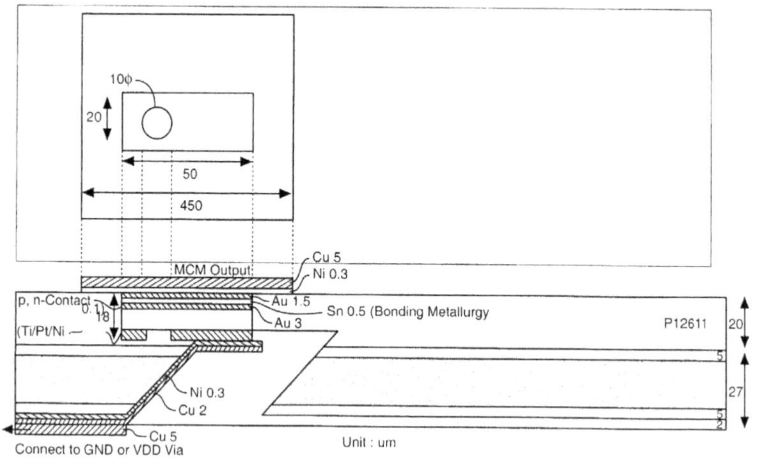

Figure 11.22 Details of a portion, where a VCSEL is optically coupled to waveguides by a metallized vertical mirror in a micro optical link module.

window of the VCSEL is orientated so that it impinges upon the metallized mirror and is reflected into the waveguide.

An exemplary fabrication process for the S-FOLM structure includes the steps of forming a first polyimide film on an Al removable substrate. After the first polymer layer is formed, contact pads and electrodes are deposited on the surface of the polymer layer. Small/thin VCSEL dies are placed onto the pad using the Au/Sn/Au metal diffusion. A second polyimide layer or other suitable polymer layer is then coated over the dies. The surface is then planarized with the chemical-mechanical planarization (CMP). An undercladding fluorinated polyimide layer is formed over the planarized wafer. A core layer is then deposited, patterned, and embedded in an overcladding coating. Vertical mirrors are formed by RIE or laser ablation. The mirror surfaces are then metalized. A polyimide layer is coated over the substrate and planarized by CMP. The Al removable substrate is removed and the first polyimide layer is removed or partially etched to make electrical contacts and/or pads and vias to the OE-Film.

As shown in Fig. 11.23, the optical connectors preferably include connector buffers to perform optical adjustment functions which facilitate communicating signals to the optical connectors. For example, a connector buffer adjusts spot size and changes the

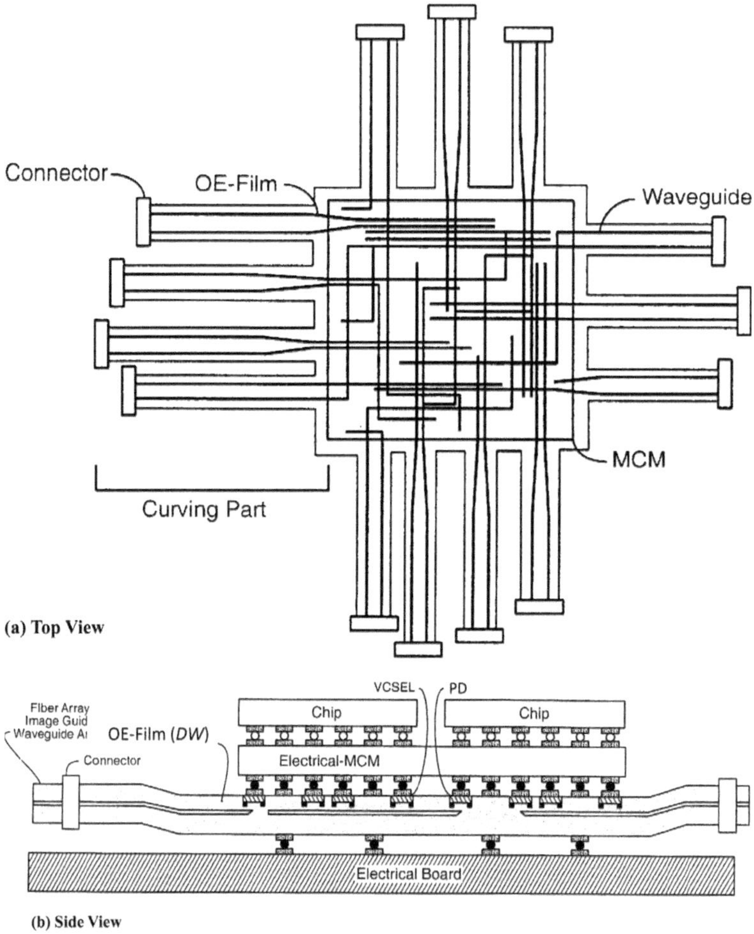

Figure 11.23 Micro optical link module having connector buffers to perform optical adjustment functions.

waveguide pitch (i.e., waveguide separation). In the connector buffer, the waveguide lengths are adjusted to minimize the signal skew. More generally, it is desirable to adjust the waveguide routes or path lengths, so that signals have the same transit time from the output to the input of the destinations. In addition, the connector buffer has curved waveguide paths so that one or more planar 1D arrays of polymer waveguides are coupled to a 2D array of waveguides (see Fig. 4.32b).

11.4 Active Optical Sheets, Boards, and Connectors

The optical tap guide described in Chapter 3 (in Fig. 3.6e) is practically enabled by active optical sheets and boards. They provide optical wiring, wherein the optical wiring section is separate and independent of electrical circuit boards, and the optical wiring section and the electrical wiring section can be freely incorporated. The active optical sheets and boards generate optical signals by probing at desired locations of voltage generation on pads, terminals, electrodes, lines, etc.

Figures 11.24a and b respectively show conceptual drawings of the transmitter part and the receiver part in an active optical

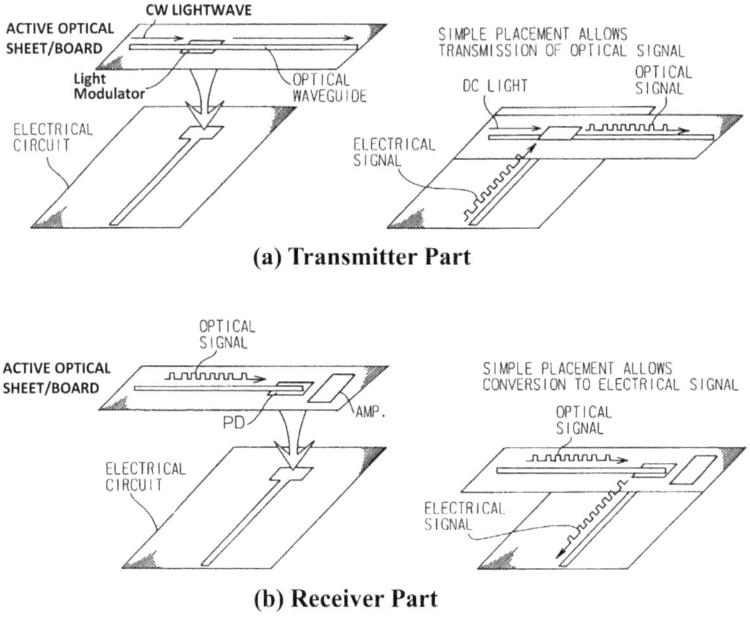

Figure 11.24 Conceptual drawings of (a) transmitter part and (b) receiver part in an active optical sheet/board for a single channel case.

sheet/board for a single channel case. When the electrode of a light modulator contacts an electrode of an electrical circuit board, the light modulator, which is driven by the voltage from the electrode of the electrical circuit board, modulates the CW lightwave to generate optical signals. The optical signals are transmitted from the transmitter part to the receiver part to be converted into electrical signals by a PD. The electrical signals come into another electrode in the same electrical circuit board or in other electronic units. Thus, simple optical interconnects are achieved. In the case that the transmitter part and the receiver part are in the same sheet, the optical signals are directly guided from the transmitter part to the receive part. It is also possible to connect both parts by an optical fiber array or waveguide array. The same functions are available when an active optical board is used instead of the active optical sheet. The sheets/boards are made of polymers such as polyimide, epoxy, silicone and acryl, glass, Si wafers, etc., and the thickness might typically be on the order of 1–1000 μm. Light modulators are typically MZMs, MRs, WPDs, etc.

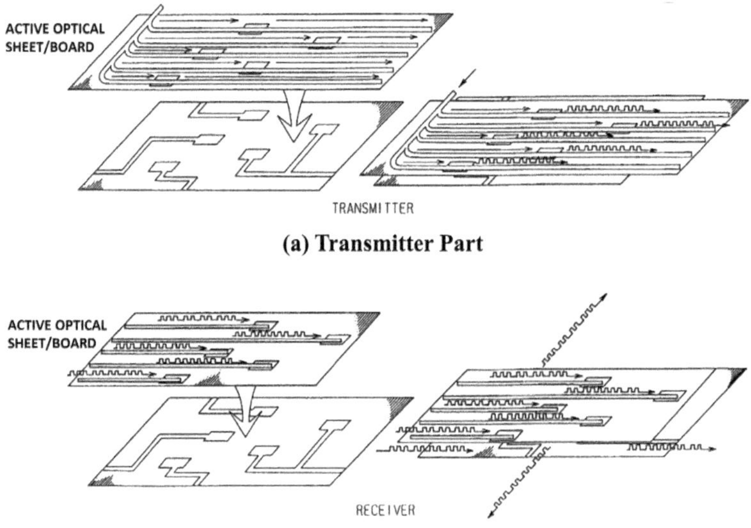

(a) Transmitter Part

(b) Receiver Part

Figure 11.25 Conceptual drawings of (a) transmitter part and (b) receiver part in an active optical sheet/board for a multichannel case.

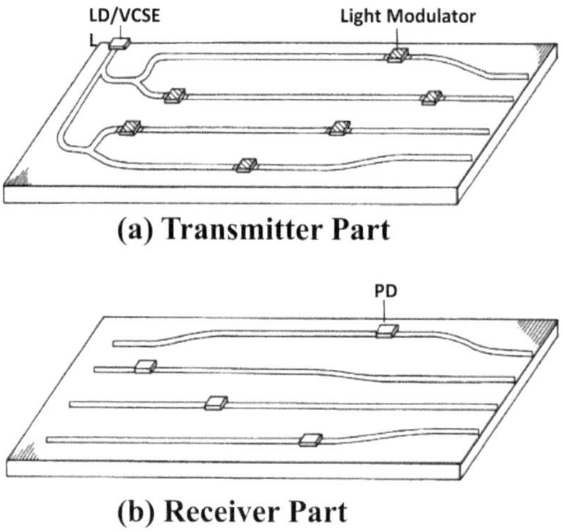

Figure 11.26 Active optical sheet/board for a multichannel case.

Figures 11.25a and b show conceptual drawings of active optical sheet/board for a multichannel case. A plurality of light modulators and PDs are respectively embedded in the transmitter part and in the receiver part. CW lightwaves emitted form an LD or VCSEL in the active optical sheet/board are branched to be guided to light modulator locations. To achieve more complicated optical interconnects, the light modulators and PDs are distributed in the same area as shown in Fig. 11.27.

Figure 11.28 shows an embodiment of connections between an active optical sheet/board and an electrical circuit board. Typically, solder bump connections are used. Connections using conductive adhesive and metal diffusion might also be used.

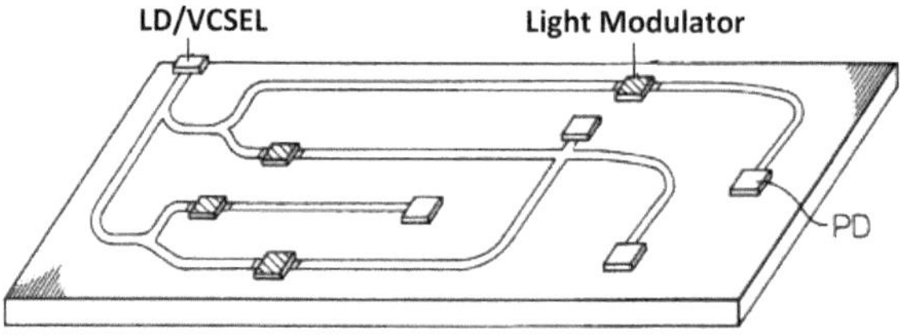

Figure 11.27 Active optical sheet/board with light modulators and PDs distributed in the same area.

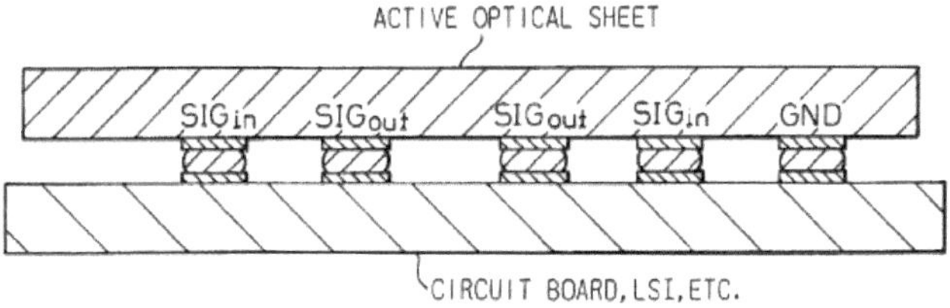

Figure 11.28 Connections between an active optical sheet/board and an electrical circuit board.

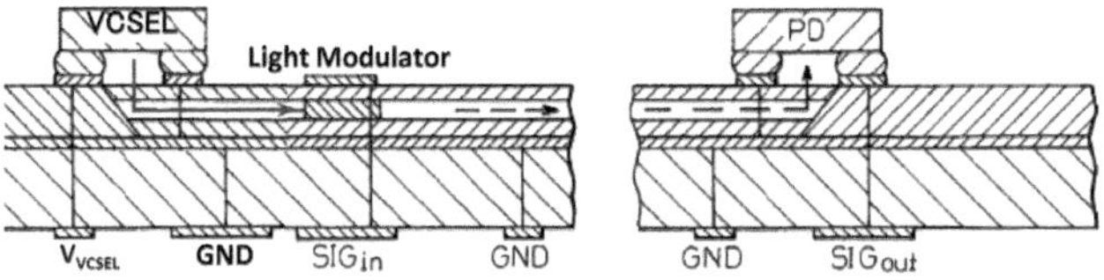

Figure 11.29 Details for transmitter and receiver parts of an active optical sheet/board.

Figure 11.29 shows details for the transmitter part and the receiver part of the active optical sheet/board. In the transmitter part, an embedded light modulator is coupled with a waveguide. The CW lightwaves emitted from a VCSEL connected to the V_{VCSEL} pad are guided to the light modulator. The CW lightwaves are modulated by the light modulator in response to voltage from the SIG_{in} pad for transmission of optical signals. In the receiver part, the optical signals transmitted through the waveguide are received at a PD, and then, electrical signals are output to the SIG_{out} pad. In addition to conventional pin PDs, small-capacity MSM devices and waveguide-type PDs can also be used. In usual cases, for PDs amplifiers and biasing voltage are necessary (they are omitted in the drawing). Sufficiently strong CW lightwave power will render the amplifier unnecessary, to allow further simplification.

Although in the preceding explanation the electrical board part is an electrical circuit board, the electrical board part can be LSIs, PCBs, MCMs, BPs, and other electronic units including peripheral devices.

Figure 11.30 shows an embodiment of an active optical connector for a parallel optical link module. The CW lightwaves emitted from a light source of a LD or VCSEL located in the active optical

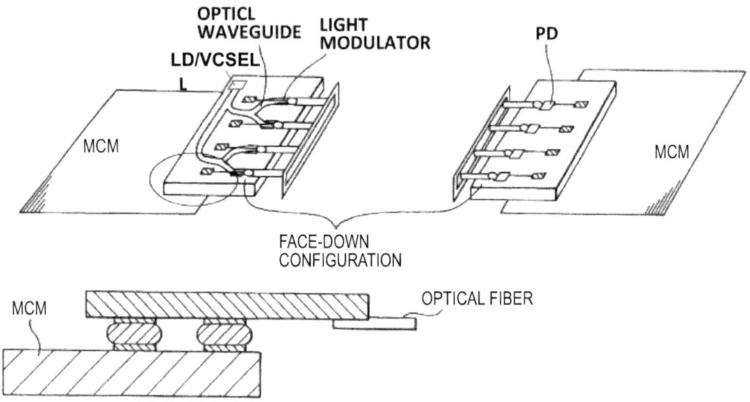

Figure 11.30 Active optical connector for a parallel optical link module.

sheet/board are branched to be guided to light modulators. The light modulators are electrically connected to the output terminals of an MCM by bump junctions. Thus, optical signals corresponding to the electrical outputs are transmitted to other electronic units through an optical fiber array or waveguide array.

References

1. T. Yoshimura, Y. Takahashi, M. Inao, M. Lee, W. Chou, S. Beilin, W.-C. Wang, J. Roman, and T. Massingill, "Systems based on opto-electronic substrates with electrical and optical interconnections and methods for making," U.S. Patent 6,343,171 B1 (2002).
2. T. Yoshimura, Y. Takahashi, M. Inao, M. Lee, W. Chou, S. Beilin, W.-C. Wang, J. Roman, and T. Massingill, "Optoelectronic substrates with electrical and optical interconnections and method for making," U.S. Patent 6,611,635 B1 (2003).
3. T. Yoshimura, Y. Takahashi, M. Inao, M. Lee, W. Chou, S. Beilin, W.-C. Wang, J. Roman, and T. Massingill, "Three-dimensional opto-electronic modules with electrical and optical interconnections and methods for making," U.S. Patent 6,690,845 B1 (2004).
4. T. Yoshimura, Y. Takahashi, J. Roman, M. T. McCormack, S. Beilin, W.-C. Wang, and M. Inao, "Optical coupling structures and the fabrication processes," U.S. Patent 6,684,007 B2 (2004).

5. T. Yoshimura, Y. Takahashi, K. Kuwabara, S. Beilin, M. Peters, W.-C. Wang, and M. Inao, "Optical reflective structures and method for making," U.S. Patent 6,706,546 B2 (2004).
6. T. Yoshimura, Y. Takahashi, J. Roman, M. T. McCormack, S. Beilin, W.-C. Wang, and M. Inao, "Single and multilayer waveguides and fabrication process," U.S. Patent 6,785,447 B2 (2004).
7. T. Yoshimura, Y. Takahashi, M. Inao, M. Lee, W. Chou, S. Beilin, W.-C. Wang, J. Roman, and T. Massingill, "Multi-layer opto-electronic substrates with electrical and optical interconnections and methods for making," U.S. Patent 6,845,184 B1 (2005).
8. T. Yoshimura, J. Roman, W.-C. Wang, M. Inao, and M. T. McCormack, "Device transfer method," U.S. Patent 6,669,801 B2 (2003).
9. T. Yoshimura, W. Sotoyama, T. Ishitsuka, K. Tsukamoto, S. Tatsuura, and K. Motoyoshi, "Active optical circuit sheet or active optical circuit board, active optical connector and optical MCM, process for fabricating optical waveguide, and devices obtained thereby," US 5,835,646 (1998).
10. T. Yoshimura, S. Tatsuura, W. Sotoyama, K. Motoyoshi, K. Tsukamoto, and T. Ishitsuka, "Optical circuit device, its manufacturing process and a multilayer optical circuit using said optical circuit device," U.S. Patent 6,338,975 (2002).
11. T. Yoshimura, T. Ishitsuka, K. Motoyoshi, S. Tatsuura, W. Sotoyama, K. Tsukamoto, T. Hayano, and A. Matsuura, "Optical circuit system and components of same," U.S. Patent 6,693,736 (2004).
12. T. Yoshimura, T. Ishitsuka, K. Motoyoshi, S. Tatsuura, W. Sotoyama, and K. Tsukamoto, "Optical circuit system and components of same technical field," U.S. Patent 6,215,585 (2001).
13. T. Yoshimura, S. Tatsuura, W. Sotoyama, K. Motoyoshi, K. Tsukamoto, and T. Ishitsuka, "Optical circuit device, its manufacturing process and a multilayer optical circuit using said optical circuit device," US 5,912,980 (1999).
14. T. Yoshimura and W. Sotoyama, "Optical device and light waveguide integrated circuit," US 5,854,868 (1998).
15. T. Yoshimura, S. Tatsuura, W. Sotoyama, K. Motoyoshi, K. Tsukamoto, and T. Ishitsuka, "Optical integrated circuit, optical circuit waveguide device and process for oriented, selective growth and formation of organic film," US 5,818,983 (1998).

Chapter 12

Future Challenges

As described in Chapter 2, core technologies that need to be developed in self-organized 3D integrated optical interconnects are as follows.

(1) Optical waveguide films with vertical mirrors and 3D optical wiring for 3D OE platforms
(2) Resource-saving all-photolithographic heterogeneous integration processes
(3) High-speed/small-size light modulators and optical switches
(4) SOLNETs for self-aligned optical couplings and vertical waveguide construction

For 1, 2, and 4, as described in Chapters 4, 5, and 7, future guidelines have been presented by theoretical predictions and experiments for proof-of-concept demonstrations. For 3, as described in section 6.3, the polymer MQD is found to be a preferable EO material for light modulators and optical switches. However, the EO effects of the polymer MQD have been predicted only theoretically by molecular orbital calculations. Therefore, it is an important future challenge to grow polymer MQDs experimentally and find future guidelines. To date, it has been found that molecular layer deposition (MLD) is promising for polymer MQD growth.

Self-Organized 3D Integrated Optical Interconnects with All-Photolithographic Heterogeneous Integration
Tetsuzo Yoshimura

ISBN 978-981-4877-04-6 (Hardcover), 978-1-003-04305-8 (eBook)
www.jennystanford.com

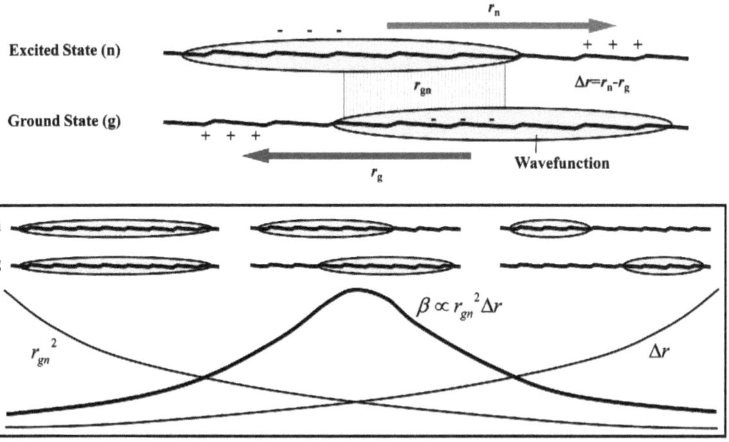

Figure 12.1 Guideline for enhancement of the Pockels effect.

Details of EO materials, including polymer MQDs as well as MLD, are described in a book published previously [1]. In the present chapter, brief reviews of the theoretical predictions of polymer MQD performance and MLD are presented.

12.1 Enhancement of the Pockels Effect by Controlling Wavefunction Shapes

In order to enhance the EO effect such as the Pockels effect and Kerr effect, electron wavefunction shapes in EO materials should be optimized. In the case of polymers, the wavefunction shapes can be optimized by controlling the molecular sequences in polymer wires. In the present section, theoretical predictions of the Pockels effect in polymer MQDs are summarized [1–4].

The Pockels effect is an EO effect originated from the second-order nonlinear optical effect. It induces refractive index changes proportional to applied electric fields. The influence of wavefunction shapes on the Pockels effect is predicted using molecular orbital calculations.

The EO coefficient for the Pockels effect in a molecule is proportional to hyperpolarizability β, which is a measure for the second-order optical nonlinearity of the molecule. In the two-level model, β is expressed as follows:

$$\beta \propto r_{gn}^2 \Delta r \tag{12.1}$$

Here, r_{gn} is the transition dipole moment between the ground state and the excited state, and Δr is the dipole moment difference between the ground state and the excited state. From Eq. 12.1, as schematically shown in Fig. 12.1, the following guidelines for the EO coefficient enhancement are derived.

(1) Promote wavefunction overlap between the ground state and the excited state, increasing r_{gn}.
(2) Promote wavefunction separation between the ground state and the excited state, increasing Δr.

For three wavefunction shapes schematically shown in Fig. 12.1, from left to right, the wavefunction separation increases while the wavefunction overlap decreases, i.e., Δr increases and r_{gn} decreases. Because Δr and r_{gn} are not independent, but are in a tradeoff relationship, the balance of the wavefunction overlap and the wavefunction separation must be considered simultaneously in optimizing β. Furthermore, Δr and r_{gn} depend on the dimensionality of the conjugated systems. Therefore, it is concluded that β is expected to have its maximum in a wavefunction shape of intermediate wavefunction separation with an appropriate conjugated system dimensionality existing somewhere between 0D and 1D, i.e., between a quantum dot condition and a quantum wire

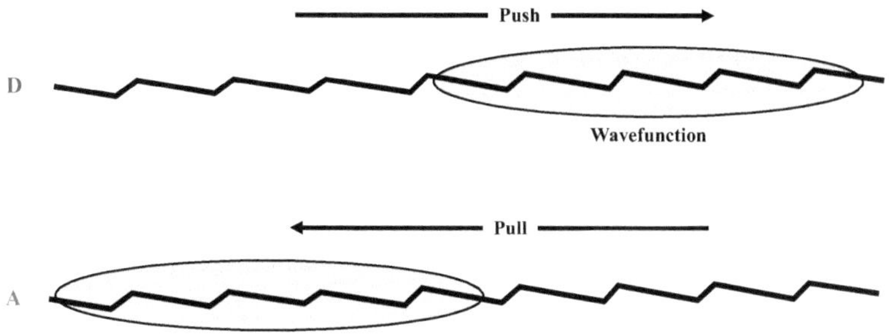

Figure 12.2 Push/pull effect on wavefunction induced by donor/acceptor groups.

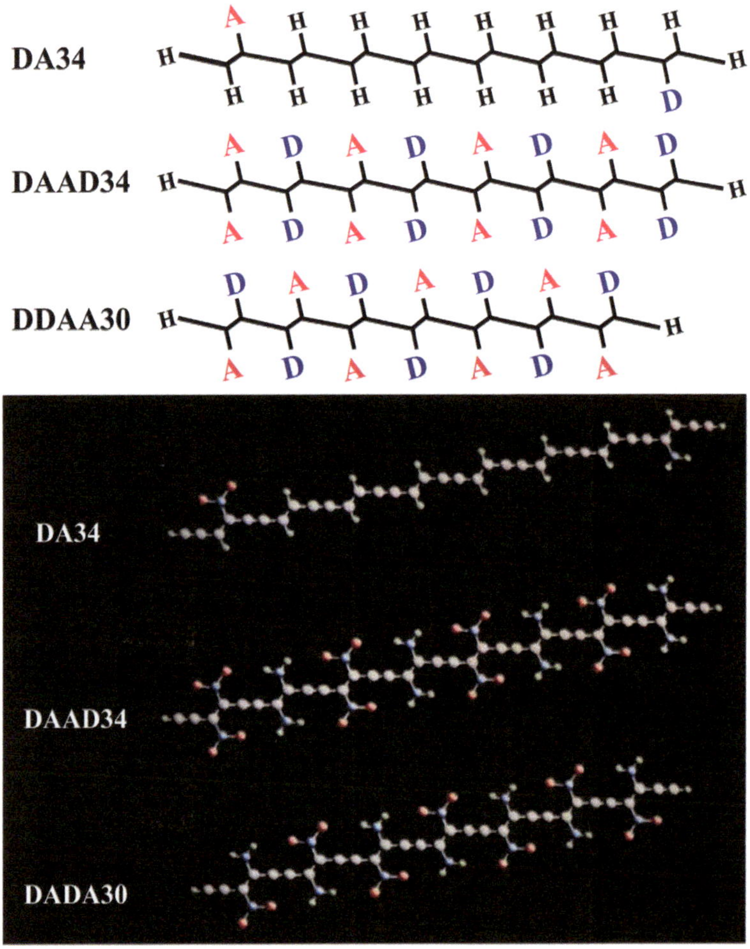

Figure 12.3 Models of conjugated polymer wires with donors and acceptors.

condition. Such optimized wavefunction conditions can be obtained by adjusting polymer wire lengths and donor/acceptor substitution sites in the wires using the push/pull effects of donors/acceptors as shown in Fig. 12.2.

Three types of models for conjugated polymer wires shown in Fig. 12.3 are considered. These molecules have poly-diacetylene (PDA) backbones. $-NH_2$ is the donor (D) and $-NO_2$ the acceptor

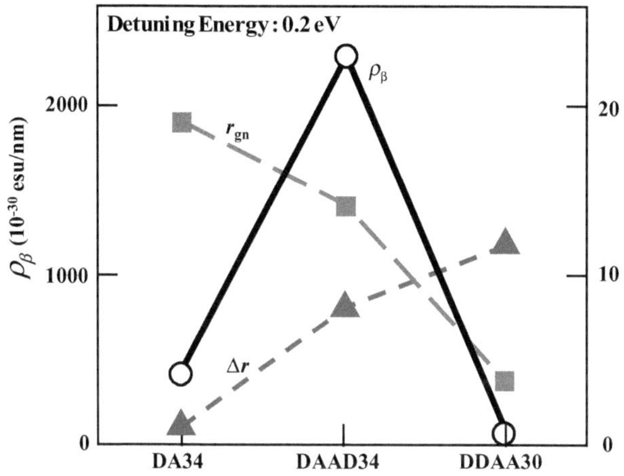

Figure 12.4 Calculated r_{gn}, Δr, and ρ_β for DA34, DAAD34, and DDAA30.

(A). DA, DAAD, and DDAA represent types of donor/acceptor substitution, and the numbers following them indicate the numbers of carbon sites N_C that correspond to molecular lengths L. For these model molecules, the relationship between β and wavefunction shapes was simulated using the molecular orbital method. Details of the simulation are described in a book and papers published elsewhere [1–4]. The calculations revealed that the wavefunction separation increases in the order of DA34, DAAD34, and DDAA30.

Figure 12.4 shows Δr, r_{gn}, and ρ_β of these model molecules for detuning energy of 0.2 eV. Here, ρ_β is β per 1 nm of molecular lengths. Δr increases and r_{gn} decreases in the order of DA34, DAAD34 and DDAA30, which makes ρ_β of DAAD34 in between maximum. This parallels to the tendency of the qualitative guideline shown in Fig. 12.1.

It is found from Fig. 12.5 that the molecular length affects ρ_β and that adjustments of the length improve ρ_β. In DAAD molecules, the ρ_β reaches maximum values near $N_C = 14 \sim 18$, corresponding to $1.4 \sim 1.8$ nm in the molecular length. Assuming the PDA wire density of 1.3×10^{14} 1/cm^2, the expected maximum EO coefficient of the DAAD is about 3000 pm/V, which is 100 times larger than the EO coefficient r_{33} of LN. It is therefore concluded that a large EO effect is expected to be obtained by controlling

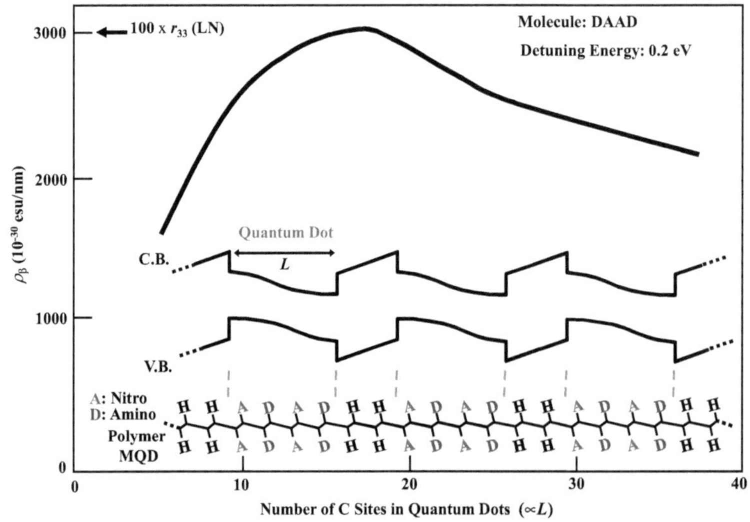

Figure 12.5 Dependence of calculated ρ_β on the number of carbon sites, which corresponds to the molecular length, *L*, for DAAD molecules. The inset represents a polymer MQD and corresponding electronic potential energy curves. DAAD18 molecules inserted in the polymer wire are regarded as quantum dots.

the wavefunction dimensionality and the separation, which can be realized by adjusting the molecular lengths and the donor/acceptor substitution sites.

In addition to the PDA backbone, other conjugated systems such as poly-azomethine (poly-AM) and poly-oxadiazole (OXD) can be the backbone for polymer wires.

Molecular orbital calculations revealed that energy gaps of DAAD molecules are smaller than those of PDA with no donor/acceptor substitution [1, 4]. Using this phenomenon, it is possible to insert DAAD molecules into a PDA backbone to construct a polymer MQD, as shown in Fig. 12.5 [1, 4]. This structure would enable the DAAD molecules, which are regarded as quantum dots, to be aligned perfectly along the wire direction. This is favorable for attaining large EO effects. In order to fabricate such molecular-order-controlled wire structures, MLD using source molecules with donors and acceptors is promising.

12.2 Molecular Layer Deposition (MLD)

Molecular layer deposition (MLD) is a monomolecular-step growth process for organic thin films with designated molecular arrangements [1, 5–11]. MLD utilizes selective chemical reactions (or electrostatic force) between different kinds of molecules. As Fig. 12.6 shows, firstly, Molecule A is provided onto a substrate surface to form a monomolecular layer of Molecule A. By switching molecules from A to B, Molecule B is connected to Molecule A. Once the surface is covered with Molecule B, the film growth is automatically terminated by the self-limiting effect similar to that in atomic layer deposition (ALD) [12]. By switching molecules in a sequence of A, B, C, D, and so on, an organic thin film with designated molecular arrangements of A/B/C/D/—is obtained. MLD is usually performed in vapor phase. It can also be performed in liquid phase [1, 8, 11].

Figure 12.7 shows an experimental demonstration of MLD using pyromellitic dianhydride (PMDA) and 4,4′-diaminodiphenyl ether (DDE) [1, 5]. When PMDA is provided onto a DDE surface, the film thickness, which is monitored by a quartz crystal microbalance, rapidly increases and then is saturated in 120 s. When the molecules are switched from PMDA to DDE, again, the film thickness rapidly

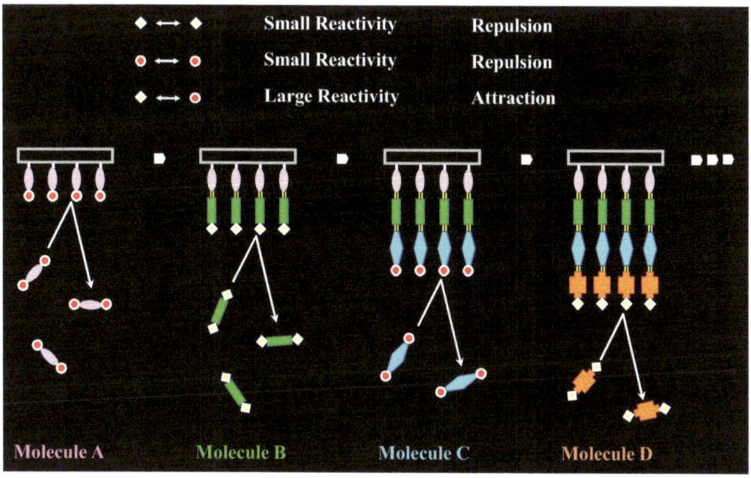

Figure 12.6 Concept of MLD.

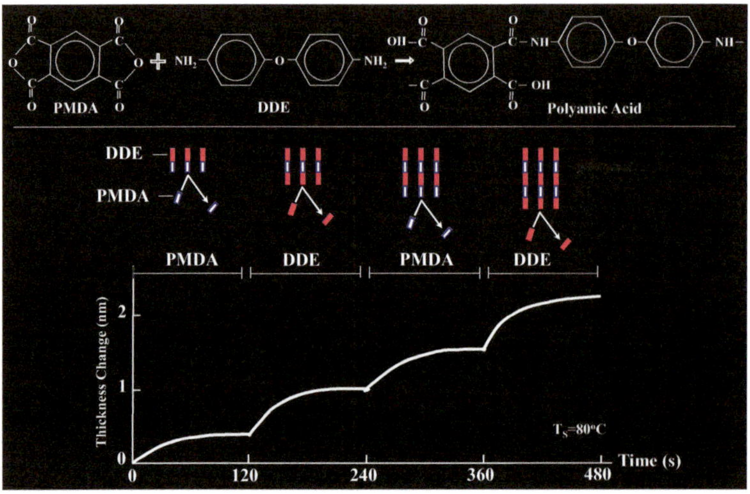

Figure 12.7 Experimental demonstration of MLD. Reprinted with permission from Yoshimura et al. [5].

increases and is saturated. By repeating the molecule switching, step-like film growth is observed. The thickness change for one growth step is close to the sizes of PMDA and DDE. These results indicate that monomolecular-step growth is performed by MLD.

Three featured capabilities of MLD are schematically depicted in Fig. 12.8.

(1) Ultrathin/conformal organic material growth: MLD enables ultrathin/conformal organic material growth on three-dimensional surfaces with arbitrary structures including beveled walls, surfaces with ultrafine patterns, particles, and porous objects.
(2) Tailored organic material growth: MLD enables tailored organic material growth with artificially controlled molecular sequences.
(3) Selective organic material growth: Utilizing patterned surface treatments, MLD enables selective organic material growth.

The capability of the ultrathin/conformal organic material growth has led to many applications. Bent et al. succeeded to form copper

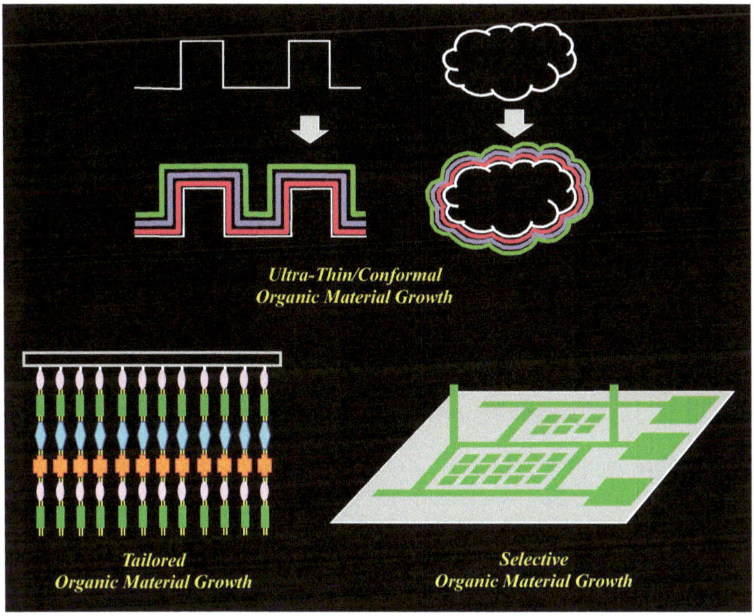

Figure 12.8 Featured capabilities of MLD.

diffusion barriers in LSIs and to deposit photoresist by MLD [13]. Weimer and Liang realized uniform polymer film coating on surfaces of nanoparticles [14]. George et al. developed a growth process for hybrid organic–inorganic polymer films [15] by combining MLD with ALD, and the films were used as gate insulators for organic TFTs [16], thin-film encapsulations for OLEDs [17], and energy technologies including batteries and thermoelectric energy devices [18]. The capability of ultrathin/conformal organic material growth is effective in depositing thin films on beveled walls for vertical mirrors/wavelength filters, in via holes, etc., in S-FOLM fabrication processes.

The capability of tailored organic material growth realizes artificial controls of the electron wavefunction shapes in organic thin films to optimize material performance or to generate new functions [1, 2, 9]. As schematically shown in Fig. 12.9, widely spread wave-functions, symmetric wavefunctions, and non-centrosymmetric

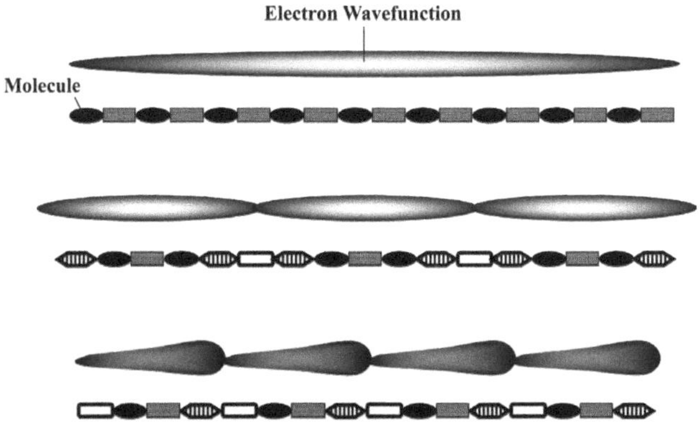

Figure 12.9 Control of electron wavefunction shapes by placing constituent molecules in designated arrangements using MLD.

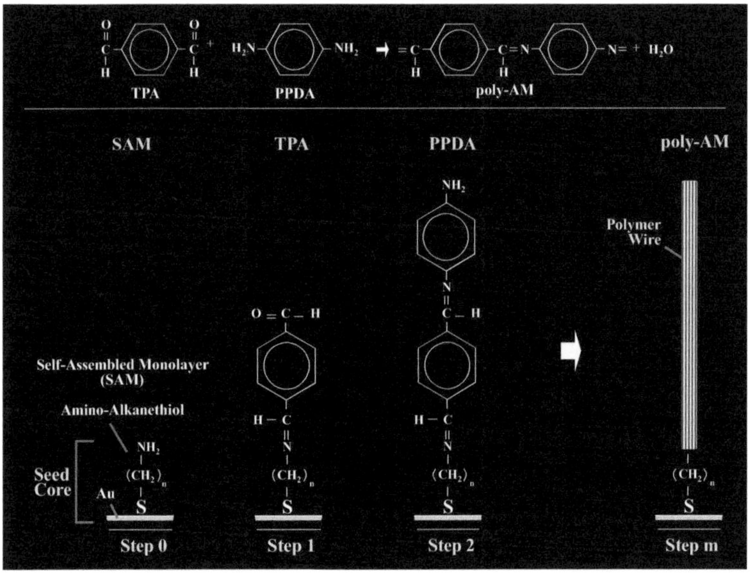

Figure 12.10 MLD process for the location/orientation-controlled growth using seed cores for selective growth. Reprinted with permission from Yoshimura et al. [7].

wavefunctions can be formed in polymer wires or molecular stacked structures by controlling molecular sequences. This capability enables us to fabricate polymer MQDs for light modulators and optical switches as described in section 12.1.

The capability of selective organic material growth is available using surface treatments [1, 7, 10, 11, 19]. Figure 12.10 shows the process for location/orientation-controlled growth using seed cores [1, 7]. First, a self-assembled monolayer (SAM) such as amino-alkanethiol with amino groups on the top is formed on a patterned Au film to fabricate a seed core. Next, terephthalaldehyde (TPA) is provided on the surface to connect it on the SAM. Then, *p*-phenylenediamine (PPDA) is provided to connect it to TPA. By repeating such growth steps, polymer wires of poly-AM can be grown nearly vertically on the seed core surface. It is expected that by distributing seed cores appropriately, as Fig. 12.11 shows, selective polymer wire growth with controlled wire orientations is achieved. Furthermore, three-dimensional polymer wire networks are constructed [1]. This approach would be promising to apply MLD to integrated photonic/electronic devices.

Selective growth can also be achieved by hydrophilic/hydrophobic patterning [1, 19]. For poly-AM film growth on a hydrophilic substrate with patterned hydrophobic regions, the film is selectively grown on the hydrophilic region, on which PPDA and TPA molecules adsorb strongly. The selective MLD can be applied to the selective deposition process of the target fabrication for SOLNETs (see Fig. 7.6) as well as the processes of S-FOLM fabrication.

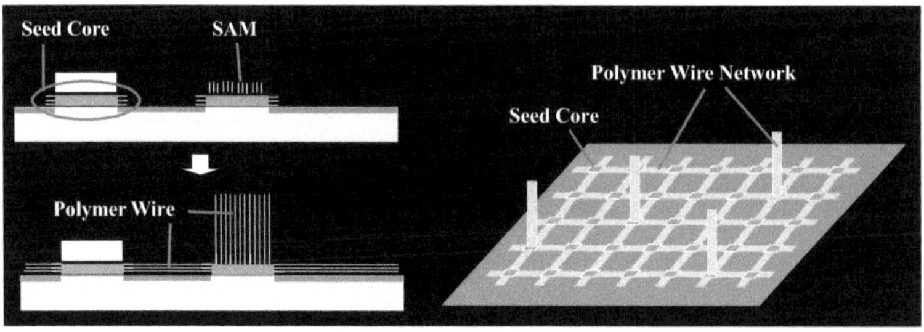

Figure 12.11 Expected structures of polymer wire network grown by the location/orientation-controlled growth by MLD.

12.3 Growth of Polymer MQDs by MLD

Figure 12.12 shows an example of the monomolecular-step growth process for polymer MQD fabrication by MLD. Three source molecules, namely, TPA, PPDA, and oxalic dihydrazide (ODH), are connected with a sequence of ODH-TPA-PPDA-TPA-ODH–. When TPA and PPDA are connected by chemical reactions, π electrons pass through the bond. When TPA and ODH are connected, on the other hand, π electrons are blocked by a series of single bonds. By using such bond characteristics, various kinds of polymer MQDs can be fabricated.

Figure 12.13 shows schematic band diagrams of poly-AM and a polymer MQD that we call 3QD structure. In poly-AM, TPA and PPDA are alternately connected. π electrons are delocalized through the polymer wire. In the 3QD polymer MQD, where three different-length quantum dots (QDs) are inserted in a single polymer wire, three source molecules are connected by MLD with a sequence of –ODH-TPA-ODH-TPA-PPDA-TPA-ODH-TPA-PPDA-TPA-PPDA-TPA-ODH– to form QDs of [OT], [OTPTP], and [OTPTPT]. The portions of ODH are barriers, and the portions consisting of TPA and PPDA are QDs.

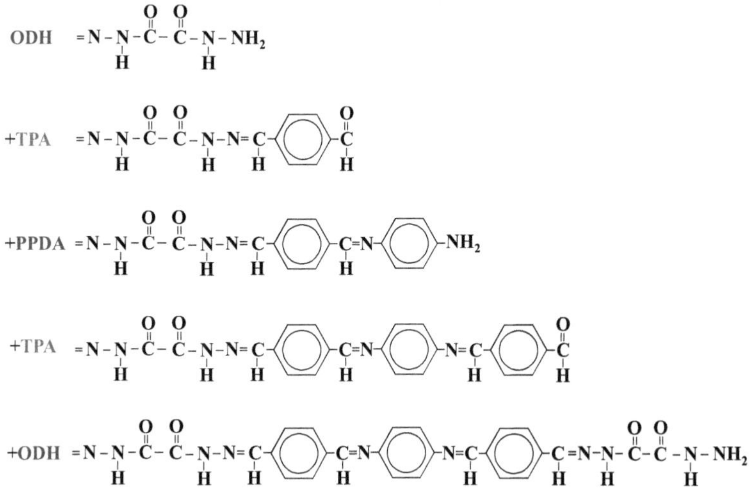

Figure 12.12 Monomolecular-step growth process for polymer MQD fabrication by MLD. Reprinted with permission from Yoshimura et al. [9].

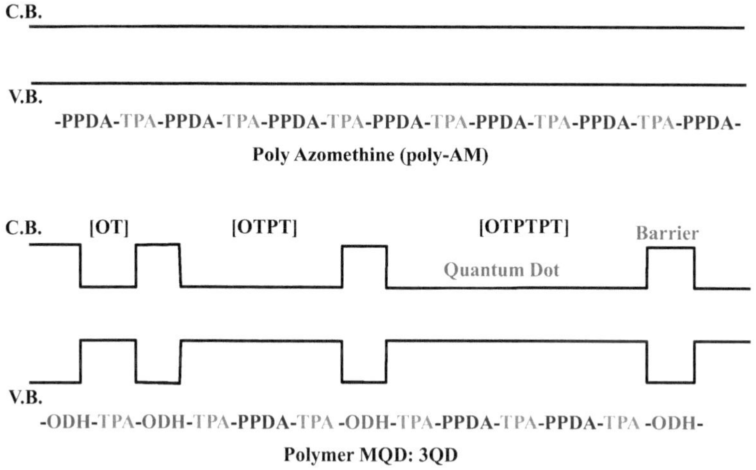

Figure 12.13 Schematic band diagrams of poly-AM and a polymer MQD that we call 3QD structure. Reprinted with permission from Yoshimura et al. [9].

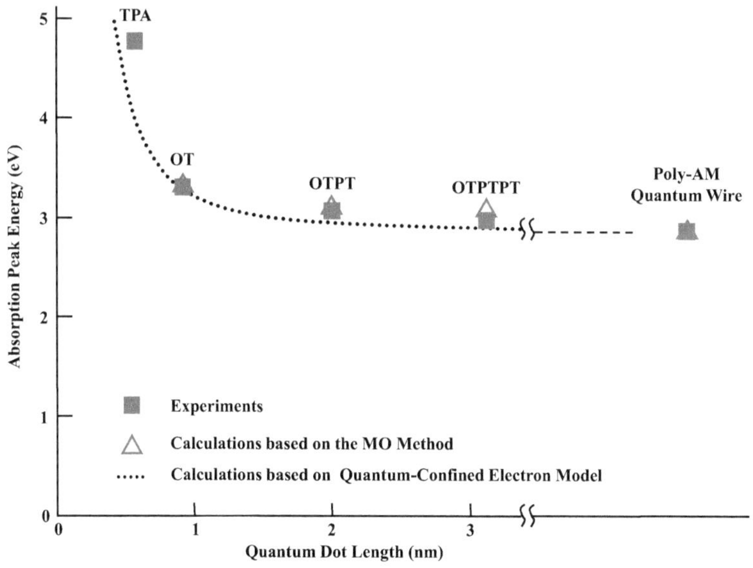

Figure 12.14 Absorption peak energy plotted as a function of the QD length. Reprinted with permission from Yoshimura et al. [9].

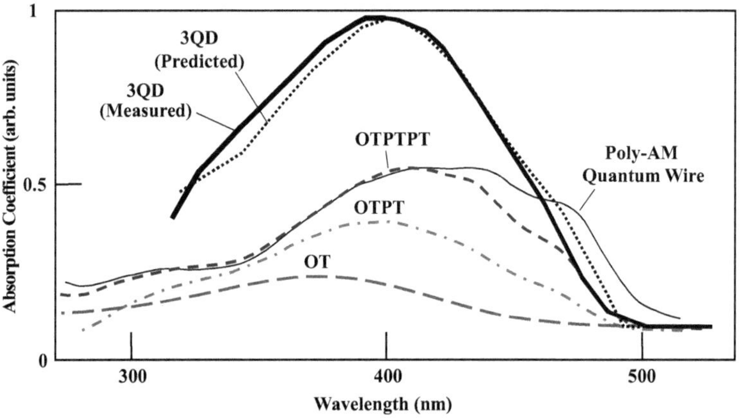

Figure 12.15 Absorption spectra of poly-AM quantum wire, OTPTP, OTPT, OT, and 3QD. Reprinted with permission from Yoshimura et al. [9].

In Fig. 12.14, the absorption peak energy is plotted as a function of the QD length [1, 9, 10, 20]. The experimental results fairly agree with the peak shift calculated by the molecular orbital method (WinMOPAC, Fujitsu), and the peak shift calculated based on the quantum-confined electron model. This suggests that the absorption peak shift is attributed to the quantum confinement effect in QDs.

As Fig. 12.15 shows, in the order of poly-AM that has no potential barriers, OTPTP, OTPT, and OT, namely, with decreasing QD lengths, the absorption peaks shift to the shorter wavelength side (higher energy side). For the 3QD polymer MQD, as shown by a solid-line curve for the measured absorption spectrum, a broad absorption band extending from ~480 to ~300 nm appears. The dotted-line curve represents the superposition of the spectra of [OT], [OTPT], and [OTPTPT]. It corresponds to the predicted absorption spectrum of the 3QD polymer MQD. The measured and predicted spectra are relatively similar, suggesting that the 3QD structure was successfully fabricated by MLD.

In order to fabricate polymer MQDs with the EO effect, source molecules having donor/acceptor groups are needed. Collaboration with researchers in chemistry is necessary to grow EO polymer MQDs in future.

References

1. T. Yoshimura, *Thin-Film Organic Photonics: Molecular Layer Deposition and Applications*, CRC/Taylor & Francis, Boca Raton, Florida (2011).
2. T. Yoshimura, "Enhancing second-order nonlinear optical properties by controlling the wave function in one-dimensional conjugated molecules," *Phys. Rev. B* **40**, 6292–6298 (1989).
3. T. Yoshimura, "Theoretically predicted influence of donors and acceptors on quadratic hyperpolarizabilities in conjugated long-chain molecules," *Appl. Phys. Lett.* **55**, 534–536 (1989).
4. T. Yoshimura, "Design of organic nonlinear optical materials for electro-optic and all-optical devices by computer simulation," *FUJITSU Sci. Tech. J.* **27**, 115–131 (1991).
5. T. Yoshimura, S. Tatsuura, and W. Sotoyama, "Polymer films formed with monolayer growth steps by molecular layer deposition," *Appl. Phys. Lett.* **59**, 482–484 (1991).
6. T. Yoshimura, E. Yano, S. Tatsuura, and W. Sotoyama, "Organic functional optical thin film, fabrication and use thereof," U.S. Patent 5,444,811 (1995).
7. T. Yoshimura, S. Ito, T. Nakayama, and K. Matsumoto, "Orientation-controlled molecule-by-molecule polymer wire growth by the carrier-gas-type organic chemical vapor deposition and the molecular layer deposition," *Appl. Phys. Lett.* **91**, 033103 (2007).
8. T. Yoshimura, H. Watanabe, and C. Yoshino, "Liquid-phase molecular layer deposition (LP-MLD): potential applications to multi-dye sensitization and cancer therapy," *J. Electrochem. Soc.* **158**, 51–55 (2011).
9. T. Yoshimura, R. Ebihara, and A. Oshima, "Polymer wires with quantum dots grown by molecular layer deposition of three source molecules for sensitized photovoltaics," *J. Vac. Sci. Technol. A* **29**, 051510 (2011).
10. T. Yoshimura and S. Ishii, "Effect of quantum dot length on the degree of electron localization in polymer wires grown by molecular layer deposition," *J. Vac. Sci. Technol. A* **31**, 031501 (2013).
11. T. Yoshimura, S. Bai, H. Tateno, and C. Yoshino, "In situ photocurrent spectra measurements during growth of three-dye-stacked structures by the liquid-phase molecular layer deposition," *J. Appl. Phys.* **122**, 015309 (2017).
12. T. Suntola, "Atomic layer epitaxy," *Mater. Sci. Rep.* **4**, 261–312 (1989).

13. H. Zhou and S. F. Bent, "Fabrication of organic interfacial layers by molecular layer deposition: present status and future opportunities," *J. Vac. Sci. Technol. A* **31**, 040801 (2013).
14. X. Liang and A. Weimer, "An overview of highly porous oxide films with tunable thickness prepared by molecular layer deposition," *Curr. Opin. Solid State Mater. Sci.* **19**, 115–125 (2015).
15. S. M. George, B. Yoon, and A. A. Dameron, "Surface chemistry for molecular layer deposition of organic and hybrid organic-inorganic polymers," *Acc. Chem. Res.* **42**, 498–508 (2009).
16. B. H. Lee, K. H. Lee, S. Im, and M. M. Sung, "Molecular layer deposition of ZrO_2-based organic–inorganic nanohybrid thin films for organic thin film transistors," *Thin Solid Films* **517**, 4056–4060 (2009).
17. D. Yu, Y. Yang, Z. Chen, Y. Tao, and Y. Liu, "Recent progress on thin-film encapsulation technologies for organic electronic devices," *Opt. Commun.* **362**, 43–49 (2016).
18. C. Ban and S. M. George, "Molecular layer deposition for surface modification of Li-ion battery electrodes," *Adv. Mater. Interface* **3**, 1600762 (2016).
19. T. Yoshimura, N. Terasawa, H. Kazama, Y. Naito, Y. Suzuki, and K. Asama, "Selective growth of conjugated polymer thin films by the vapor deposition polymerization," *Thin Solid Films* **497**, 182–184 (2006).
20. T. Yoshimura, S. Tatsuura, W. Sotoyama, A. Matsuura, and T. Hayano, "Quantum wire and dot formation by chemical vapor deposition and molecular layer deposition of one dimensional conjugated polymer," *Appl. Phys. Lett.* **60**, 268–270 (1992).

Epilogue

In January 2009, I had two offers to write books on different subjects. One of them was from Jenny Stanford Publishing for a book titled *Optical Electronics: Self-Organized Integration and Applications*. I was afraid that it might be very hard to complete two books at the same time. So, I asked my wife, Yoriko, "How do you think about writing two books at the same time in English?" In reply, she said, "I know that you have accomplished everything you once told me." Her answer made me accept the two offers.

During the preparation of these books, Yoriko passed away, on September 27, 2009. She continued to receive encouragement from her friends of Joshi Gakuin High School and Musashino Academia Musicae, friends in Morimura-Gakuen, students of her piano school, and her relatives and family for five months before September 27.

In 2017, the Director and Publisher of Jenny Stanford Publishing, kindly gave me another offer to write a book on optical interconnects. I believe that this fortunate chance is a result of Yoriko's help. I thank Yoriko for her sincere support and encouragement since 1977, and have placed her photographs in the epilogue as a memorial to her. By the way, the cover of this book was designed by our daughter, Naoko Yoshimura.

Most of the concepts described in the present book were produced during a period I was in Fujitsu Computer Packaging Technologies Inc. (FCPT), Zanker Road, San Jose, California, from 1997 to 2000. Every day, I drove beside Intel on San Tomas/Montague Express Way to commute to FCPT while listening to KFRC 99.7 Oldies brought by Cammy Blackstone in cool winds from fully open car windows. I enjoyed exciting experiences with friends at FCPT. The California Feeling that "Silicon Valley is like Arcadia" brought various new ideas to me.

Yoriko Yoshimura,
California, 2000

From the left: Yoriko, Chikako, Naoko, and Tetsuzo Yoshimura, New York, 1998

I should note how students in my laboratory made great contributions to the present book. The students, whose names can be found in the references of every chapter, promoted the research to get interesting and important results with their excellent effort and passion. I am sincerely proud of them.

Yoriko and I hope that *Self-Organized 3D Integrated Optical Interconnects* will contribute to the R&D of high-performance computing and switching systems as well as solar energy conversion systems in future.

Tetsuzo Yoshimura
September 27, 2020

Index